From Additive Manufacturing to 3D/4D Printing 1

To Laurent and Denis, who allowed me to dedicate myself to additive manufacturing.

Series Editor
Jean-Charles Pomerol

From Additive Manufacturing to 3D/4D Printing 1

From Concepts to Achievements

Jean-Claude André

iSTE

WILEY

First published 2017 in Great Britain and the United States by ISTE Ltd and John Wiley & Sons, Inc.

ISTE Ltd
27-37 St George's Road
London SW19 4EU
UK

www.iste.co.uk

John Wiley & Sons, Inc.
111 River Street
Hoboken, NJ 07030
USA

www.wiley.com

Library of Congress Control Number: 2017950122

British Library Cataloguing-in-Publication Data
A CIP record for this book is available from the British Library
ISBN 978-1-78630-119-2

Contents

Acknowledgments

Sincere thanks to 3A, BeAm, 3DCeram, 3D Systems, Fives, Nanoscribe, Poietis, Prodways, XtreeE, Alex (alias Alexandre Martel, co-founder of 3D Natives.com) and Laurent Dupont, head of the Lorraine Fab Living Lab® for their effective cooperation, and particularly for graciously providing certain images.

Prototype part (3DCeram, 2017), reproduction of the Église de Bonsecours in Nancy (LRGP, 1994) and metallic part (3A – Applications Additives Avancées, 2017)

Foreword

The evocative expression "3D printing" has been overtaken in everyday speech by the expression generally preferred by scientists and engineers, "additive manufacturing". In both cases, it is a matter of manufacturing objects in successive layers, and soon every workshop and every school will have a 3D printer and engage in additive manufacturing. Self-service workshops known as fab-labs already offer users the possibility to create their own objects. However, the adventure is not over, as "4D" is coming up over the horizon with materials that evolve over time, not to mention "bio-printing", which aims to create organs to be used to repair the living. Furthermore, the 3D printing of tomorrow, which will be performed without layers, threatens to make the term "additive manufacturing" obsolete, thereby making it possible to return to the initial concept of 3D printing. Whatever the case may be, we are faced with not only a very active and booming world, but also a complex world that calls on numerous skills in physics, engineering, chemistry of materials and mechanics with a resolutely multidisciplinary and convergent approach.

To understand the origin of the ideas in additive manufacturing/3D printing, learn about the current state of what is known and explore the developments to come, what could be better than to ask one of the inventors of the technology and one of the first French patent holders in the field, Jean-Claude André, to share his knowledge with us? This led to the idea of this 3-volume edition that I am pleased to present; a work that is both erudite and prospective, as its intention is to start at the genesis of the ideas that led to additive manufacturing to anticipate the impact and future of still "additive" technologies and, beyond this, to encourage reflection on the interactions between science and society of today and tomorrow.

If the first patents date back to 1984, an era where lasers, photo-materials and computer-aided design had already been mastered, was the idea of additive manufacturing completely disruptive as would be said today. What was creative was to put all of this knowledge together to come to something entirely new. Nevertheless, approval for the concept of additive manufacturing came rather quickly. It is on this basis that other additive methodologies, currently many of them with very specific niches, could be developed. These range from prototype and industrial parts to art, variable spatial scales – from the decameter to the nanometer –, from the inert to the living, from industrial organizations to very delocalized forms of manufacturing, etc.

On the basis of these works with varied applicative and societal spectrums, some of which are in the process of becoming stabilized, others to be invented, the principles of additive manufacturing can serve as an example, even as a "laboratory" to better understand how the interactions between research and society can (and must) develop, whether this is through new scientific concepts and the associated concepts of creativity, interdisciplinary scientific and technological operations, the popularization of public research, links with society in terms of the creation of new markets and jobs, and also forms of responsibility and ethics.

Throughout these three volumes, the author would like to invite you to reflect on the circuits between the applications that pose new scientific questions and prior research which opens the door to new applications or new products. The more we progress in the field of new niches, the more previously unasked scientific questions are considered, questions whose answers (if they exist) are supported and encouraged by public authorities and industry, which are gaining awareness of an immense industrial and/or medical market, as is the case for bio-printing. From dream to reality, scientists are often in the position to anticipate the length of the path; however, a dynamic is created. This leads to cultural changes and changes in practices, particularly concerning the importance of creativity, sharing enthusiasm for research, openness with others, the multiplying (and sometimes inhibiting) effect of public actors, on the one hand, and the economic world, on the other, as this work illustrates wonderfully.

This saga of additive manufacturing, told by one of its inventors, teaches us that creativity alone does not suffice; it is necessary to have a good dose of perseverance as well, and it is, of course, necessary to keep moving after the first failures. In addition, this shows us that sometimes the research structures and the environment are not entirely receptive to innovation, even when success comes relatively quickly.

Jean-Claude André also explains with great enthusiasm how we give shape to an idea to feed our intuition, which in turn increases creativity. On the whole, these three volumes provide a wealth of information on additive manufacturing, and additionally, they illustrate and encourage veritable reflection on the task of a researcher and research structures, as well as the role of creativity in research, and finally, they invite us to rethink and reinforce the relations between science and society.

Jean-Charles POMEROL
President of the Incubateur AGORANOV
and the ISTE Editions Scientific Committee

Preface

"We have too often forgotten that specialists are created from amateurs, just as soldiers are made from civilians". [LAT 07]

"In France, strangely enough, it is not those used to sailing the seas, the specialists of the real and tangible, who are asked for advice guiding the flagship, but the members of a caste who stay at port and who, for the most part, have only purely theoretic knowledge of the sea". [BEI 12]

"Technology has taken on a new breadth and organization. Here, I am searching for its specific structure, and I have noticed that it exists as a system, in other words, as an organized whole". [ELL 04]

"Those in the organization who have ideas to do things otherwise or better are divided into two categories: those who do not dare and those who dare. Those who do not dare understand very well the risks and the importance of new ideas, but they are paralyzed by risk taking and the fear of displeasing. Having never tried anything, they have not known failure and are thus unharmed by reproach [...], they are quitters. Those who dare, the innovators, move forward by challenging conventional ideas, organizations, and sometimes procedures. They stir up fears and a lack of understanding and are truly criticized...". [PHI 12]

"Science has largely renounced an interdisciplinary vision allowing the merits of different results to be faced". [THO 83]

"Theory is when everything is known and nothing works. Practice is when everything works and no one knows why. Here, we have united theory and practices: nothing works… and no one knows why!". [EIN 07]

"These creatures of man [machines] are exacting. They are now reacting on their creators, making them like themselves. They want well-trained humans; they are gradually wiping out the differences between men, fitting them into their own orderly functioning, into the uniformity of their own regimes". [VAL 57]

"Speaking of discipline is designating the scientific activity as a particular form of the division of labor in the social world". [FAB 06]

"The imagination is brilliant in that it produces images that enlarge reality and really invent it". [GUÉ 15]

"In cultural terms, no enterprise is built with dreams alone and none without. Action, if it is to be successful, is by necessity guided by practical circumstances. But the goal of any action is defined, implicitly or explicitly, by the deep nature of the human being, his dreams, his vision of life, his culture. The dynamics of life, the challenge of risk and uncertainty, today require from us a new creative effort leading to the reconstruction and to the re-conquest of the notion of progress, which the philosophies and the ideologies of certainty have shuttered almost to the point of destruction". [GIA 90]

"Researching is inventing the world, it is setting new rules of functioning for an ephemeral world. Not like the tyrants who also invent a new world for themselves, but impose it upon others. The researcher does not recreate the world, but rather unravels it to make it. He/She imagines one, then compares it with the real world to clarify it and not to exhaust it. Researching is an endless quest". [ROS 01]

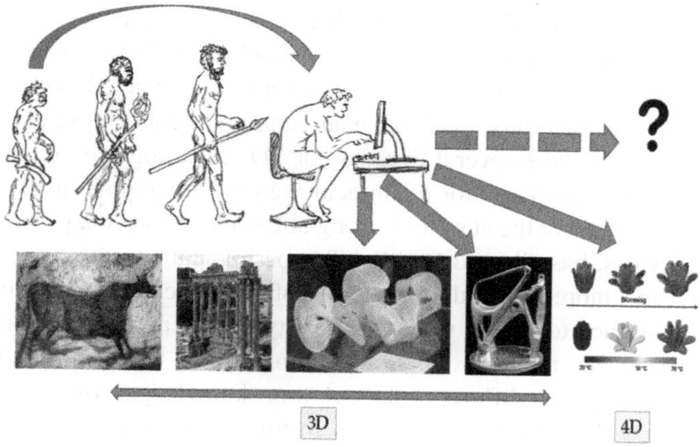

Figure 1. *From additive manufacturing to 3D/4D printing*

This book (in three volumes) is the result of a demand that has been repeated countless times for different reasons, notably among these, of the oversight and the reminder of the oversight to cite a French school that in 1984 succeeded to patent the first additive manufacturing process, stereolithography, several weeks before the Americans (who were working on the same subject, without either party knowing it). However, at the same time, thirty or so years later, it is a history lesson that can be told about a process concept, tossed out in France, without any malice of course, by "clairvoyant hierarchists", the explosion of the research team who felt their future was blocked and an American technical-economic development which has today led to several books and more than 50,000 scientific publications on additive manufacturing, because consequent applicative markets exist with profitable enterprises (and also because there is an immense attraction field around this subject that conditions the actions of a great number of researchers).

So why have we entitled these three volumes "From Additive Manufacturing to 3D/4D Printing"? First, it was about locally bringing material and/or energy to perform a transformation (e.g. from a powder to a solid or from a liquid to a solid). The expression "additive" then takes on its true meaning. But for a short time now, researchers have been developing (or working on) new processes that allow this change to be avoided through the additions mentioned at the start of this paragraph. It thus becomes possible to create an object in one go. Moreover, the use of so-called "smart" materials authorizes the introduction of a complementary parameter, i.e. time or functionality. The 4D aspect is thereby introduced.

The first volume on additive manufacturing is strongly linked to the existence of an effective economic market, one that is already significant, stemming from technological research in the engineering sciences connected to an essential component, that of materials (and of manipulating them to prepare them for manufacturing). It will take several decades for 3D technology to emerge and find its place as a robust technology for manufacturing objects in quite diverse domains. This situation, linked from the start to a strong attractiveness on the part of industrial R and R&D services, has allowed for "field" experimentation with competent users who are more and more demanding in terms of manufacturing qualities (without seeking in this preface to define what this quality, a true portmanteau, represents). Mastery by users, on the one hand, and competition between the bearers of knowledge pertaining to different 3D printing knowledge, on the other, are translated into new demands to be satisfied. In this framework, this demand has in fact made up one of the driving forces of incremental research, a "technology pull" described in Volume 2 (at least as much as is known (or published)).

A solution is good if and only if the concept, its demonstration with the right people, a culture of industrial innovation, and time and finances effectively come together. Maybe at that time, in 1984, there was a closed system of opinion and self-centered management that had not even thought of a possible debate on futuristic technological openings. This conformity to a manufacturing follower style of thinking was more and more often considered to be obsolete. But there was also, beyond socio-economic milieus, an incredible viscosity with many scientists: the most common attitude was not openness to other explicative schemata, but in the majority of cases, the ignorance and/or refusal to accept their existence. Tricks that only imperfectly fit into our ethics as researchers (at the time) must be made and likely developed.

According to estimation methods, the revenue from additive manufacturing lies somewhere between 5 and 40 billion euros (we could think that this is an estimation of the number of protesters in a claim by the police or trade unions!). Some speak of a revolution and others imagine senseless promises (which, according to Audétat [AUD 15], could put every emerging sector in danger); in short, things are booming at present with seven main stabilized technologies and a new kind of governance (Jeremy Rifkin's "makers"). This appreciative placement of the normalizers into categories is indeed rather artificial. Beyond a recent manufacturing technique that associates computer science and matter, 3D printing, with cheaper and cheaper home machines (down to a few hundred euros), constitutes a paradigm shift that impacts product design (which can even be defined, thanks to "open-source" systems), creation (from heavy industry to one's "garage"), consumption and the business models that result from them (from market activity, a new handicraft and DIY (Do-It-Yourself) to counterfeiting).

In fact, the progression rates are always in the double figures (between 20 and 40% per year), which leads some to believe that the additive manufacturing processes will continue to evolve for a long time to become a widespread technology, as they increasingly occupy ever-new applicative niches, quashing the other manufacturing methods that made up the skeleton of 20th Century industrial manufacturing. But what do tens of billions of euros per year represent for the world relative to France's "small" debt amounting to 2 trillion euros? It is therefore difficult to project a future which leads to a possible hegemony of additive manufacturing; besides, it would be more interesting to explore how intelligent synergies can be implemented with technology that emerged long before 1984. Yet, as is resurfaced in Volume 2, there are spaces, still relatively empty, where an attempt is made to challenge the very concept of adding material to processes.

The early 21st Century is marked by the "hegemonic" presence of the digital transition with the technological and practical complements of additive manufacturing processes likely to affect Western society in a quick and profound way. "In the face of radical innovation markets, where the first arrivers can acquire decisive, dominant positions and make the passage of other markets and the economic actors in place disappear, keeping a distance and watching things happen can lead to considerable social and economic costs" [FRA 17]. To go beyond this already uncertain space and become involved in disruptive innovations implies taking risks, thus accepting potential failure, facing their possible negative consequences, and being capable of learning all the lessons this teaches. "If we do not proactively incorporate innovation, this will end up being imposed all the same, in an even more disruptive manner" [FRA 17]. In short, it may be useful to anticipate.

In roughly a century, the number of researchers in Europe has gone from a few thousand to a few million, and despite some disturbances, this trend is continuing. Research activities have been the subject of reassuring discourse on the researcher's independence, on the one hand, and on the other, of a certain programming of research with the aim of achieving goals: security (before the fall of the "Iron Curtain", for example) and economic developments (from mass production with ECSC projects to information and communication sciences and technologies) participating in different forms of competition from France and the European Union.

On this basis, the stereotypical image of the scientist, responsible for the truth and good, is still part of the idealized image, which often positions him/her very highly in relation to a social reality of which he/she only has an imperfect mastery. The will to achieve the best "research efficiency" has led to the promotion of rather mono-disciplinary processes that are easier to manage from "peers", referents of a discipline. On the one hand, in-depth scientific study is maintained by actors from the same field provided that the guarantee of excellence is defined and respected; on the other, for the State, it is easier to realize international comparisons discipline after discipline. Indeed, and this is necessary to remember, without really noticing it, we have gone from a limited worldwide scientific elite to mass research (with tens of thousands of scientific journals) which represents a characteristic that is not discussed by developed nations: research must indeed allow society to respond to the great challenges that loom today: employment, progress, security, global warming, health and quality of life, sustainable development, etc.

Without seeking to speak of two worlds exploring different paradigms, one of in-depth study, the other of responding to social demand (even its anticipation), for this aim would be too limited, rather we look at evolutions translated by a research program that takes account of the different and sometimes antagonistic imperatives (see Volumes 2 and 3). This situation actually shows, at least in part, that the researcher is an element of society who is not independent, even if forms of "grand isolation" have long protected him. But, in the European Charter for Researchers signed by France at the CNRS (National Center for Scientific Research) in 2005, a reminder is given that "Researchers should focus their research for the good of mankind and to expand the frontiers of scientific knowledge, while enjoying the freedom of thought and expression, and the freedom to identify methods by which problems are solved, according to recognised ethical principles and practices."

Without this having been noticed by most of the research actors financed by the State, even if the notion of good is not easily defined (in any case, it does not simply mean the absence of evil), this sentence is a reminder of the role of research centers as a social (or societal) actor, implying new approaches like functioning through interdisciplinary projects and strategic reflections negotiated by stakeholders, stemming from a new prospective work. Considering their importance for the development of citizens' quality of life, research associated with technology is an element that is really starting to be discussed. Indeed, it has participated in the "natural" evolution of things and technological progress has long allowed man to be free from a number of material constraints. In this framework, the rhythm of implementing research results has been greatly modified and complicated, thanks to a more and more frequent coming-and-going between "manufacturing" and research, and thanks to a hybridization of technologies as well as, to a lesser extent, modes of research action (added value from the ability to interact). To work in the new economy of knowledge with greater partnership, there is a need for better reflection on creativity, innovation, and the societal impact of scientific and technological activities. So then, in today's context of growing co-constructed and contractual research actions, must/can we break away from the "researcher's temptation of innocence", of the consoling illusion of "neutral" science, or of the simple transfer of responsibility to the deciders/financers? It will be understood that these are somewhat the stakes of the current evolutions/revolutions applicable to additive manufacturing, and particularly to its future.

With the concept of informed matter, there must be a possibility to modify the shape of objects in time (4D printing), to print living matter (bio-printing), etc. It is thus conceivable to come closer to life by flirting with its possible prolongation! This questioning, like 3D printing pushed to its limits (nanomanufacturing, micro-fluidics, electronics and robotics) associated with other domains, does not correspond to an economic market present today, but instead, if researchers, breaking with the traditions of incremental innovation, succeed (thanks to a bit of creativity and epistemic exploration), immense markets (relative to the "modest" market today amounting to 10 billion euros per year) should open up. The illustrative example of bio-printing which could correspond to a market worth several hundred billion euros per year is a great demonstration of the stakes linked to research concerning initial findings, presented in Volume 3.

If it is necessary to put some of this enthusiasm into perspective, the "classic" additive manufacturing technologies, which have already successfully demonstrated their numerous capacities of industrial development, offer application fields, some of which are very recent and possible, thanks, in particular, to disciplinary research, enabling existing manufacturing processes to be improved. This concentration on a clearly identified objective, process–material optimization, has limited more creative research leading to weaker programming and support for "divergent" researchers, whose numbers, for various reasons, are rather limited in the world of research. Nevertheless, these new applications called 4D printing, bio-printing, 5D printing, etc. result from more complex interdisciplinary activities that, if they succeed, could open markets, no longer in the 4-digit range (billions of euros around the world), but in all likelihood in the 5- or 6-digit range!

There are thus (at least) two types of challenge in additive manufacturing, one is the realization of 3D pieces which contribute a (the most) crucial input relative to the more traditional manufacturing techniques (prototyping, foundry, soldering, etc.) and the other is more prospective on openings in new fields with renewed approaches (and with the associated difficulties). For this reason, with the publisher (ISTE), there was a wish to present the 3D domain in three parts, one with validated scientific and technological bases (certainly with potential redundancies relative to other works on this subject) and the others based on a field of possibilities that offers new epistemological questions, terrible risk-taking, but considerable stakes.

In the first three volumes, it was actually about writing two open "scenarios" that were slowly constructed within a framework, but without a very strict preliminary plan, the scenarios in which the elements were to be introduced and discussed would be spread in an *a priori* graded manner. Each chapter has some degree of autonomy, which can be translated by possible repetitions (as few as possible, however), with a "history" that is progressively fed thanks to the in-depth reading of hundreds (thousands?) of publications, numerous times meandering through and delving into beautiful ideas and scientific meetings for debates, sometimes with success. The gray literature has been a vital source for what is happening in the field at times, which explains the numerous references to the websites in some chapters.

In Volumes 1 and 2, the reader is sensitively placed within the "summary table of disciplines" published in 1829 by Auguste Comte with an "institutional" organization for scientific disciplines, enabling incremental research and development in additive manufacturing. In Volume 3, the idea is to place the reader in a less programmable mode of functioning, with a recursive, systematic and self-organizing character of knowledge, a better willfulness in processes, which sets it apart from the first two, yet it is nevertheless complementary (because it is still constructed using what is known). However, a bit of naivety and/or ignorance may allow for progress to be made in the domain by tackling new paths of creation from a small amount of scientific and technical knowledge in a less "professional" manner, but full of enthusiasm towards a new world to be explored.

An intentional artifact (linked to the engineer and/or designer's work) may be considered a means of connecting an "internal" environment, the substance, the functioning, and the organization of the artifact itself and an external environment, the surroundings in which it is implemented. If the two environments are compatible, the artifact responds to the specifications. As underlined by H.A. Simon [SIM 04] in another framework, the knowledge of an artifact as an additive manufacturing machine "benefits from an advantage on the knowledge of nature, for it is based on valid, previous foundations whose ends will be perverted with a certain dose of new willingness to give projects intelligibility and openings on society." This notion can also be found within the facts in the three works, but with different interdisciplinary openings.

In Volume 3, for the researcher who studies the behaviors associated with the intrusion of temporal aspects and functionality in additive manufacturing, the systems operate for sufficiently long, entirely determined times. But, like "self-organization" phenomena, they can become very off-balance and sensitive to factors considered to be negligible near equilibrium. This is the intrinsic activity of the increasingly complex system, with an increasingly nonlinear behavior, which determines how it is possible to describe its relationship to the environment, which thus generates the type of intelligibility that will be pertinent to understand its possible stories. It is thus not only a matter of an applicative field with its constraints, but also of a theoretical domain to be approached and interrogated in order to resolve the end/means equation in a robust way so as to achieve it.

It will be understood that the epistemological foundations of the reflections in Volume 3 are based on the complexity paradigm, where interdisciplinarity is projected as one of the means of study. The disciplinary approach is too often divided, fragmentary and linear, hence a master idea aiming to know how to percolate through disciplinary borders so that the complexity paradigm can truly spread, notably because the recomposition of thought categories can no longer be based on borders and disciplinary subjects, but on boundary subjects based on the creative, the divergent, who, having no fear of recursiveness, hope to legitimately respond to the great risks society must face.

This change in delivering research for a more systematic approach does not hope to be the indicator of a field of scientific disciplines that, hoping to keep its power, loses its authority, even if current societal issues still cannot handle constructive forms of subordination well. It aims for a real, responsible integration of activities open towards society, bearers of meaning, allowing new research in additive manufacturing to be made to emerge as credible scientific evidence of movements that are materializing.

The evocation of different attractors of disruptive innovation in 3D manufacturing is the focus of Volume 3, in addition to its scientific and technical aspects. The author uses his experiences in this volume to recreate a bit of the history of new additive manufacturing processes, which could, in case of success, invade our daily lives in some years. It is in the spirit of creating a history, and interiorizing it by trying with the time and means available to re-establish them with a personal vision, with the risk of committing mistakes, of having failed with a promising idea. But this is the price to pay.

In the three volumes on the subject of additive manufacturing, it is shown that in relation to almost every problem, there is in fact a creative avant-garde with low inertia: this is carried out by groups of divergent researchers working in practice on the problem at hand. Then there are all the followers, who will structure the "paradigm" and engage it only in forms of conservatism authorizing research to improve processes or materials ("programmable" research). It will take years, even decades, for this paradigm to change positions – often with shoves (linked to the work of the creative by following information provided by the avant-garde). "Paths must be transformed into roads, the ground leveled, etc., so that the landscape will transform significantly until it becomes the main group's parking place" was written by L. Fleck in 1935. Could this context, in terms of research, be adapted to economic development? These characteristics of considering time, and its management, are the elements to be taken into consideration in a process of spatial and temporal transformation of matter that displays significant advantages.

Thus, beyond scientific aspects, indispensible techniques will be discussed to examine how the edifice of additive manufacturing was and is being built through its cultural filters and filters of understanding and interpretation. Anticipating the future of the field of additive manufacturing in the larger sense, to be in a position to prepare ourselves, is considered one of the keys for the long-term durability and competition of companies. This imperative to think of the future, to add to this divergent thought to create new devices for creating objects with the adapted material, devices that are functional, adaptive, "smart", etc., today seems even more significant considering the instability of the environment, the speed of evolution and the generalization of uncertainty. In such a context, research locations must be "offerers" of concepts, of their demonstration to anticipate the productive industrial future, not to mention the technological, economic and governance systems in which, on shorter and shorter reference times, companies evolve (undergoing nonlinear dynamics, splits and breaks). This mission is not only meant for individual researchers, but also for everything around them: research units, their administration and also (and above all else) the proactivity of economic milieus.

In terms of tomorrow and the future, can we not foresee new means of creating objects? At present, we have mastered synthesis, the way in which the objects are constructed. But we could also ask ourselves if it wouldn't be possible to develop systems in which we could give objects an intentionality, thus giving it the choice to look for itself for what changes it needs to make, thus moving onto self-organization with the selection of necessary elements that it would extract from a "bank" for the edification of the final object. This would go beyond the 4D printing that tackles the functional and evolutionary assembly of materials that should be able to come together to create an upgradeable object and that could be made easier through "programmable matter": "Programmable materials and objects that are themselves created would thus make assembly factors and heavy installation procedures superfluous… Robotization, the heart of progress in 20th Century productivity, could thus be integrated into the products themselves, with, as can be imagined, some ethical problems to be taken into consideration" [FRA 17]. Let us thus dream together of this future. The process attempted in these volumes therefore aims to try to question a present (it is impossible to know if this present will likely be able to achieve all its goals) and to determine the conceptual elements that could lead to an original future with access to new applicative niches by exploiting revisited paradigms.

Beyond the exhaustion of the reserves and consequences, it is also the way in which we understand scientific policy to be carried out by taking into consideration different world actors that should evolve to stimulate this nascent domain. In the reflection these books are aiming to create in its readers, it will likely be a matter of proposing changes to be undergone, which correspond to the conceptual displacement of the economy allowed by technology towards a new economy of creativity making a better effort to consider social, economic, organizational, geopolitical, even emerging environmental constraints. It is a form of "design thinking" that is thus to be considered. A reflection on the processes to help the integration of societal data, far from its disciplinary culture, would probably also be projectable (if only on the organizational aspects). In the end, it would be a matter of demonstration, through changes negotiated with the responsible authorities (some of whom are mute), leading to better exploration of the complexity, which can be done well, if not better, maybe with less equipment, but otherwise in a context of social and/or socio-economic demand that it would be advantageous to anticipate, if not follow. The paradigm shift would then take place thanks to scientific initiatives, which are marginal today, which remain aporetic in the paradigm in crisis, and which should be muted in a new scientific era, less framed, applied to 3D printing.

These three volumes can serve to think about the future in the domain that remains exciting for the author after more than 30 years since his 1984 patent, so that we can again find its place concerning its abilities of industrial creation and development in an ever more competitive environment. 3D, 4D, even 5D technologies constitute a path of promotion (among others that stem from the author's competence) of this desire for renewal.

NOTES.–

– For these three volumes, the search for the greatest possible number of specific or general visions concerning the subject of additive manufacturing, which can help the reader, has led to the presentation of the bibliography chapter by chapter and in alphabetical order. In fact, it was almost impossible to classify the bibliography through the numbering of entries.

– Some repetitions in the chapters of these three volumes may exist in an attempt to give them certain coherence and to provide them some degree of autonomy.

Jean-Claude ANDRÉ
Research Director at CNRS
August 2017

Bibliography

[AUD 15] AUDÉTAT M., *Sciences et technologies émergentes: pourquoi tant de promesses?*, Herrmann, Paris, 2015.

[BEI 12] BEIGBEDER C., *Puisque c'est impossible, faisons-le*, J.C. Lattès, Paris, 2012.

[EIN 07] EINSTEIN A., quoted by Debonneuil M., *L'espoir économique: vers la révolution du quaternaire*, Bourin, Paris, 2007.

[ELL 04] ELLUL J., *Le système technicien*, Le cherche midi, Paris, 2004.

[FAB 06] FABIANI J.L., "A quoi sert la notion de discipline?" in BOUTIER J., PASSERON J.C., REVEL J. (eds), *Qu'est-ce qu'une discipline?*, EHESS, Paris, 2006.

[FRA 17] FRANCE STRATÉGIE, "2017/2027 – Répondre à l'innovation disruptive – Actions critiques", available at: http://www.strategie.gouv.fr/publications/20172027-repondre-linnovation-disruptive-actions-critiques, 2017.

[GIA 90] GIARINI O., STAHEL W.R., *Les limites du certain: affronter les risques dans une nouvelle économie de service*, Presses Polytechniques et Universitaires Romandes, Lausanne, 1990.

[GUÉ 15] GUÉRIN M., *La croyance de A à Z; un des plus grands mystères de la philosophie*, Encre marine, Paris, 2015.

[LAT 07] LATOUR B., *L'espoir de Pandore; pour une version réaliste de l'activité scientifique*, La Découverte, Paris, 2007.

[PHI 12] PHILIPPE J., "L'innovation managériale, comment innover dans l'univers bancaire?", in EUROGROUP CONSULTING, L'art du management de L'innovation dans le service public, Eurogroup, Paris, 2012.

[ROS 01] ROSE J., *Profession quasi-chercheur*, L'Harmattan, Paris, 2001.

[SIM 04] SIMON H.A., *Les sciences de l'artificiel*, Folio-Essais, Paris, 2004.

[THO 83] THOM R., *Paraboles et catastrophes*, Champs Science, Paris, 1983.

[VAL 57] VALÉRY P., *Œuvres complètes*, La Pléiade, Paris, 1957.

Introduction

"The emerging potential for using 3D printing is illuminating some of the inefficiencies in mass production: the stockpiling of components and parts, the large amount of working capital required for such stockpiling, profligate waste of materials, and of course the expense of employing large numbers of people… It continues and accelerates a longer-term trend towards de-dematerialization of manufacturing goods – a trend that has already kept the total tonnage of global goods constant over the past half century, even as their value has increased more than three-fold". [GOR 13, TAY 14]

"You need have a lot of ideas to have good ones". ([PAU 14], quoted by [BRY 14])

"This could be called the Sirius paradox: it is futile to remain outside reality and to cultivate a haughty view certain of a good analysis. However, upon drawing too close to reality, it becomes even denser and seems to saturate the horizon of micromeanings that end up clouding our eyes and making any analysis impossible". [MEI 14]

"There are also artificial fogs that we take to be fortifications; they constitute the primary obstacles, for they set up a barrier at the very root of any process of alternative imagination". [VIV 12]

"All progress in speed is celebrated by the media as a success and adopted as such by the public, but experience shows that the more time we gain, the less we have. The faster we go, the more hassled we are". [ELL 88]

"Researchers' imaginations coupled with smart capitalism has barely any limits as soon as they start dreaming of an alleged world freeing us from the possible technological maximum of material and bodily constraints that chains us to this world. It will nevertheless be easy to show that this liberation dialectically turns again to servitude". [HUN 15]

"Progress, more than a staunch figure, is that agitated fever, this continuous transport that works on our entire society, which leaves it neither rest nor calm nor joy. The treatment necessary for this evil is unbeknownst to us". [TÖP 35]

"The novelty of modern times therefore has a very unique accent. It is not just about a consciousness of the time that distinguishes itself from the past and institutes a boundary between 'here' and 'today,' but rather a rupture that, by rejecting the authority of the models, takes away all of their value as good examples: all antecedence is declared over and obsolete". [BLU 99]

"There are thus no "induction rules" for a general application, thanks to which hypotheses or theories could mechanically be derived or inferred through empirical data. To go from data to theory, the creative work of the imagination is necessary". [HAM 14]

[Technology]: "The sum of all the tools and know-how used as an intermediary between Man and the real, namely to learn about the state of the real". [SER 97]

I.1. Introduction

The large-scale digitization of the world around us offers opportunities for innovation that create so many levels for businesses to grow and become more competitive [CIG 17]. "The event undoes the temporality required to invent another time, from which another world, another view is unveiled" [DUF 11]. 3D printing, which makes use of this possibility, has been around since approximately 1984, but it has only been understood and developed within society for approximately the past ten years. Its possible origins (at least for the author) can be found in reading comics by André Franquin (Spirou from the publisher "Petits Formats", 1975) and Hergé [TIN 14]. For various reasons, the editors consulted did not understand the interest of this transition between dream and realization, which will not allow the readers of these three volumes to connect creativity and innovation.

In 1984, in a common unit of the CNRS with the Lorraine University in Nancy (which has become one of the components of the LRGP-UMR 7274 CNRS-UL for the "Réactions et Génie des Procédés" (Reactions and Process Engineering), established at the ENSIC (a school in Chemical Engineering at the Lorraine University – UL)), the author of this work and two colleagues from the former CILAS (Alain Le Méhauté and Olivier de Witte) filed the first patent in additive manufacturing, their so-called laser stereolithography, three weeks before C. Hull from the USA [AND 84, HUL 84]. Figure I.1 presents the first page of this patent with the second by Hull for comparison.

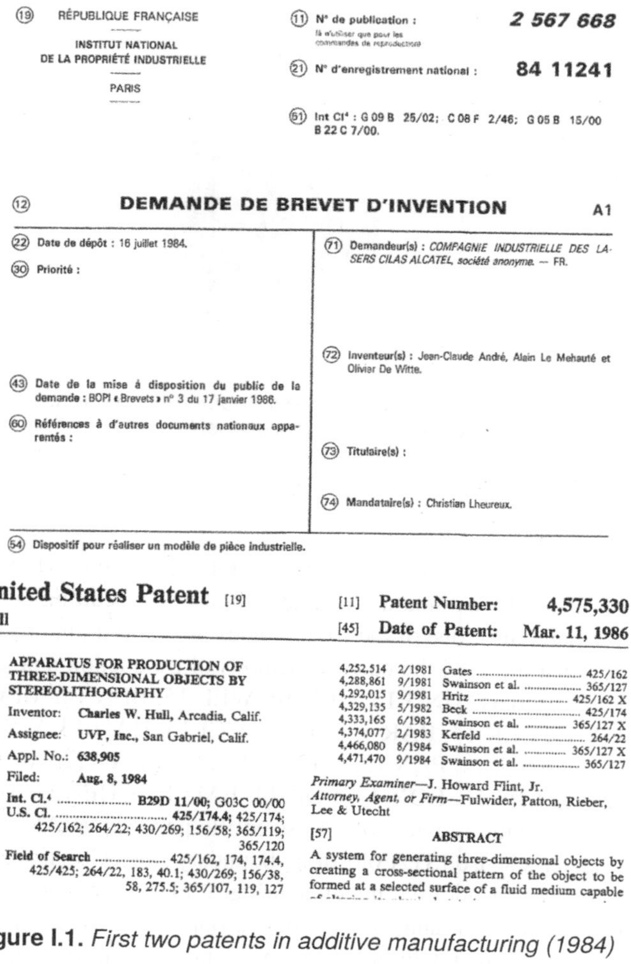

(19) RÉPUBLIQUE FRANÇAISE

INSTITUT NATIONAL DE LA PROPRIÉTÉ INDUSTRIELLE

PARIS

(11) N° de publication : **2 567 668**

(21) N° d'enregistrement national : **84 11241**

(51) Int Cl⁴ : G 09 B 25/02; C 08 F 2/46; G 05 B 15/00 B 22 C 7/00.

(12) **DEMANDE DE BREVET D'INVENTION** A1

(22) Date de dépôt : 16 juillet 1984.

(30) Priorité :

(43) Date de la mise à disposition du public de la demande : BOPI « Brevets » n° 3 du 17 janvier 1986.

(60) Références à d'autres documents nationaux apparentés :

(71) Demandeur(s) : *COMPAGNIE INDUSTRIELLE DES LASERS CILAS ALCATEL, société anonyme.* — FR.

(72) Inventeur(s) : Jean-Claude André, Alain Le Mehauté et Olivier De Witte.

(73) Titulaire(s) :

(74) Mandataire(s) : Christian Lheureux.

(54) Dispositif pour réaliser un modèle de pièce industrielle.

United States Patent [19]

Hull

[11] **Patent Number: 4,575,330**

[45] **Date of Patent: Mar. 11, 1986**

[54] **APPARATUS FOR PRODUCTION OF THREE-DIMENSIONAL OBJECTS BY STEREOLITHOGRAPHY**

[75] Inventor: **Charles W. Hull**, Arcadia, Calif.

[73] Assignee: **UVP, Inc.**, San Gabriel, Calif.

[21] Appl. No.: **638,905**

[22] Filed: **Aug. 8, 1984**

[51] **Int. Cl.⁴** **B29D 11/00**; G03C 00/00

[52] **U.S. Cl.** **425/174.4**; 425/174; 425/162; 264/22; 430/269; 156/58; 365/119; 365/120

[58] **Field of Search** 425/162, 174, 174.4, 425/425; 264/22, 183, 40.1; 430/269; 156/38, 58, 275.5; 365/107, 119, 127

4,252,514	2/1981	Gates	425/162
4,288,861	9/1981	Swainson et al.	365/127
4,292,015	9/1981	Hritz	425/162 X
4,329,135	5/1982	Beck	425/174
4,333,165	6/1982	Swainson et al.	365/127 X
4,374,077	2/1983	Kerfeld	264/22
4,466,080	8/1984	Swainson et al.	365/127 X
4,471,470	9/1984	Swainson et al.	365/127

Primary Examiner—J. Howard Flint, Jr.
Attorney, Agent, or Firm—Fulwider, Patton, Rieber, Lee & Utecht

[57] **ABSTRACT**

A system for generating three-dimensional objects by creating a cross-sectional pattern of the object to be formed at a selected surface of a fluid medium capable

Figure I.1. *First two patents in additive manufacturing (1984)*

"3D printing (see http://play.tojsiab.com/bUV0ckUtdFBBRWsz) actually encompasses a whole series of processes that share the fabrication of objects through the deposit of successive, extremely fine layers of material [or more simply additions like melted wires], which are gradually solidified by an energy source (laser, for example) […]." It is defined by AFNOR (The French National Organization for Normalization) [AFN 11] as "the set of processes allowing the layer-by-layer fabrication, through the addition of matter, of a physical object from a digital object." It allows several specific axes to be explored, such as those presented in Figure I.2 from Conner *et al.* [CON 14], the axes that will be defined and explained in greater detail in the current document.

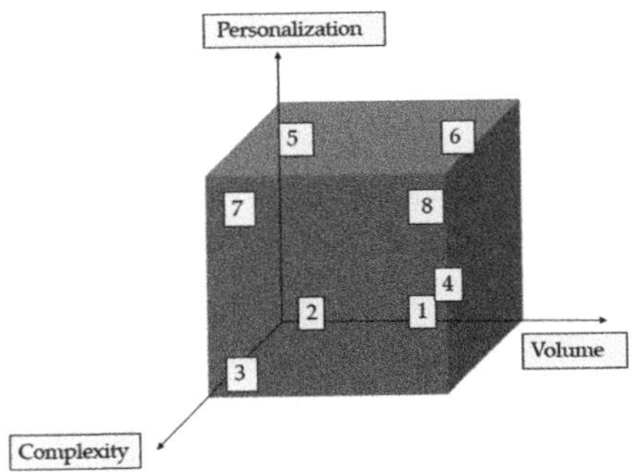

Figure I.2. *Intervention axes of additive manufacturing (1: Mass production; 2: Manufacturing of few parts; 3: Complexity pertinence; 4: Mass production complexity; 5: Art; Individual Personalization; 6: Mass personalization; 7: Handmade and Fab-Lab products; 8: Total manufacturing freedom)*

"3D is thus considered – with mobile Internet, the Internet of Things (IoT), cloud computing, big data, the automation of intellectual trades, cutting-edge robotics, or even advanced materials – one of the technologies linked to digital technology likely to profoundly transform […] the current methods of production and, as a result, the current economic models" [ING 15, GEB 14, WEL 15, BLA 14, MER 13, TEC 15, MOR 14, ACA 15, BER 14, BAR 15, LIP 14, CCI 15, HAU 14, WIN 12, HOS 14, GIB 15, SCU 15b, HLA 14, PAN 13, FOR 15, KAZ 14, RAT 12, ACA 16, KOH 16, WHI 13, KEA 15, UPS 16, ROY 13, FOR 14, FAD 15, CHE 15, COT 14, SIR 16, HAG 15, GAO 15 (with a review of 348 scientific papers)], [BAU 16], etc. It is in Star and Griesemer's sense [STA 08] of a "boundary-object" that we talk about 3D printing, 3D copying or additive manufacturing. ("Plastic enough to be adapted to the local needs and constraints of various groups that use it, all while being sufficiently robust to maintain a common identity from one site to another" [ING 15].)

The current document does not, however, cover everything within this extremely broad field. It attempts to limit itself, above all, to the components' "processes" and "materials", their reactions, but these being in connection to the applications, expansions beyond this "border" are inevitable (and slightly desirable to the author). We must not forget that Gartner placed additive manufacturing among the top ten strategic priorities in developed countries for 2016 [ALE 15]. Innovations in the field should enable industrial developments in space, transport, energy, military materials and medicine. According to this author, an increase in more than 60% per year for businesses is expected until 2019. There are thus questions we will deal with in this work to discuss differences between promises and reality.

I.2. Historical reminder of 2D1/2 processes [AND 15, AND 16, LAV 15]

Although a great deal of progress has been made in the field of computer science, aspects of the actuator and material formatting are still a bit inconsistent at producing prototype parts, personalized shapes with artistic or biomedical applications in mind, etc. There is thus room for development for processes producing active parts in a considerable worldwide market. Other less futuristic applications are already being implemented using rustic processes to produce passive parts in very diverse fields.

These realizations of objects without a machine tool (the one that removes matter) are developing more and more from so-called "additive manufacturing" processes. The process using light, developed in Nancy in the 1980s, was based on knowing the coordinates of the object to be created, saved in a computer piloting galvanometric mirrors and sequencing the movement of the light to transform a liquid oligomer (a resin) into a solid through the desired polymerization of a layer, "voxel" by "voxel" (see Figure I.3).

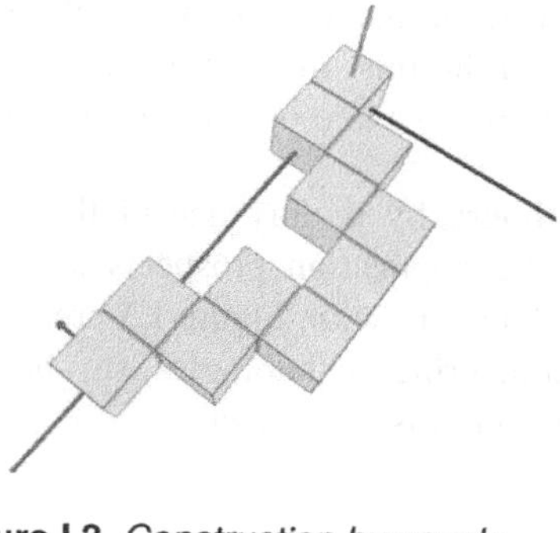

Figure I.3. *Construction by voxels*

The addition of a second layer, then a third, etc. thus allowed a prototype part to be created, just as a mason does when building a wall. This basis is still a foundational concept for additive manufacturing processes (see Figure I.4, from [AND 94]).

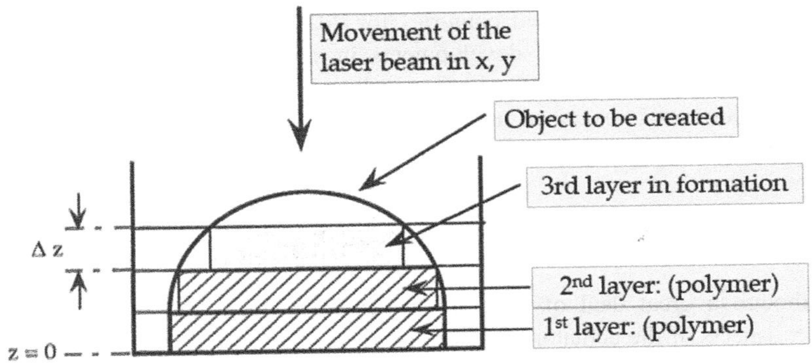

Figure I.4. *Historic schema of the principle of creating an object layer by layer*

Different problems associated with this process have gradually been solved in the refinement of the process; they will be mentioned with others connected to the description of different additive manufacturing technologies. This "3D copying" process using a quickly broadcast digital file was immediately attractive for socio-economic environments wishing to create less costly, complex or prototype pieces impossible to be otherwise produced. However, few managers from CILAS and CNRS believed in this when the operation started to take shape, which means that the know-how that could have been acquired, at least in France, is practically no longer present on the national territory. The example of an innovation from research (which represents a significant market today) will allow the problems of technology transfer to be mentioned in this first volume, and it may also allow for the consideration of propositions to avoid or at least limit them.

At present, stereolithography is only one of the seven additive manufacturing technologies with specific application prospects and limitations that will also be described in this book. To begin with, Figure I.5 presents some differences between various additive manufacturing technologies (to be compared with subtractive manufacturing, where material is taken away).

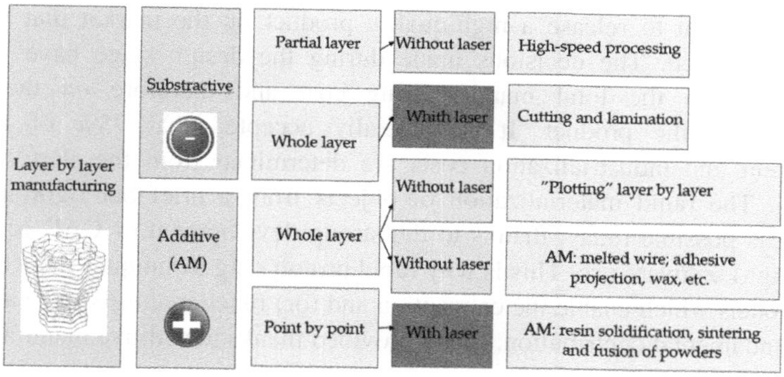

Figure I.5. *Different manufacturing technologies (according to [AFP 15]); AM: Additive Manufacturing*

This rather recent manufacturing technology (approximately 30 years old) is defined in Figure I.6 [DIR 14]. Currently, it has a definition according to the French norm NF-E 67-001: "All of the process allowing a physical object to be manufactured from a digital object, layer by layer, through the addition of matter."

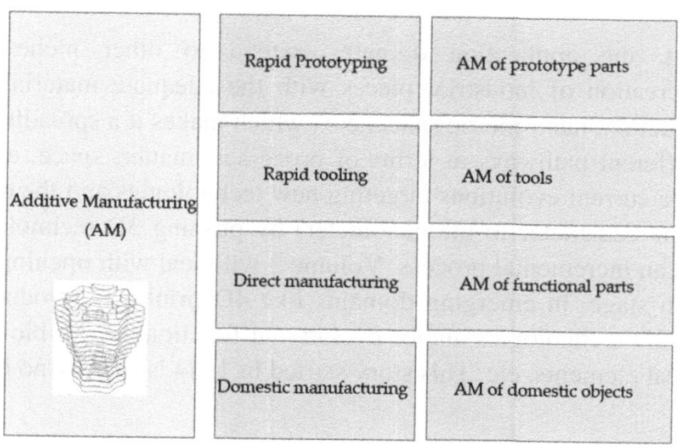

Figure I.6. *Principles of additive manufacturing*

If there is one thing engineers are well aware of, it is the fact that they will be led to modify or revisit the conception of their prototype during production. Additive manufacturing gives full freedom to rethink a part and innovate right away, with no delay or major financial consequences. This freedom presents numerous advantages: reduction of production costs, improvement of product quality, diversification of models, and ultimately, an increase in revenue [TRA 15].

It is important to release a high-quality product on the market that has been perfectly approved. The decisions made during the design stage have profound repercussions on the total manufacturing cost and therefore on the overall profitability of the product. It is generally accepted that 75% of the total development and industrialization costs are determined from the start of design [AFP 15]. The rapid materialization of objects from a brief but rigorous digital definition is possible today, thanks to the strong development of CAD (computer-aided design) software use. This is why rapid prototyping techniques, which quickly lead to models which enable the conception and (or) functional tests to be approved, make a fundamental contribution, easily provided thanks to additive manufacturing.

With an unbroken digital chain, rapid prototyping, through related methods and means, allows all of the actors affiliated with the product to:

– test alternative solutions as early as possible;

– approve technical and (or) technological solutions reserved for the future product: appearance, ergonomics, processes, etc.;

– work in a parallel and coordinated way in a context of collective and simultaneous engineering to optimize the product.

At present, the application domains extend to other niches: art, home applications, creation of industrial pieces with the adequate material, prosthetics, repairs, construction, nano-metric space, etc., which makes it a spreading technology that covers different pathways in terms of processes, matter, space, etc. Volume 2 will present the current evolutions targeting new technologies and the exploration of space (from the decameter to the nanometer) by pushing 3D technologies to their limits through an incremental process. Volume 3 will deal with openings, essentially in the research stage, in emerging domains like 4D printing (introduction of time elements into 3D technologies and/or evolutional functionalities), bio-printing with living biological elements, etc. This story started in 1984 is thus by no means over.

I.3. Framing the application market

Over the last two centuries, the needs and expectations of the manufactured parts on the market have greatly evolved with a considerable increase in manufactured objects and radical evolutions in the methods of fabrication [RUF 14]. Industrial revolutions and world wars are the trigger elements for the progressive passage from artisanal production to mass production. The former depends on a qualified workforce using general purpose machinery to make the product demanded by the client unit-by-unit with the materials available. Mass production, on the contrary, responds to a demand greater than the supply. It is based on the realization of a restricted range of products, manufactured in high volumes, using dedicated means of production. Henry Ford and Louis Renault were, in the United States of America and France respectively, the first to apply this kind of production to the automobile industry [MAR 99].

The economic crisis period following the "Trente Glorieuses" (the 30-year post-war boom) then saw the progressive emergence of mass personalization that responds to the context of a supply now superior to the demand [PIN 99]. To support the market, it became necessary to propose products likely to respond better to the client's different expectations. "This multiplication of models will be made by moving towards the development of 'personalized' products allowing the client to select the combination that he/she likes the best. Such an approach allows businesses to maintain mass production, no longer at the level of products, but at that of their components. However, it requires reconsideration of the way the product industrialization cycle is approached" [RUF 14].

Since the 2000s, this kind of production has been evolving with the development of computer tools and new information and communication technologies, which reduces not only the delivery time for objects but also the possibility of realizing objects, i.e. customized objects. "In the current context of market saturation associated with the rarefaction of raw materials, these advances push the concept of mass personalization to its climax. In other words, to perfectly personalized production" [RUF 14, HU 11]. Figure I.7 shows these different changes in the ways of producing material consumer goods.

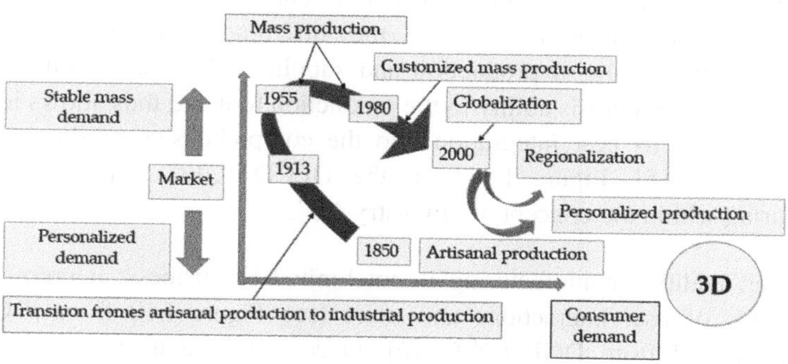

Figure I.7. *Evolution in the methods of producing manufactured products*

The current market is worth several billion euros per year, for some, even in the range of tens of billions, based on 2D1/2 processes where the construction of the object takes place voxel by voxel (the elementary volume), with an imposed spatial nonlinearity (the layer), then, more generally, through the addition of matter whose nature changes in the space. This rapid technological evolution since 1984, which will be discussed in this first volume, has been more and more often a reference for new innovation processes [HAR 15, GEB 14, KIE 15, BER 12, DEL 15, PRI 14, ROW 12, GAR 12, MER 13, ELW 14, VAE 13, MEL 12, PIE 13, WEN 08, BOU 14, 3D 15, MAR 15, D'AV 15, DAR 15, ZHA 15a, GEI 17], etc. The interest of additive manufacturing is that it allows the limitations of traditional manufacturing to be overcome so as to do better.

After the steam engine, electricity and automation, the "robolution" (revolution brought about by robotization) including additive manufacturing is the fourth industrial revolution (see [FRA 75, MOU 15]). The concept of Industry 4.0 corresponds to a new way to organize the means of production: the goal is to implement the so-called "smart" factories capable of greater adaptability in production and capable of more efficiently allocating resources, thereby opening the way to a new industrial revolution [VAL 16]. After automation, this new revolution would be founded on the smart factory, characterized by an interconnection not only between the machines and systems within the production sites, but also between them and the exterior (clients, partners and other production sites). Industry 4.0 corresponds in some way to the digitization of the factory (see Table I.1 inspired by [SCH 15]) which represents the temporal evolution of ruptures (Kondratieff cycle) in industry and Figure I.8 from Pau [PAU 14] which explains the displacement of the different forms of activities. Through reference to the Internet of Objects, cyber-physical systems and additive manufacturing, i.e. the virtual networks serving to control physical objects, the smart factory is characterized by continuous and instantaneous communication between the different tools and workstations integrated into the chains of production and supplies. The use of communicating sensors adds a self-diagnosis ability to the production tool and thus allows its remote control as well as its best integration into the comprehensive productive system [WIK 15i, HIN 15]. Figure I.9 from the OECD [OEC 17] places additive manufacturing within the concept of "Industry 4.0".

However, additive manufacturing is not limited to business; it presents much broader fields of use, interactions and users (see Figure I.10 from [DGE 16]). Table I.2, created from DGE [DEG 16], places additive manufacturing among France's national priorities, with prospective features summarized in Table I.3. From a historic point of view, Table I.4, taken from Rayna and Striukova [RAY 16], shows the stages; the transitions between the different developments in the technology from the creation of the concept to industrial and home applications. It underscores rather long temporalities in the order of human generations (which was not the case of the transistor, but which fits rather well into the classical dynamics of innovation). Table I.5, from the Office of the Deputy Assistant Secretary of the Army [OFF 16] shows the perception that Americans have of this recent technology's position.

Typicality	Period	Technological domains
1.0	> 1780	Steam engine, charcoal and beginning of industrialization
2.0	> 1870	Division of labor, electricity, automobile and mass production
3.0	> 1970	Electronics, automation, information and communication technologies, Internet
4.0	> 2000?	Internet of Things, cyber-physical systems, etc.

Table I.1. *Kondratieff cycle*

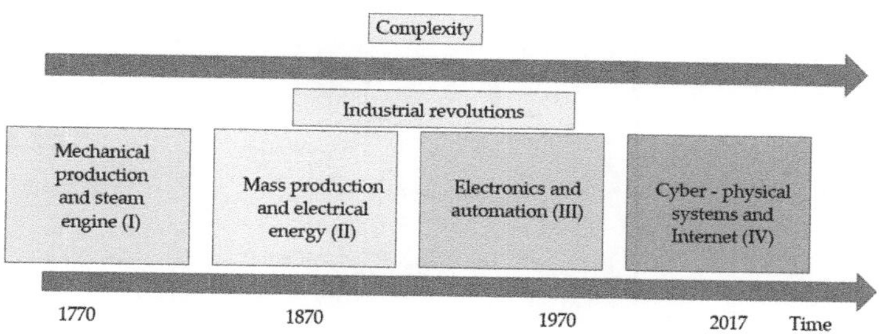

Figure I.8. *Technological changes: from industry 1.0 to 4.0*

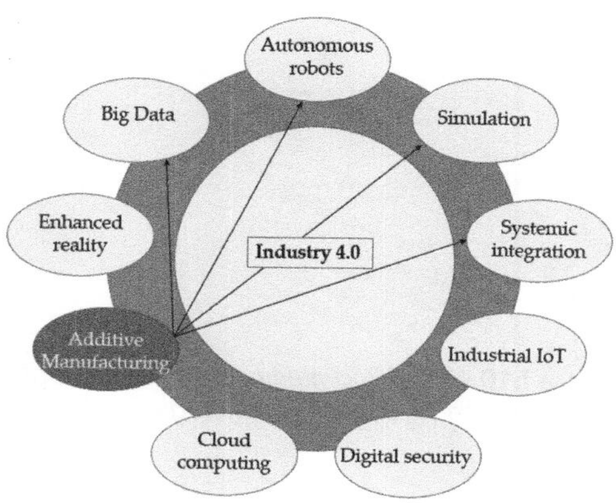

Figure I.9. *Main components of "Industry 4.0" (IoT – Internet of Things)*

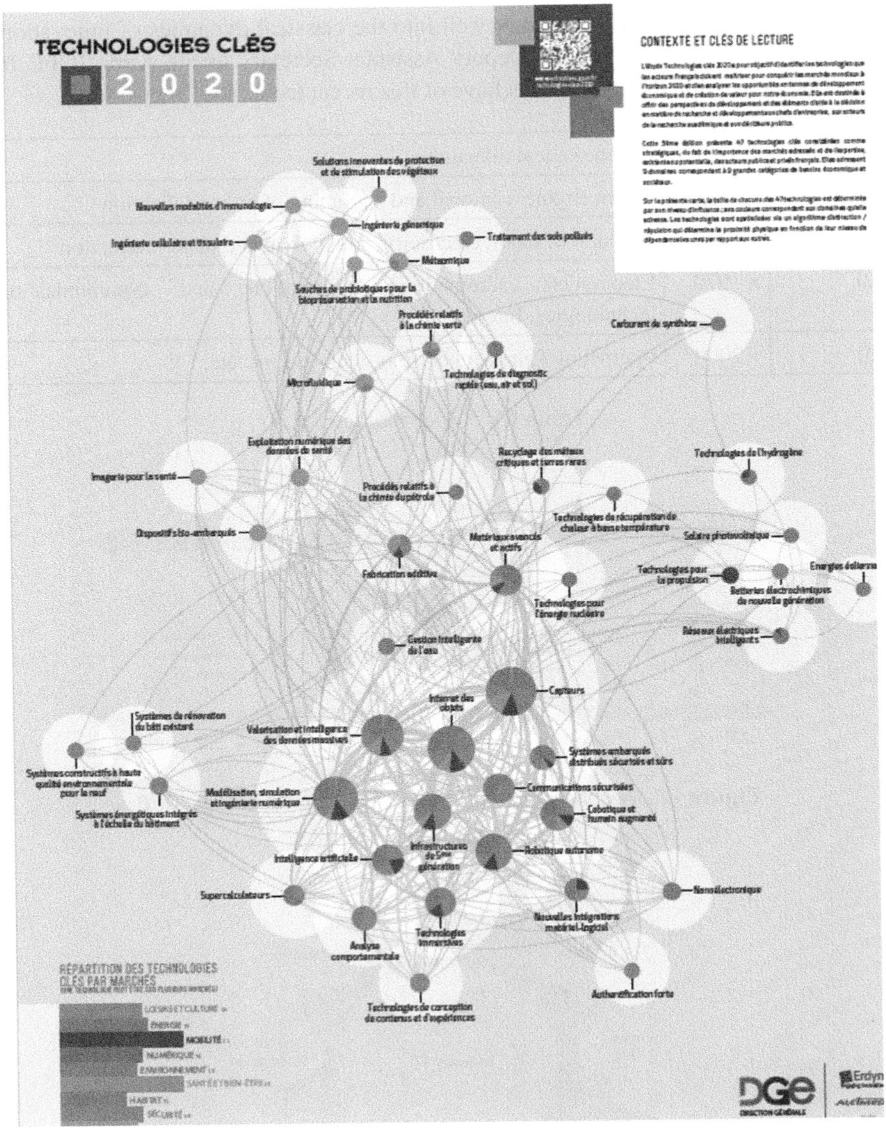

Figure I.10. *2020 Priorities of Key Technologies in France (French: "Technologies clés")*

However, as the SWOT analysis published by the same organization shows, even if France was a pioneer in the field, the current situation is relatively concerning, as indicated by the synthesis of this analysis, presented in Table I.2.

Strength	Weaknesses
High demand in space and aeronautics Demand in the luxury sector Good academic competences Outsourcing leaders present	Modest number of French manufacturers Few raw material producers
Opportunities	Threats
Large number of applicable fields Democratization of technology Return of manufacturing to national soil	Buy-back of French start-ups by foreign groups Counterfeiting

Table I.2. *SWOT analysis corresponding to additive manufacturing*

Subjects	Comments
Improvement of production speed, object size and reliability	Domain where creativity must be expressed in the field of manufacturing processes to make production faster: variable voxel dimension, new energy localization, etc. Exploration of scales: from decameters to nanometers Coupling of spatial resolution and production velocity Object again, respect for instructions concerning spatial parameters, robustness, etc.
Materials	Numerous materials are already used (polymers, resins, metals, metal alloys, ceramics, etc.). However, the range of materials should be widened to respect industrial demand Coupling of materials and processes Multi-material 3D printing Smart and adaptive materials
Bio-printing	Revolution in medicine aiming to print biological tissues
New niches	Micro-fluidics, sensors and actuators, 3D electronics, bio-printing, space, etc.

Table I.3. *Some prospective elements concerning additive manufacturing*

Adoption	Start	Design	Instrumentation	Manufacturing	Distribution
Rapid prototyping	Early 1990				
"Rapid" tooling	Late 1990				
Direct manufacturing	Late 2000				
Home manufacturing	Early 2010				

Table I.4. *Stages of the adoption of 3D printing technologies*

Concerning the prospective aspects, the recent report from the PwC (2016) on industry 4.0 sheds light on the fact that the respondents believe that in 2025, automobiles will start to be manufactured using 3D printing (84.4%) and that 5% of consumer products will be produced with this technique (81.1%). Anything seems to be imaginable. Another survey conducted by Boston Consulting Group (2016) presents the primary advantages associated with industry 4.0 (including additive manufacturing) in Figure I.11. A number of attractive features of additive manufacturing can be seen in this figure.

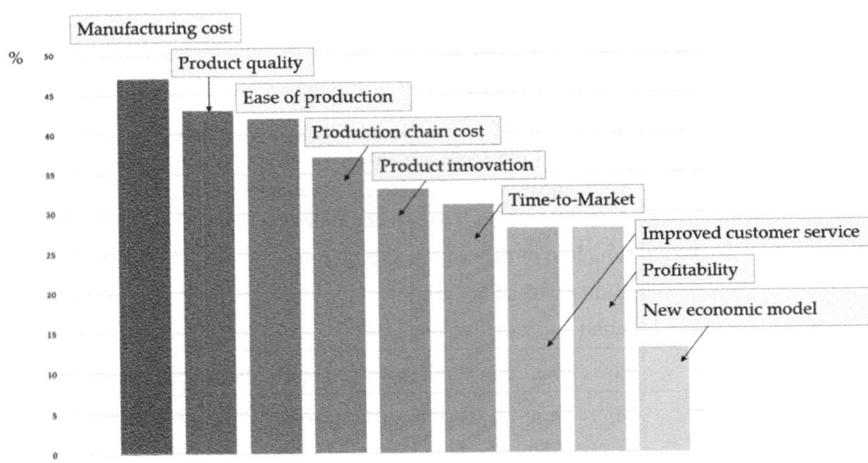

Figure I.11. *Results of Boston Consulting Group's survey on the attractiveness of devices associated with industry 4.0 (sum of first-, second- and third-priority responses for those surveyed)*

Framework	Position of additive manufacturing	Comments
Technological competition	2nd/24	Located between robotics and autonomous systems and social autonomization; the bio-printing component included in medical advances is in sixth place
Importance of the technology in society's activities	8th/24	The bio-printing component included in medical advances is in third place; the social autonomization (first) and robotics and autonomous systems and social systems (second) aspects are voted on; this result should be connected to the still modest revenue from 3D technologies. It is worth mentioning that in this survey, education is in the last place.

Table I.5. *Perceived impact of additive manufacturing according to the Office of the Deputy Assistant Secretary of the Army [OFF 16]*

New technologies are being developed with five principal concerns presented and involved in more prospective aspects:

– showing that additive manufacturing belongs to the industrial process, enabling the "time-to-market" to be reduced;

– finding new technological outlets: from toys for children (Mattel) to printing biological organs;

– innovating to maintain the dynamics and to last: incremental approach to new processes and especially materials and multi-materials; breakaway approach in revisiting the concept (example of 4D printing; see Volume 3);

– performing concept tests and communicating; the initial demonstrators are developed in light structures with a modest investment, but strong creativity;

– manufacturing final products in the most limited number of stages possible.

These different dimensions are dealt with in the present work with, in some cases, extensions in terms of modeling and research.

Beyond R&D aspects that are not the target of Sculpteo's survey [SCU 16a], by taking one of the examples presented in their report, the authors give an indication of the "great economic masses" for some of these pillars; as the application target does not cause any considerable changes in the results, only one example is presented in Figure I.12 and it is considered to be representative of the others.

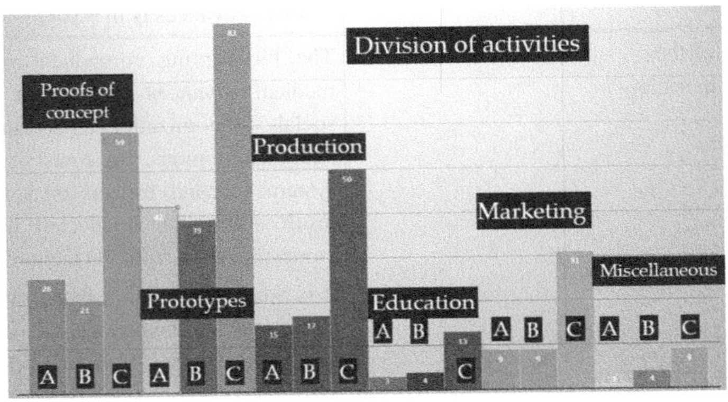

Figure I.12. *Division for the "High Tech" rubric of 3D activities according to Sculpteo [SCU 16a] – (A: High-Tech Activities; B: Totals; C: "Expert user" activities). For a color version of the figure, see www.iste.co.uk/andre/printing1.zip*

I.4. A transition to "acceptance"

The McKinsey Global Institute [MCK 17] introduces the general principle of moving from the concept to industrialization, presented in Figure I.13 in a rather linear approach (which suffices to pursue the subject). First, this figure gives the first priority to a technical potentiality before going through classical stages of prototyping and economic analyses to tackle the market. Its adoption therefore raises classical questions of attractiveness (or even acceptance), fields that fall under psychological aspects of perception in a technicist culture. In the same report, the importance of automation aspects in the large fields of activity is the object of the authors' attention, insofar as automation can be a determining element in economic development. Figure I.14 considers two domains, that of "management" and that of "manufacturing". This comparison highlights the very large importance of automation for the latter (average value of 60% according to the authors' criteria, as opposed to just 30% for management). Additive manufacturing, with its relationship between fabrication and information technologies, fits into this framework and then responds to the manufacturing industries' principles of development.

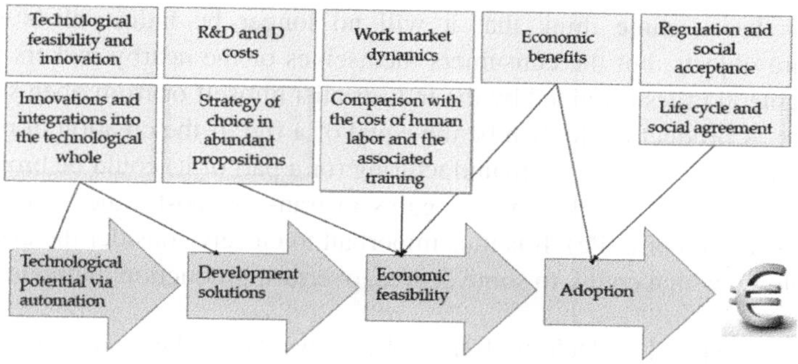

Figure I.13. *Factors influencing technological innovation*

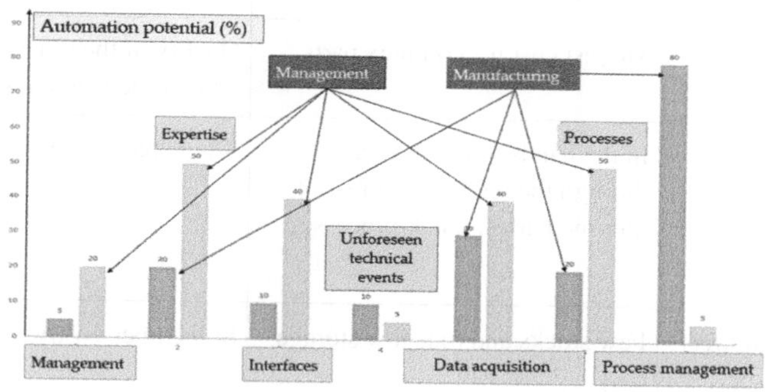

Figure I.14. *Importance of "Automation" aspects in the "Management" and "Manufacturing" branches*

3D can challenge the concept of a logistic chain and it could have numerous consequences: lean manufacturing, reduced storage, transport, energies and associated costs (as long as objects can be produced with the appropriate material). According to Branche [BRA 15], "the entirety of this chain even becomes obsolete. The owners of 3D printers need only model their part or retrieve a 3D file to manufacture their own object, thereby short-circuiting all the usual intermediaries. It is in this way that the 'do it yourself' (DIY) philosophy, open sourcing, i.e. sharing information, connected to the generalization of digital technologies including additive manufacturing, has given rise to the Makers Movement and Fab-Labs, or 3D manufacturing workshops."

On this basis, some think that it will no longer be industrialists who will manufacture objects, but the consumers themselves or the nearby makers thanks to downloadable 3D files provided by the industrialist himself or from open sources on the Internet. A product could then be the equal of a file. If the creation of parts from "good" material is possible, the manufacturing (or a part of it) could be brought back to national/regional soil, allowing for gains in transport costs (then limited to the digital transmission of a file). It is thus important to closely consider the evolution of 3D technology, which could, to some degree, overturn production methods.

Table I.6, inspired by Deloitte [DEL 14], shows the applications that have come from research laboratories today.

	Advantages	Difficulties
Rapid prototyping	Modest cost for complex parts	Quality of the objects created with home machines
Supply and innovation times	In the hands of the designer through the reduced number of intermediaries; quick processes	Single parts for the essential
Reduction of indirect costs	Few stocks and thus low storage costs	Sometimes costly materials
Mass personalization	Possibility for single-use or some pieces adapted to a specific use	No collective manufacturing
Mass production	Not yet possible	Not yet possible
Materials	Large range	Not so many multi-materials
Environmental aspects	Less waste	Not many HSE studies (Hygiene, Security and Environment)

Table I.6. *Advantages and difficulties of using additive manufacturing processes*

Furthermore, a technology cannot be developed against its users, hence a need to appreciate the desirability of a process based on factual aspects and impressions. By acceptance, we consider here the potential attraction of the novelty for citizens. In the field of convergence exploration, the knowledge of risks (regardless of their their origins and consequences on man and the environment) is not stabilized. From a factual point of view, the decisions that should be based on the use of the precautionary principle can be translated by:

– a temporary (or definitive) restriction;

– a restriction on use;

– targeted information;

– specific further research on the risks/dangers for researchers and the general population;

– "simple" monitoring;

– an irresponsible expectation.

"The existence of unknown and uncontrolled risks, implicit in the argument of 'absence of zero risk' is not constitutive of reality, but inherent to a deficit not only of knowledge, but also of coherent concepts in relation to the technological ambitions pursued" [POU 06]. In a situation of uncertainty, several criteria for the realization and impact of the research program have already been foreseen by Chevassus-au-Louis [CHE 07]. They have been recently completed by Porter and [YOU 09, KIM 13, KIM 01], and, very recently, by Tchiehe and Gauthier [TCH 17]. On the basis of these works, mapping (see Figure I.15) has been proposed to several students of the LRGP (the author's lab) and the ENSIC (the engineering school where the lab is located) on the following 12 criteria:

Framework of a technology's attractiveness

– attractiveness/self-image;

– ease of use;

– cognitive aspects to be implemented (aid for intellectual activity, games, etc.);

– cost;

– efficiency ("price/quality" ratio);

– positive perception of the technology.

Risk "consistency"

– risk observability (short time);

– analogy with other risks;

– non-reducibility of the risk (possibility or impossibility of taking preventive measures);

– severity (gravity potential; concerned population);

– plausibility (potential knowledge of the effects);

– irreversibility of the effects on man (e.g. cancers).

By attributing a value to each of these considerations (from 1 to 5), it is thereby possible to situate a "center of gravity" for the possible risk/the attractiveness of a technology on this dodecagon, enabling us to attempt to understand issues that can affect society. For this analysis, which could lead to the appropriate measures, there is a possibility of implementing adapted deliberation processes (which, however, go beyond the subject of this work) internally (possibly with outside support).

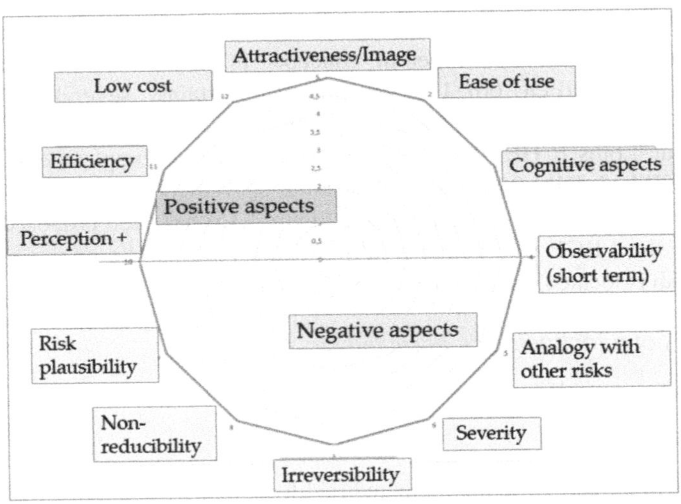

Figure I.15. *Mapping the 12 criteria connected to the attractiveness of a technology and the associated risks*

After having defined what additive manufacturing represents (which they already knew) in less than an hour, it was possible to create the map presented in Figure I.16 demonstrating a very strong attractiveness of additive manufacturing devices for highly educated people who had never created objects through 3D printing methods. They were also asked about other technologies: the NBIC convergence (Nano, Bio, Info and Cognition) and ICT (Information and Communication Technologies); these results will be presented in Chapter 2.

However, according to Castelfranchi [CAS 13], quoted by Peyré – Tartaruga *et al.* [TAR 16], the deployment of a technology should partially be separated from its desirability and the understanding of its application potential for the public; indeed, this can demonstrate a wait-and-see attitude *vis-à-vis* its development.

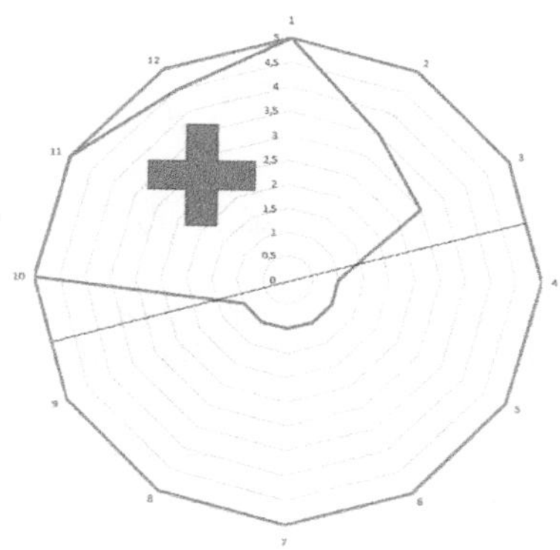

Figure I.16. *Perception of the high attractiveness of 3D technologies for students*

I.5. Societal impact of additive manufacturing

In terms of organization in enterprise, 3D manufacturing processes change the deal, as shown in Figure I.17 from Cotteleer, Trouton, and Dobner [COT 16]. It is a matter of integrating the management of the flow of digital data into enterprise for different missions (and making them compatible with one another): object design, its adaptation (reverse engineering), modeling of the part's manufacturing, manufacturing plan, quality assurance, etc. These "intellectual" operations must serve the effective manufacturing of the object, its later processing and functionality monitoring.

According to another report from Deloitte University Press [DEL 16], the "exponential" development of digital technologies provides additive manufacturing with specific opportunities by simplifying the interactions between the designer/the user and the digital system. This evolution is presented in principle in Table I.7. However, one question that will not be dealt with in any of the chapters of Volume 1 concerns the effective mastery of designing a 3D object: is it extracted from a databank or is it totally conceived? The position of the cursor between these two extremes is important in a process of individual autonomization of a technology.

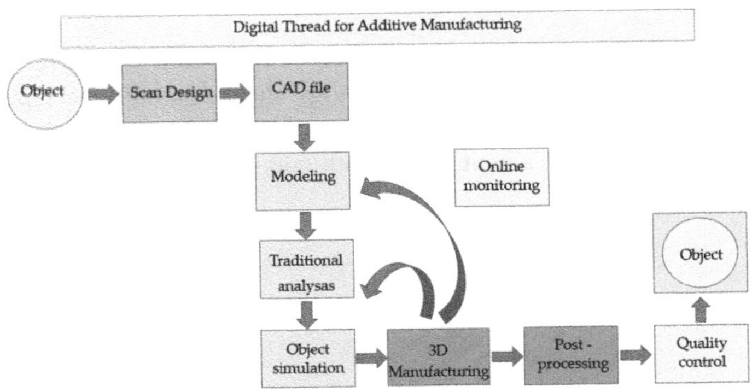

Figure I.17. *The 3D enterprise, according to Deloitte: from the designed object to its creation (according to [COT 16])*

Nature of the man–machine interface	Comments
Origin (3 relational steps)	Man–hardware–software–machine interface requiring programming skills and going through an alphanumeric keyboard (previously going through the step of punch cards or teletype!)
Progress 1 (2 steps)	Hidden hardware; word recognition; touch screens; intuitive command
Progress 2 (1 step)	Augmented and/or virtual reality; integration of man–system relations
Progress 3 (total integration)	Intuitive interactions; movement recognition; empathic "programming"

Table I.7. *Evolution of man–machine interfaces*

The integration of additive manufacturing technologies in the ecosystem of innovation has recently been studied by PwC [PWC 16a, PWC 16b]; it is the object of a useful complement to the framing presented in Table I.6 (see also Table I.8), which indeed places 3D printing at the center of current and future devices as a recognized manufacturing method.

Period	Framework	Comments
> 1990	Emergence of digitization	Ubiquitous digital products: computers, digital telephones **Conceptualization and proofs of concept of additive manufacturing; first enterprises involved in 3D machining** Web/Internet platforms Process automation Digital industrial standardization Placing "business models" online
2000–2015	"Disruptive" digital technologies	Cell phones Digitized and automated industrial processes (including **additive manufacturing**) Digital functions Users interfaces; shared access Cloud computing Sensors, Internet of Things, digital video
2016–2020	Industry 4.0	Horizontal and digital integrations along the value chain New "business-to-business" enterprises Real-time data analysis in relation to users Reinforcement of the power of computers in processes **Digital manufacturing (including industrial-scale additive manufacturing)**
> 2020	Digital ecosystem	Integrated and flexible networks along the value chain **Digital manufacturing (including bio-printing)** Virtualized processes Digitized supplies Virtual user interfaces "Complete" spread of the Internet of Things Smart algorithms (aided deciding, adaptive processes, robotics, etc.

Table I.8. *Position of additive manufacturing in industrial devices*

Independently of this aspect, does 3D printing condemn giant factories and their production chains in the long term? The managers of Local Motors are convinced of this [GÉN 16]. This car manufacturer's development relies on less than 4000 m^2 microfactories installed near consumers. In September 2014, Local Motors printed a prototype of the car christened Strati in 44 hours for the body and frame. A traditional car is made up of 20,000 different parts. The manufacturer was to commercialize its first mass-produced vehicles in 2016.

The example of the Strati and its industrial model based on 3D printing can overturn entire sections of industry. It is gaining ground for various reasons:

– Political: in 2013, Barack Obama maintained that the mastery of this technology was strategic and that it would allow the United States of America to become reindustrialized;

– Technological: single-stage manufacturing of very complex parts, personalization of the object and reduced consumption of materials (in traditional processing, 100 kg of raw materials are needed to produce a 15-kg part, while additive manufacturing would only need 15 kg);

– Re-conception: manufacturing a complex assembly into a single part manifests itself through a gain in weight. This is significant in a domain where the consumption of energy is an important parameter: satellites, avionics, automobile, etc.

– Modest maintenance: no more need to store spare parts because they could be created through online additive manufacturing; only the manufacturing materials would need to be stored;

– Re-localized production: to the users' neighborhood (from a 3D machine in a workspace to one owned by a washing machine repair man!), or in microfactories (example of Local Motors), or with the "makers" or part providers (accessible 3D files and on-site production);

– Logistical simplification: only the materials used in additive manufacturing must be transported;

– Personalization: case of prostheses, art, jewelry, clothing, shoes, etc. that use a software database that can be adapted to the client's demand;

– Reduction of transport costs: the extension of 3D printing and the relocalization of certain products closer to consumers could reduce world air cargo traffic and maritime cargo activity;

– Legal problems: a legal battle will ensue, with an industrial property right that is mishandled. The democratization of 3D printers could allow millions of individuals, and also malicious competitors, to clone all or some commercial products at a low cost;

– But at the same time, for Obaton [OBA 16, OBA 15], the die has not yet been cast: "Thus, we confirm, for example, that at the level of the powder used, this degrades over time and during reuse of the unused material. This has a direct impact on the properties of the piece created. It is therefore important to correctly measure these elements and to understand how the properties of the powder evolve and impact the creation of the pieces produced."

These different elements with a societal nature will be the subject of a chapter questioning the consideration of this technology in the panoply of manufacturing devices with its advantages and limitations.

In one of their last volumes, Wohlers Associates [WOH 14] reported that the 3D printing market, including the sale of 3D printers, consumables and associated services around the world, represented more than US $3 billion in 2013. This figure, variable according to the authors, is associated with an annual 34.9% per year growth, i.e. the fastest annual change in 17 years. From 2011 to 2013, the annual growth rates had already increased by 32.3%. As indicated schematically in Figure I.18 [MIZ 15, PIN 16, CON 14], today, this rather recent generic technology (approximately the past 30 years), defined by DIRECCTE [DIR 14], allows for the realization of small series of complex parts, otherwise inaccessible to classical methods for the same time and cost, thanks to its performances (see also [WEL 15, MEL 14]). Furthermore, at the end of the IT chain, this process, without (too much) direct human intervention, authorizes a specific relationship to the object (the same kind of relationship as the one that exists between a citizen and his information and communication systems).

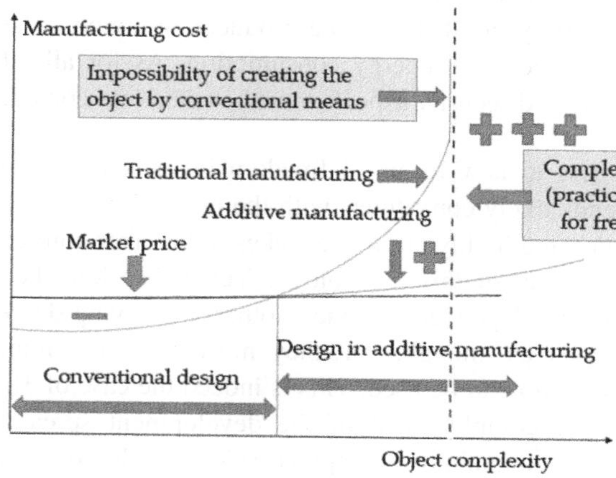

Figure I.18. *Interests and limitations of 3D printing relative to the conception of an object (the sign – represents the advantageous domain for traditional conception and the sign + for conception in additive manufacturing that becomes +++ when the object cannot be created by conventional means)*

According to Dodziuk [DOD 16], the pertinence (without contest) of 3D technologies can be found in a certain number of situations (even in terms of sustainable development):

– very complex shapes (example of injection nozzles, assembly parts for aviation with less materials and a cost reduced by 75%);

– mobile parts (example of catenanes, which are interlocked chemical structures);

– hollow objects (example in aeronautics, allowing a gain in transport weight);

– personalized implants and creation of personalized surgical tools;

– prostheses;

– bio-printing (evolving);

– creativity and art;

– societal change.

Thus, GAO [GAO 15] places additive manufacturing among the technologies that will form the foundation of sustainable development, in the sense that it is possible to avoid material losses. This data is all the more true as a large portion of unused materials are (at least partially) directly recyclable. Although this aspect is uncontestable, it is nevertheless necessary to consider that 3D printing joins a global process which links information and communication technologies (with materials with a shelf life of a few years that must be produced and whose dangerous waste must be managed) and electrical energy consumption. As for all other processes, environmental, hygiene and security aspects must be taken into consideration.

Nevertheless, access to new forms of freedom in the creation of objects with extremely variable ends likely constitutes, with the possibility of creating otherwise unfeasible parts, major and highly attractive stakes to be taken into consideration in the azimuthal development of this technology. Yet, it has long been known that different technologies are dependent on one another, to varying degrees, of course. Gille [GIL 78] reminds us, though, that they must have a minimum coherence among themselves to be able to be used. This is indeed the case of 3D technologies, which are a material accomplishment of the development spreading into every industrial and domestic domain of computer science. Moreover, the term "3D printing" reminds us of office printing associated with the most individual computers.

At the same time, access to "open-source" technologies, making numerous patents available in the public domain (after 20 years of industrial protection), and development via the Internet of information and exchange networks, has created another form of freedom connected to the manufacturing personal objects at home or in a Fab-Lab. The change in the cultures linked to this phenomenon is considered by some to be a form of organizational revolution in 21st Century society, because it is likely to change the methods of manufacturing by bringing them closer to the user.

But simultaneously, according to Mousion [MOU 15a], the question of intellectual property (reminiscent of pirated copies of music) begs to be asked in global democratization. He writes: "Indeed, the legislator has foreseen limitations in the exercise of intellectual property rights. These limitations are restrictively listed in the intellectual property code." Concerning the author's rights, French article L.122-5 provides a list of situations in which the author cannot forbid representations, copies or reproductions of his work. The exception for private copy listed in the second paragraph of this article L.122 is only reserved if three conditions are all fulfilled. The copy:

– Is realized using a lawful source, i.e. an original work acquired legally by the copyist;

– Is reserved for strictly private use;

– Does not violate the normal exploitation of the work and does not harm the legitimate interests of the right holder.

"Consequently, tomorrow if you scan a work bought by a neighbor or reproduce it at home, the three conditions will not be fulfilled and you can be prosecuted for counterfeiting, naturally on condition that the object's right holder is aware of the act of reproduction, which could happen if you put the digital version of the work that you scanned on the Internet. New technologies like 3D printing simply contribute more flexibility and speed to the act of copying. It is up to us to protect ourselves with a legal device, but also with other technologies. Far from being outdated, let's say that we are fighting an endless battle that must therefore not be abandoned" [MOU 15]. Let us note, without seeking to excuse these methods, that the worldwide counterfeit market is worth 140 billion euros per year [OEC 17] (as opposed to 5–10 billion for the total additive manufacturing market); however, vigilance is necessary.

This general societal context, which is an important element in a technology's growth dynamics, will also be the subject of a presentation. Thus, this first volume will not only cover technical and scientific aspects, but will also attempt to extend this reflection to other dimensions like those of the creativity associated with a high demand on society's part for original process/material pairs better responding to the demand, that of successful innovation, respect for regulations, analysis of societal changes, and when it is necessary, ethical aspects. These dimensions are integrated into the general concept of additive manufacturing, which is one of the forms of the notion of convergence defined in Figure I.19 as the will to extract disjointed knowledge from means, facilitating industrial applications through the integration of disciplinary scientific knowledge, which results in a coupling between scientific, economic and social aspects, as expressed in Figure I.20 [BUT 16].

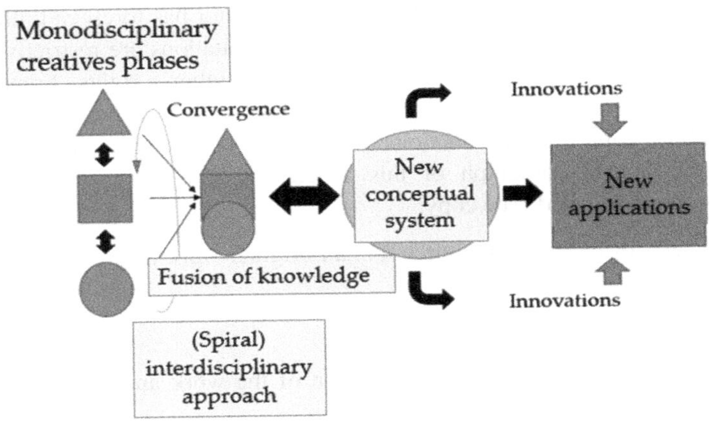

Figure I.19. *Principles of convergence/divergence*

According to Wikipedia (https://en.wikipedia.org/wiki/Technological_convergence), the term "technological convergence" aims to integrate separately developed components/concepts into a technical system (which will then be called "convergent"). The increased fertilization of technological convergence would be the source of innovations (see, for example, the GPS, [LAU 05]). According to AFTU [AFT 06], it is indeed an interdisciplinary approach. The expression "convergent technology" indicates the somewhat fortuitous encounter of innovations in the fields of microelectronics, bioinformatics, nanotechnologies and cognitive science. Technology progresses to the borders of scientific knowledge under the pressure of powerful, sometimes speculative, commercial or political interests. This statement is confirmed by Cornu [COR 08, COR 09], who writes, "Does technology, beyond its undeniable progress, help support the dream to make applications acceptable by the greater masses or does it rather give some fans of science-fiction legitimacy to influence research programs? Whatever the case may be, a 'convergence of interests' has allowed people to gather around this inspiring program." 3D technologies are indeed part of this program!

Additive manufacturing thus results from a coupling of scientific, technological, economic and social aspects, as expressed in Figure I.20 for a successful innovation [BUT 16]. It is within the articulation and integration between these different partially disjointed components that it is possible to judge whether or not technologies are becoming "mature". The signs are promising.

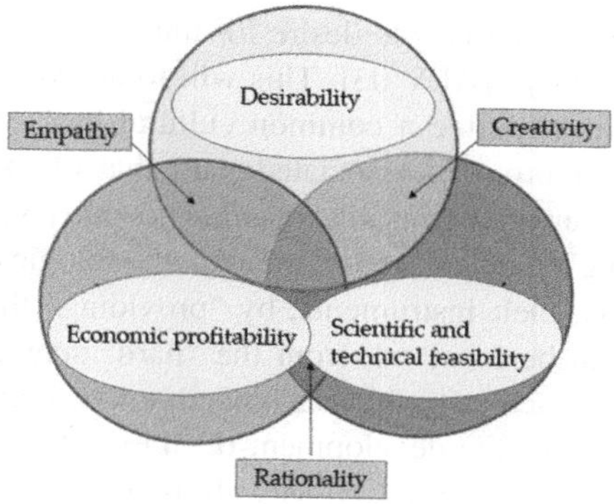

Figure I.20. *Reminders about the founding principles of a successful innovation*

The communities where a single disciplinary passion is shared are innovative, but they sometimes have a tendency to withdraw into themselves. Additive manufacturing's luck, from a scientific point of view, lies in its ability (to be obliged to) develop an interdisciplinary, heuristic, temporary and open research space, authorizing a more collective functioning of the creation and exploitation of new concepts, because the demand is present. Thus, through association with scientific disciplines, on matters of public utility, discontinuities should be made to emerge which lead to recompositions and scientific adaptations. These are the components that must constitute the specificity of the development in the field of 3 or 4D printing through new approaches, other links between man and artifacts, between workers/users and production/artifact systems. This is why we must take into consideration the way in which the objects studied by scientists and technicians return to the social world. This is a central question that strongly ties research to the economy, to the social and political, and also places additive manufacturing processes in direct and responsible connection with society. Currently, it is a matter of encouraging the demand for better trust between all the partners involved in disciplinary hybridization. Then, propositions are made in the three volumes so that researchers will further consider innovations as ends and less as means (in particular for financing).

To do this, all the scientific and technical means must be exploited to satisfy an intentional research objective on "objects/devices", be they material or not. "Either the researcher treats the object partially by ignoring or silencing all the links that inevitably exist between this partial view and the global biases about the object that cannot be avoided, or it places its global biases at the heart of scientific interrogation: not as interesting elements to be looked at, but as the motors of a dynamic that then takes the inevitable desire for unity as a footboard towards what becomes a demand for unity" [ALV 03]. This will to approach the real, with all its difficulties [NIC 96], must define a common cultural basis, that of modeling and integration (systematic approach), associated with that of experimentation, at the same time that there is an indispensable opening to other scientific and technical domains, by "borrowing" their concepts, their methodologies (permanent learning and blending principles), their instruments, by "providing" them subjects of study, etc. Every discipline can be involved, from the "hard" sciences to the human and social sciences. Thus to reach an ambitious objective which gathers living scientific and technological forces for the development of 3D printing, we must be clear on our subsidiarity (common cultural basis) as well as on the relational approaches to be envisioned which allow for the desired interdisciplinary opening to be developed. Indeed, the disciplinary reconciliation can favor the innovation produced at the interface due to unexpected rapprochements, the project method, the exploration aiming at different ends for the same instrument, etc. [VIN 07]. This book perhaps provides the chance to encourage "brainstorming", as well as an opportunity to take into consideration the difficulties of interdisciplinarity and thus the efforts to be supported in action taken by the decision makers so that this creative indiscipline may develop.

In an attempt to conclude this introduction, all of the factors concerned in Volume 1 are presented in Figure I.21.

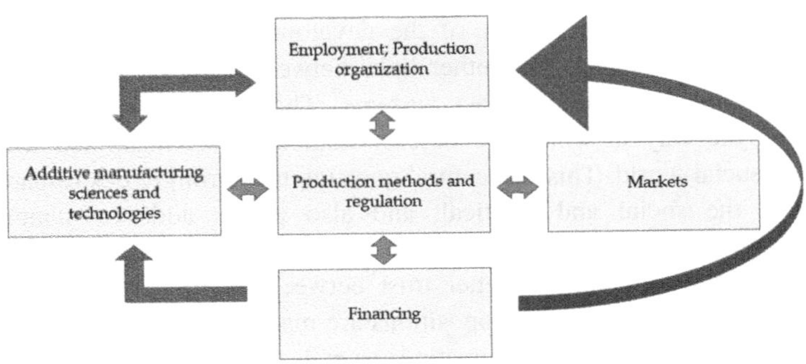

Figure I.21. *Interdependency associated with the development of additive manufacturing*

Thus, the aim of Volume 1 is to try to understand how societies (industry, public) accept engagement in new processes, strategies and behaviors associated with innovation in additive manufacturing (and vice versa). It stems from this analysis of a process emerging from nonlinear, recursive and organized aspects that could be studied as complex systems. The analysis of these elements presented throughout the different chapters of Volumes 1 to 3 sheds light on several elements that deserve attention:

– the nonlinearity connected to the interrelations between technological devices and society, with positive and negative control loops, change rates over time;

– the production of a "new order" where additive manufacturing is recognized as legitimate in the field of object production, be these industrial or domestic objects, with the development of forms of scientific, technical and social autonomy;

– the emergence of additive manufacturing implies new processes and reciprocal actions between stakeholders, possible organizational ruptures associated with new manufacturing methods;

– technological development implies interdisciplinary concerns in investment activities for enhancement in known niches and the exploration of new applicative niches (see Volumes 2 and 3);

– in these looming future developments, where several scientific disciplines are called together to mutualize their knowledge and their ways of functioning (explored in detail in Volume 3), the success of a certain number of operations with considerable economic potential (as new technologies must be created, applied to new fields) is reached by supporting creativity, and also through a better mastery of the complexity of the phenomena in question.

In this sense, the subject of 3D printing (or additive manufacturing), in the movement that it creates in its unsteady development, somewhat constitutes a "laboratory" of study of the operations of research, R&D, design or user companies' strategies and those of the public. This is a component that comes from the description of the "simple" technical or scientific act, but it seemed important to exploit the context associated with 3D printing to analyze and propose progress paths for the reinforcement of this technology's impact, paths which could possibly see applications in other areas. The debate introduced on this occasion is to know whether the standardizations of technical knowledge and research's methods of functioning have an impact on innovative and entrepreneurial behaviors in the field of 3D manufacturing because this standardization of behaviors can be fed with a counterpoint: the promotion of creativity and individual identity [RUA 17].

Bibliography

[3D 15a] 3D NATIVES, "Des maisons de 200 m^2 imprimées en 3D pour 3500 €", available at: http://www.3dnatives.com/habitations-impression-3d-shanghai/, 2015.

[3D 15b] 3D NATIVES, "L'Université de Sheffield développe l'impression 3D à grande vitesse", available at: http://www.3dnatives.com/sheffield-impression-3d-vitesse-04062015/, 2015.

[3D 15c] 3D NATIVES, "La fabrication additive source d'économie pour l'aéronautique!", available at: http://www.3dnatives.com/additive-economie-aeronautique-10062015/, 2015.

[3D 15d] 3D NATIVES, "Opel utilise l'impression 3D pour réduire ses coûts de production", available at: http://www.3dnatives.com/opel-impression-3d-production25112015/, 2015.

[3D 15e] 3D NATIVES, "Toshiba dévoile un premier prototype d'une imprimante 3D métal", available at: http://www.3dnatives.com/toshiba-imprimante-3d-metal-14122015/, 2015.

[ACA 15] ACADEMIE DES TECHNOLOGIES, "Les enjeux stratégiques de la fabrication additive: positionnement de l'Académie des Technologies", available at: http://academie-technologies-prod.s3.amazonaws.com/2015/06/18/08/31/49/22/Avis_de_l_AT_sur_Fabrication_additive_juin_2015_1_.pdf, 2015.

[ACA 16] ACATECH, "Industrie 4.0: International Benchmark, Options for the Future and Recommendations for Manufacturing Research", available at: http://www.acatech.de/fileadmin/user_upload/Baumstruktur_nach_Website/Acatech/root/de/Publikationen/Sonderpublikationen/INBENZHAP_E_web.pdf, 2016.

[AFN 11] AFNOR, NF E 67-001, Fabrication additive – Vocabulaire – Paris, France, October 2011.

[AFP 15] AFPR, "Guide à la mise en place du prototypage rapide", available at: http://code80.net/RP2/content/sourcing/guide/info/contexte.htm, 2015.

[AFT 06] AFTU (Association Pour Une Fondation Travail-Université), "De nouveaux enjeux pour la politique scientifique", Technologies convergentes, évaluations divergentes, Education Permanente, available at: http://www.ftu.be/documents/ep/EP-24.pdf, 2006.

[ALE 15] ALEX, "Gartner places 3D printing among the top 10 strategic technology trends for 2016", available at: http://www.3ders.org/articles/20151224-gartner-places-3d-printing-among-the-top-ten-strategic-technology-trends-for-2016.html, 2015.

[ALV 03] ALVAREZ-PEREYRE F., "L'exigence interdisciplinaire", *MSH*, Paris, France, 2003.

[AND 84] ANDRE J.C., LE MEHAUTE A., DE WITTE O., Dispositif pour réaliser un modèle de pièce industrielle, French patent no. 84 11 241, 16 July 1984.

[AND 94] ANDRE J.C., CORBEL S., *Stéréo-photolithographie Laser*, Polytechnica, Paris, France, 1994.

[AND 15] ANDRE J.C., "N'est pas Prométhée ou Mary Shelley qui veut: une petite histoire de la stéréo-photo-lithographie…", *Actualité Chimique*, vols 397–398, pp. 99–101, 2015.

[AND 16] ANDRE J.C., "Désordres dans la troisième dimension; un cheminement entre formation et interdisciplinarité", *VARIA*, vol. 5, pp. 62–72, 2016.

[BAR 15] BARLIER C., BERNARD A., *Fabrication additive: du prototypage rapide à l'impression 3D*, Dunod, Paris, France, 2015.

[BAU 16] BAUMERS M., DICKENS P., TUCK C. *et al.*, "The cost of additive manufacturing: machine productivity, economies of scale and technology-push", *Technological Forecasting & Social Change*, vol. 102, pp. 193–201, 2016.

[BER 12] BERMAN B., "3D printing: the new industrial revolution", *Business Horizons*, vol. 55, pp. 155–162, 2012.

[BER 14] BERCHON M., LUYT B., *L'impression 3D*, Eyrolles, Paris, France, 2014.

[BLA 14] BLACKMAGIC3D, "The 8 best 3D printing stories of 2014", available at: http://blog.blackmagic3d.com/8-best-3d-printing-stories-2014/, 2014.

[BLU 99] BLUMENBERG H., *La légitimité des temps modernes*, Gallimard, Paris, France, 1999.

[BOS 16] BOSTON CONSULTING GROUP, "Sprinting to Value in Industry 4.0", available at: https://www.bcgperspectives.com/Images/BCG-Sprinting-to-Value-Industry-4-Dec-2016_tcm80-217083.pdf, 2016.

[BOU 14] BOUFFARON P., "Impression 3D: Les prémisses d'une nouvelle (r)évolution industrielle", available at: http://sf.france-science.org/wp-content/uploads/2014/11/SMM14_025.pdf, 2014.

[BRA 15] BRANCHE P.M., "Impression 3D; la fin annoncée des transports", available at: http://www.voxlog.fr/dossier/13_2/logistique-et-transport-limpression-3d-remet-en-cause-la-supply-chain, 2015.

[BRY 14] BRYNJOFSSON E., MCAFEE A., *Le deuxième âge de la machine; travail et prospérité à l'heure de la révolution technologique*, O. Jacob, Paris, France, 2014.

[BUT 16] BUTTIN N., SAFFRE B., *Activer l'économie circulaire*, Eyrolles, Paris, France, 2016.

[CCI 15] CCI PARIS-IDF/CONSEIL GENERAL DE L'ARMEMENT, "L'impression 3D: porte d'entrée dans l'industrie du 21ème siècle", available at: http://www.adverbe.com/wp-content/uploads/2015/09/rapport-impression3d.pdf, 2015.

[CHE 15] CHEN D., HEYER S., IBBOTSON S. *et al.*, "Direct digital manufacturing: definition, evolution, and sustainability implications", *Journal of Cleaner Production*, vol. 107, pp. 615–625, 2015.

[CHE 07] CHEVASSUS-AU-LOUIS B., *L'analyse des risques: l'expert, le décideur et le citoyen*, Quae, Paris, France, 2007.

[CIG 17] CIGREF, "Cercle innovation: L'usine du furur", available at: http://www.cigref.fr/wp/wp-content/uploads/2016/12/CIGREF-Cercle-Innovation-Usine-du-futur-2017.pdf, 2017.

[CON 14] CONNER B.P., MANOGHARAN G.P., MARTOF A.N. *et al.*, "Making sense of 3-D printing: creating a map of additive manufacturing products and services", *Additive Manufacturing*, vols 1-4, pp. 64–76, 2014.

[COR 08] CORNU J.M., *Nouvelles technologies, nouvelles pensées? La convergence des NBIC*, FYP, Limoges, France, available at: http://alltrends.over-blog.net/article-25668522.html, 2008.

[COR 09] CORNU J.M., quoted by Manach J.M., "La convergence des sciences (1/3): Un choix de société", available at: http://www.Internetactu.net/2009/05/25/la-convergence-des-sciences-13-un-choix-de-societe/, 2009.

[COT 14] COTTELEER M.J., "The 3D opportunity primer - The basics of additive manufacturing", available at: https://dupress.deloitte.com/dup-us-en/focus/3d-opportunity/the-3d-opportunity-primer-the-basics-of-additive-manufacturing.html, 2014.

[COT 16] COTTELEER M.J., TROUTON S., DOBNER E., "3D opportunity and the digital thread; Additive manufacturing ties it all together", available at: https://dupress.deloitte.com/dup-us-en/focus/3d-opportunity/3d-printing-digital-thread-in-manufacturing.html, 2016.

[DAR 15a] DARPA, "Boosting confidence in new manufacturing technology", available at: http://www.darpa.mil/NewsEvents/Releases/2015/05/29.aspx, 2015.

[DAR 15b] DARPA NEWS, "Boosting confidence in new manufacturing technology", available at: http://www.doncio.navy.mil/chips/ArticleDetails.aspx?ID=6426, 2015.

[D'AV 15] D'AVENI R., "The 3D Printing revolution", available at: https://hbr.org/2015/05/the-3-d-printing-revolution, 2015.

[DEL 14] DELOITTE, "Du nouveau dans le secteur manufacturier: les répercussions de l'impression tridimensionnelle", available at: http://www2.deloitte.com/content/dam/Deloitte/ca/Documents/insights-and-issues/ca-fr-perspectives-et-enjeux-daffaires-du-nouveau-dans-le-secteur-manufacturier.pdf, 2014.

[DEL 15] DELOITTE, "The future of manufacturing; making things in a changing world", available at: http://d2mtr37y39tpbu.cloudfront.net/wp-content/uploads/2015/03/DUP-954_Future-of-manufacturing_vFINAL_3.31.15.pdf, 2015.

[DEL 16] DELOITTE UNIVERSITY PRESS, "Tech Trends 2016; innovating in the digital era", available at: http://d27n205l7rookf.cloudfront.net/wp-content/uploads/2016/03/DUP_TechTrends2016.pdf, 2016.

[DGE 16] DGE, "Technologies clés 2020", available at: http://www.entreprises.gouv.fr/files/files/directions_services/politique-et-enjeux/innovation/technologies-cles-2020/technologies-cles-2020-poster.pdf, 2016.

[DIR 14] DIRECCTE, "L'impression 3D: Etat des lieux et perspectives", available at: http://www.centre.direccte.gouv.fr/IMG/pdf/L_impression_3D_etat_des_lieux_et_perspectives_-_DIRECCTE_Centre_en_partenariat_avec_CCI_Centre_-_10_decembre_2014-2.pdf, 2014.

[DOD 16] DODZIUK H., "The added value: 3D printing brings much more than cheaper manufacturing", available at: http://3dprintingindustry.com/2016/02/23/the-added-value-3d-printing-carries-much-more-than-cheaper-manufacturing/?utm_source=3D+Printing+Industry+Update&utm_medium=email&utm_campaign=9a0e5caa42-RSS_EMAIL_CAMPAIGN&utm_term=0_695d5c73dc-9a0e5caa42-64583257, 2016.

[DUF 11] DUFOURMENTELLE A., *Eloge du risque*, Payot, Paris, France, 2011.

[ELL 88] ELLUL J., *Le bluff technologique*, Hachette, Paris, France, 1988.

[ELW 14] ELWANY A., "Making 3D futures reality", *Industrial Engineer* vol. 9, pp. 32–35, 2014.

[FAD 15] FADHEL N.F., CROWDER R.M., WILLS G.W., "Provenance in the additive manufacturing process", *IFAC-Papers on Line*, vol. 48, no. 3, pp. 2345–2350, 2015.

[FOR 14] FORD S., "Additive manufacturing technology: potential implications for U.S. manufacturing competitiveness", *Journal of International Commerce and Economics*, available at: http://www.usitc.gov/journals, 2014.

[FOR 15] FORMLABS, *Design, Adoption, and Readiness: The Future of 3D Printing and Additive Manufacturing*, e-Book, available at: http://www.liebertpub.com/lpages/3dp-formlabs-learning-center/178/, 2015.

[FRA 75] FRANQUIN A., *Spirou et les hommes-bulles/les petits formats*, Dupuis, Marcinelles, Belgium, 1975.

[GAO 15a] GAO, "3D Printing: opportunities, challenges and policy; implications of additive manufacturing", available at: http://www.gao.gov/assets/680/670960.pdf, 2015.

[GAO 15b] GAO, "Defense additive manufacturing", available at: http://www.gao.gov/assets/680/673099.pdf, 2015.

[GAO 15c] GAO W., ZHANG Y., RAMANUJAN D. *et al.*, "The status, challenges, and future of additive manufacturing in engineering", *Computer-Aided Design*, vol. 69, pp. 65–89, 2015.

[GAR 12] GARTNER, "Gartner's 2012 Hype cycles for emerging technologies identifier "Tipping points" technologies that will unlock long-awaited technology scenarios", available at: http://www.gartner.com/newsroom/id/2124315, 2012.

[GEB 14] GEBLER M., SCHOOT-UITERKAMP A.J.M., VISSER C., "A global sustainability perspective on 3D printing technologies", *Energy Policy*, vol. 74, pp. 158–167, 2014.

[GEI 17] GEISSBAUER R., WUNDERLIN J., LEHR J., "The future of spare parts is 3D – A look at the challenges and opportunities of 3D printing", available at: http://www.strategyand.pwc.com/media/file/The-future-of-spare-parts-is-3D.pdf, 2017.

[GEN 16] GENEAU D., CAULIER S., "Le 3D imprime sa marque sur l'industrie; Construire localement, à moindre coût, avec un service après-vente amélioré", available at: http://www.ledevoir.com/economie/actualites-economiques/465475/le-3d-imprime-sa-marque-sur-l-industrie-construire-localement-a-moindre-cout-avec-un-service-apres-vente-ameliore, 2016.

[GIB 15] GIBSON I., ROSEN D., STUCKER B., *Additive Manufacturing Technologies: 3d Printing, Rapid Prototyping and Direct Digital Manufacturing*, Springer, New York, 2015.

[GIL 78] GILLE B. (ed.), *Histoire des techniques*, La Pléiade, Paris, France, 1978.

[GOR 13] GORE A., *The Future; Six Drivers of Global Change*, Random House, New York, 2013.

[HAG 15] HAGUE R., "The third industrial revolution – considering the impact of additive manufacturing", available at: https://www.eversheds-sutherland.com/documents/sectors/diversified-industrials/DI-Additive-Manufacturing-Brochure-23-06-14-Screen.pdf, 2015.

[HAM 14] HAMPEL C., *Eléments d'épistémologie*, Armand Colin, Paris, France, 2014.

[HAR 15] HARRISON M., "Small innovative company growth: barriers, best practices and big ideas" available at: https://www.sba.gov/sites/default/files/advocacy/FINAL_Innovation_Report.pdf 56pp., 2015.

[HAU 14] HAUSMAN K.K., HORNE R., *L'impression 3D pour les nuls*, First Interactive, Paris, France, 2014.

[HIN 15] HINKS J., "5 things you should know about industry 4.0", available at: http://www.techradar.com/news/world-of-tech/future-tech/5-things-you-should-know-about-industry-4-0-1289534, 2015.

[HLA 14] HLAVIN M., "3D printing: the next industrial revolution", *Appliance Design*, vol. 62, pp. 22–23, 2014.

[HOS 14] HOSKINS S., *Le guide de l'impression 3D pour artisans et designers*, Marabout, Paris, France, 2014.

[HU 11] HU S.J., KO J., WEYAND L. *et al.*, "Assembly system design and operations for product variety", *CIRP Annals - Manufacturing Technology*, vol. 60, pp. 715–733, 2011.

[HUL 84] HULL C., Method for production of three-dimensional objects by stereolithography, US Patent 5,762,856, 1 August 1984.

[HUN 15] HUNYADI M., *La tyrannie des modes de vie; sur le paradoxe moral de notre temps*, Le bord de l'eau, Lormont, France, 2015.

[ING 15] INGELAERE R., "Innovation technologique et performance industrielle: l'exemple de l'impression 3D", available at: http://www.lecese.fr/sites/default/files/pdf/Avis/2015/2015_07_innovations_techno_impression_3d_.pdf, 2015.

[KAZ 14] KAZIUNAS A., *Pratique de l'impression 3D*, Eyrolles, Paris, France, e-book, available at: https://books.google.fr/books?isbn=2212270631, 2014.

[KEA 15] KEARNEY A.T., "3D Printing: A Manufacturing Revolution", available at: https://www.atkearney.com/documents/10192/5992684/3D+Printing+A+Manufacturing+Revolution.pdf/bf8f5c00-69c4-4909-858a-423e3b94bba3, 2015.

[KIE 15] KIETZMANN K., PITT L., BERTHON P., "Disruption, decisions and destinations: enter the age of 3D printing and additive manufacturing", *Business Horizons*, vol. 58, pp. 209–215, 2015.

[KIM 15] KIM H.C., "Acceptability Engineering: the Study of user Acceptance of Innovative Technologies", *Journal of Applied Research and Technology*, vol. 13, pp. 230–237, 2015.

[KOH 01] KOHLHUBER M., KAGE M., KARG M., *Additive Fertigung*, available at: http://www.acatech.de/fileadmin/user_upload/Baumstruktur_nach_Website/Acatech/root/de/Publikationen/Kooperationspublikationen/3Akad_Stellungnahme_AdditiveFertigung_2016_web.pdf, 2016

[LAU 05] LAURENT L., "Nanotechnologies et technologies convergentes: quel avenir à vingt ans?", *Le Banquet*, available at: http://www.revue-lebanquet.com/nanotechnologies-et-technologies-convergentes-quel-avenir-a-vingt-ans/, 2005.

[LAV 15] LAVIGNE C., "J.C. André, A. Le Méhauté, O. de Witte: history of the digital sculpture tools", *What Things May Come: The 3rd International 3D Print Sculpture Exhibition*, Georgetown, Texas, USA, 26–27 February 2015.

[LIP 14] LIPSON H., KURMAN M., *Impression 3D: la prochaine révolution*, Wiley & First Interactive, Paris, France, 2014.

[MAD 17] MADEELI, "Fabrication Additive: Enjeux et Impacts en Occitanie Pyrénées-Méditerranée", available at: http://www.madeeli.fr/wp-content/uploads/2017/01/Fabrication-Additive-en-Occitanie-Etude-2016.pdf, 2017.

[MAR 99] MARTY C., *Industrialisation des produits mécaniques*, Hermès, Paris, France, 1999.

[MAR 15a] MARION F., "Le papier devient lumière", available at: http://www.up-magazine.info/index.php?option=com_content&view=article&id=4016:le-papier-devient-lumiere&catid=81:methodo&Itemid=483, 2015.

[MAR 15b] MARION F., "Pourquoi les startups françaises ne deviennent-elles pas des géants?", available at: http://www.up-magazine.info/index.php?option=com_content&view=article&id=4976:pourquoi-les-startups-francaises-ne-deviennent-elles-pas-des-geants&catid=103:analyses&Itemid=542&utm_source=newsletter_123&utm_medium=email&utm_campaign=up-nl-vnew, 2015.

[MCK 17] MCKINSEY GLOBAL INSTITUTE, "A future that works: automation, employment and productivity" available at: http://www.mckinsey.com/~/media/McKinsey/Global%20Themes/Digital%20Disruption/Harnessing%20automation%20for%20a%20future%20that%20works/MGI-A-future-that-works-Full-report.ashx, 2017.

[MEI 14] MEIER D., "Regarder le monde de façon interdisciplinaire: l'espace du possible", in LEMAY V., DARBELLAY F. (eds), *L'interdisciplinarité racontée; chercher hors frontières, vivre l'interdisciplinarité*, P. Lang, Bern, Switzerland, 2014.

[MEL 12] MELCHELS F.P.W., "Celebrating three decades of stereolithography", *Virtual and Physical Prototyping*, vol. 7, pp. 173–175, 2012.

[MEL 14] MELLOR S., HAO L., ZHANG D., "Additive manufacturing: a framework for implementation", *International Journal of Production Economics*, vol. 149, pp. 194–201, 2014.

[MER 13] MERTZ L., "A new world for 3D printing offer", *IEEE Pulse*, vol. 4, pp. 12–14, November–December 2013.

[MIZ 15] MIZERET J., "Les technologies de fabrication additive pour la créativité, le prototypage, la fabrication", available at: http://www.swissmem.ch/fr/presentations-fabrication-additive.html, 2015.

[MOR 14] MORA A., "L'impression 3D: L'industrie du futur est en marche", *Management*, vol. 217, pp. 64–71., March 2014.

[MOU 15a] MOUSSION A., "Impression 3D: quel impact sur la propriété intellectuelle? L'INPI répond dans une interview", available at: http://www.priximprimante3d.com/loi/, 2015.

[MOU 15b] MOUSSION A., "Un vol de médicaments dissimulé grâce à l'impression 3D", available at: http://www.priximprimante3d.com/conteneur/, 2015.

[NIC 96] NICOLESCU B., *La transdisciplinarité; manifeste*, Du Rocher, Monaco, 1996.

[NOR 04] NORDMANN A., Converging technologies: shaping the future of European societies, European Commission, Brussels, Belgium, 2004.

[OBA 15] OBATON A.F., BERNARD A., TAILLANDIER G. *et al.*, "Fabrication additive: Etat de l'art et besoins métrologiques engendrés", *Revue Française de Métrologie*, vol. 37, pp. 21–36, 2015.

[OBA 16] OBATON A.F., "La maîtrise du processus de fabrication additive est un véritable enjeu pour les industriels", available at: http://www.usinenouvelle.com/article/tribune-la-maitrise-du-processus-de-fabrication-additive-est-un-veritable-enjeu-pour-les-industriels.N385082, 2016.

[OEC 17] OECD, "La contrefaçon touche un cinquième des téléphones mobiles expédiés dans le monde", available at: http://www.oecd.org/fr/presse/la-contrefacon-touche-un-cinquieme-des-telephones-mobiles-expedies-dans-le-monde.htm, 2017.

[OEC 17] OECD, Key issues for digital transformation in the G20; Report prepared for a joint G20 German Presidency/OECD conference, available at: http://www.oecd.org/G20/key-issues-for-digital-transformation-in-the-G20.pdf?utm_source=Adestra&utm_medium=email&utm_content=Read%20the%20OECD%27s%20report%20for%20the%20G20%E2%80%A6&utm_campaign=OECD%20Science%2C%20Technology%20and%20Innovation%20News%20Jan17&utm_term=demo, 2017.

[OFF 16] OFFICE OF THE DEPUTY ASSISTANT SECRETARY OF THE ARMY (RESEARCH & TECHNOLOGY), Emerging Science and Technology Trends: 2016–2045 – A Synthesis of Leading Forecasts, available at: http://www.defenseinnovationmarketplace.mil/resources/2016_SciTechReport_16June2016.pdf, 2016.

[PAN 13] PANETTA K., "Need something? Print-it: 3D printing in engineering", *Electronic Component News*, vol. 57, pp. 16–19, 2013.

[PAU 14] PAU W., "IoT: 3C's of Internet of Things (aka Welcome to Industry 4.0!)", available at: https://www.linkedin.com/pulse/20141104130144-5544582-iot-3c-s-of-Internet-of-things-aka-welcome-to-industry-4-0 and http://waynepau.blogspot.ca/2014/11/iot-3cs-of-Internet-of-things-aka.html, 2014.

[PEY 17] PEYRÉ TARTARUGA I.G., CAZAROTTO R.T., BACKX MARTINS C.T. *et al.*, "Innovation and public understanding of science: possibility of new indicators for the analysis of public attitudes to science, technology and innovation", available at: https://mpra.ub.uni-muenchen.de/76288/1/MPRA_paper_76262.pdf, 2017.

[PIE 13] PIERCE A., "3D printing: from the simple to the complex", *Technology Today*, vol. 73, pp. 8–9, 2013.

[PIN 16] PINKERTON A.J., "Lasers in additive manufacturing", *Optics & Laser Technology*, vol. 78, pp. 25–32, 2016.

[POR 09] PORTER A.L., YOUTIE J., "How interdisciplinary is nanotechnology?", *J. Nanoparticle Research*, vol. 11, pp. 1023–1041, 2009.

[POU 06] POUTEAU S., "Biotechnologies: la vie en morceaux et l'éthique en péril", *Journal International de Bioéthique*, vol. 17, pp. 15–27, 2006.

[PRI 14] PRINCE J.D., "3D printing: an industrial revolution", *Journal of Electronic Resources in Medical Librairies*, vol. 11, 39-45, 2014.

[PWC 16a] PwC, "Industry 4.0: opportunities and challenges for consumer product and retail companies", available at: http://www.strategyand.pwc.com/media/file/Industry-4.0-RC.pdf, 2016.

[PWC 16b] PwC, "The future of industries: Bringing down the walls", available at: http://www.pwc.com/gx/en/industries/industrial-manufacturing/publications/pwc-cips-future-of-industries.pdf, 2016.

[RAT 12] RATTO M., REE R., "Materializing information: 3D printing and social change", *First Monday*, vol. 17, available at: http://firstmonday.org/ojs/index.php/fm/article/view/3968/3273, 2012.

[RAY 16] RAYNA T., STRIUKOVA L., "From rapid prototyping to home fabrication: how 3D printing is changing business model innovation", *Technological Forecasting and Social Change*, vol. 102, pp. 214–224, 2016.

[ROW 12] ROWE-PRICE T., "A brief history of 3D printing", available at: http://individual.troweprice.com/public/Retail/Planning-&-Research/Connections/3D-Printing/Infographic, 2012.

[ROY 13] ROYAL ACADEMY OF ENGINEERING, "Additive manufacturing: opportunities and constraints – A summary of a roundtable forum held on 23 May 2013 hosted by the Royal Academy of Engineering", available at: http://www.raeng.org.uk/publications/reports/additive-manufacturing, 2013.

[RUA 17] RUANO-BORBALAN J.C., "La 'grande standardisation' et les transformations du savoir dans les sociétés technoscientifiques", *Innovations*, vol. 1, pp. 17–41, 2017.

[RUF 14] RUFER S., Quel est l'impact de l'impression 3d sur la supply chain?, Research paper, Université Paris-Dauphine, available at: http://www.memoireonline.com/04/15/9049/Quel-est-l-impact-de-l-impression-3D-sur-la-supply-chain-.html, 2014.

[SCH 15] SCHWAB K., "What is the fourth industrial revolution?", available at: https://agenda.weforum.org/2015/12/the-fourth-industrial-revolution-what-it-means-and-how-to-respond/, 2015.

[SCU 15a] SCULPTEO, "Your guide to the 4 areas where 3D printing is unstoppable", available at: http://www.sculpteo.com/en/get/ebook/unstoppable_3D_printing/, 2015.

[SCU 15b] SCULPTEO,"3D printing and the tiny house movement in the US", available at: http://www.sculpteo.com/blog/2015/08/11/3d-printing-and-the-tiny-house-movement-in-the-us/?utm_medium=email&utm_campaign=Newsletter+12082015+US&utm_content=Newsletter+12082015+US+CID_66d4c457b681ed42b4dd366d94b6d4ce&utm_source=Campaign%20Newsletter, 2015.

[SCU 15c] SCULPTEO, "Apprendre l'impression 3D", Elephorm, available at: www.elephorm.com, 2015.

[SCU 16a] SCULPTEO, "The state of 3D printing 2015 vs 2016", available at: http://www.sculpteo.com/blog/2016/02/17/the-state-of-3d-printing-2015-vs-2016/?utm_medium=email&utm_campaign=Newsletter%202172016%20US&utm_content=Newsletter%202172016%20US+CID_6038000b1c0963ea3c725ccb68a200a5&utm_source=Campaign%20Newsletter&utm_term=Predictions%20for%20the%20State%20of%203D%20Printing%20Report%202016, 2016.

[SCU 16b] SCULPTEO, "The state of 3D printing", available at: http://www.sculpteo.com/static/0.30.0-64/download/report/Sculpteo_State_of_3D_Printing.pdf, 2016.

[SIR 16] SIRICHAKWAL I., CONNER B., "Implications of Additive Manufacturing for Spare Parts Inventory", *3D Printing and Additive Manufacturing*, vol. 3, pp. 56–63, 2016.

[SER 97] SERRES M., FAROUKI N., *Sciences en techniques – Dictionnaire des sciences*, Flammarion, Paris, France, 1997.

[STA 08] STAR S.L., GRIESEMER J.R., "Ecologie institutionelle, traductions et objets-frontières: des amatures et des professionnels au musée de zoologie vertébrée de Berkely 1907–1939," in LAHIRE B., ROSENTAL C. (eds), *La cognition au prisme des sciences socials*, Editions des Archives Contemporaines, Paris, France, 2008.

[TAY 14] TAYLOR S., "Trend evolution: 3D printing trends", available at: http://3dprintingindustry.com/2014/02/24/trend-evolution-3d-printing-trends-part-1/, 2014.

[TCH 17] TCHIEHE D.N., GAUTHIER F., "Classification of risk acceptability and risk tolerability factors in occupational health and safety", *Safety Science,* vol. 92, 138–147, 2017.

[TEC 15] TECHNOPOLIS, "Regional Innovation Monitor Plus 2015 – Mapping advanced manufacturing networks and exploring new business opportunities", available at: https://ec.europa.eu/growth/tools-databases/regional-innovation-monitor/sites/default/files/report/RIM%20Plus_Mapping%20Advanced%20Manufacturing%20Networks_Thematic%20Paper.pdf, 2015.

[TIN 14] TINTIN.COM, "Tintin en impression 3D2", available at: https://fr.tintin.com/news/index/rub/100/id/4168/0/tintin-en-impression-3d, 2014.

[TÖP 35] TÖPFFER R., Quoted by Bouveresse J. (2017), *Le mythe moderne du progrès*, Agone, Marseille, France, 1835.

[TRA 15] TRANCHARD S., ROJAS V., "L'avenir en fabrication 3D", available at: http://www.iso.org/iso/fr/home/news_index/news_archive/news.htm?refid=Ref1956, 2015.

[UPS 16] UPS, "3D Printing: The Next Revolution in Industrial Manufacturing - New research from UPS and the consumer technology association (CTA)", available at: https://www.ups.com/media/en/3D_Printing_executive_summary.pdf, 2016.

[VAE 13] VAEZI M., SEITZ H., YANG S., "A review on 3D micro-additive manufacturing technologies", *International Journal of Advanced Manufacturing Technology*, vol. 67, pp. 1721–1754, 2013.

[VAL 16] VALERO J., "Industry 4.0: The future is here", available at: http://www.euractiv.com/section/innovation-industry/linksdossier/industry-4-0-the-future-of-manufacturing-is-here/#ea-accordion-further-reading, 2016.

[VIN 07] VINCK D., *Sciences et société; sociologie du travail scientifique*, A. Colin, Paris, France, 2007.

[VIV 12] VIVERET P., *La cause humaine; du bon usage de la fin du monde*, Les liens qui libèrent, Paris, France, 2012.

[WEL 15] WELLER C., KLEER R., PILLER F., "Economic implications of 3D printing: market structure models in light of additive manufacturing revisited", *International Journal of Production Economics*, vol. 164, pp. 43–56, 2015.

[WEN 08] WENDEL B., RIETZEL D., KUHNLEIN F. *et al.*, "Additive processing of polymers", *Macromolecular Materials and Engineering*, vol. 293, pp. 799–809, 2008.

[WIK 15] WIKIPEDIA, "Industrie 4.0", available at: https://fr.wikipedia.org/wiki/Industrie_4.0, 2015.

[WIN 12] WINNAN C.D., *3D Printing: The Next Technology Gold Rush*, Amazon, New York, USA, 2012.

[WHI 13] WHITE G., LYNSKY D., "Economic analysis of additive manufacturing for final products: an industrial approach", Swanson School of Engineering, University of Pittsburgh, pp. 3039–3050, 2013.

[WOH 14] WOHLERS ASSOCIATE, "3D printing and additive manufacturing state of the industry", available at: http://wohlersassociates.com/index.html, 2014.

PART 1

From Spectacular Applications to the Economic Market of Additive Manufacturing

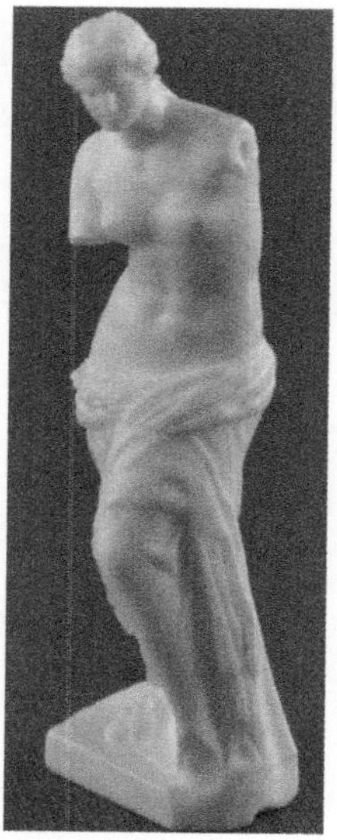

Ceramic part created by stereolithography (photography kindly provided by 3D Ceram)

Introduction to Part 1

"Fashion always combines a taste for imitation and a taste for change, conformism and individualism, an aspiration to melt into a social group and a desire to stand out, if only in the small details". [SIM 03]

"It is thus important to study the nature of the final clientele's expectations and their criteria for evaluating the quality of the service". [ROU 03]

"Technologies […] not only produce instruments that transform our lives, they modify the reality around us, reorganize social life, the movement of which has been racing since the Industrial Revolution". [KLE 11]

"Technology is then the projection or rather reflection surface on which an image appears, the characteristics of which send their image back to the human genius, to the development of sciences whose machines and processes are applications, to a socio-economic organization that determines its risks, even to the accomplishment of a fate sealed by a metaphysical 'choice' that founds the identity of the West". [TIN 94]

"We learn from laws and the way of speaking about our slaves to better manage them. And thus, without even noticing it, we renounce our human qualities, our own laws. We dehumanize ourselves, we adopt the style of our technical slaves". [GHE 74]

"Arthur Koestler was not far off when he described the behavior of certain scientific creators in comparing them to sleep-walkers, stumbling over major discoveries. But it is still necessary for the moment to be favorable and the mind prepared, and possibly for luck to come to your aid". [DE 12]

"No authority decides what a good idea is". [FEY 98]

"It is understood that France has always preferred a precise preparation of operations described by its past experiences – and the strict respect for the structured hierarchical instructions that result from them – to the awareness of changes and new realities, thus to the detriment of innovation". [SER 11]

"The ability to imagine, or the imagination, consists of nothing more than the power that the soul has to form itself from the images of objects by producing a change in the fibers of this part of the brain, which can be called the primary part as it responds to every part of our body". (Malebranche [MAL 75] quoted by Roquette [ROQ 07])

"Innovation is a novelty or significant change that is made in the political government against the use & rule of its constitution". (Diderot & D'Alembert Encyclopédie [DID 74] according to Godin [GOD 15])

"Science, despite being factually plural, struggles to be experienced as a place for the expression of true pluralism". [COU 15]

"It is possible to find exceptional situations for which this cloud of data has a 'very special' shape, different from what is 'normally' expected. The search for these 'divergent' situations must lead to interesting information about the internal structure of the unknown". [THO 83]

In the imagination of researchers and engineers at the beginning, 3D printing/additive manufacturing is one of the methods of so-called "morphogenetic engineering" [DOU 13, MC 15, DOU 08, BOU 11, BOU 06, MUR 07, THO 89], which authorizes the production of material objects "through thought" (or based on mental conception) using a computer and a physical interface: the 3D printer.

For some, it would be rather natural to proceed with a work dedicated to additive manufacturing from the beginning, that is, by speaking of the origin of this concept and its evolution over time. However, it seemed more interesting to the author to show, through the use of a few examples, the potential of 3D techniques, from pratical realizations in different applicative fields to socio-economic modifications. Access to new forms of freedom in the creation of objects with very diverse ends likely constitutes, with the possibility of creating infeasible parts alternatively, strongly attractive major stakes to be taken into consideration in the all-around development of technology.

This bias thus allows value to be given to an emerging technology that still has not found all of its limits, to try to make the reader dream about new applicative fields, and to record the technology through a collective imaginary in new societal habits, all while reflecting on some constraints. It also allows factual information on the development of technological innovation in additive manufacturing to be added.

For Zeese and Flowers [ZEE 13], who based themselves on an older work from Moyer [MOY 87], the emergence of a technology follows eight steps. Established at the time to protest against nuclear power in an in-depth and scientific way, in 1987 the "Movement Action Plan" engaged reflection on our knowledge of the following elements, which must also be considered concerning additive manufacturing (while this is a matter of an attractive potential, as opposed to the problems of peacefully managing nuclear waste):

– Phases 1–3: start of the technology, information to industrial environments, and first communication to the public based on the fact that the innovation produces objects that broaden the field of possibilities, creates a new reality, even properly invents it [GUÉ 15];

– Phase 4: measurement of the technology's attractiveness with potential openings; support from originals (the "makers" who will be mentioned below) who only imperfectly join the traditional and still current economic system; according to Chomsky and Herman [CHO 08], "the system perfectly adjusts to a certain degree of dissidence", illustrating the fact that it is not monolithic and is open to innovation, but that this coexistence of ideas nevertheless remains boxed in; creation of a financially credible demand by playing on a form of economic system "tectonics";

– Phases 5–6: creation of "consistency" within the proposition (and thus of a widened market) with future forms of colonization; densification of the present moment for the profitability and appropriation of propositions, even prescriptions proposed by the "system" [TRA 15] allowing investment in the virtual world; "An impression is thus had that civilization is something imposed on a reluctant majority by a minority having understood how to take control of the means of power and coercion" [FRE 27].

– Phases 7–8: comforting and media-friendly exploitation of success thanks to spectacular results and a large number of promises; transfer from the production sphere to the consumption sphere (particularly with performing tools/machines, only a small proportion of whose capacities we use) with the risk of no longer managing to distinguish what comes from its will and what is provided by the web.

Over more than 30 years, since the first patent, this is indeed what has been observed: from slightly confidential, even laborious initial phases, to an almost considerable and constantly growing economic market because the citizens, in their perception, believe that it is possible to rediscover some freedom of action, to make our ways of seeing the world and influencing our social behaviors evolve. 3D technologies, with their mythogenic character, can create living tissues or organs, made on demand, parts of what Lipovetsky [LIP 06] calls emotional consumption. For some, this is a path to reinvest in the artisanal nature of former times, yet, connected to the Internet with his/her computer. Should we then follow Ferrari [FER 15] on this aspect of things when he writes, "Technology hardly ever appears to be the product of conscious human efforts aiming to increase material power anymore; it seems rather to be a large-scale biological event throughout which the internal structures of the human body are transported more and more into the world surrounding man?"

Flower (Prodways, 2017 – Copyright ©Prodways ©Hamilton de Oliveira)

1

Some Significant Examples

"But the fundamental nonsense, nonsense itself, model nonsense, is the way in which Joseph considers progress, not only as a means, but as the goal, as the only goal of happiness. In this way, he pursues his goal, for behind progress is always found more; in this way, he does not enjoy, pleasure being indefinitely postponed; in this way, he scorns the past, he disdains the present that is great, he waits for the future that is always before him". [TÖP 35]

"As for a skeptical mind, it can be tempted to conclude [...] that the true goal of progress, if there is one, has finally become clear: everything happens as if it was in fact nothing more than the continuation of progress itself". [BOU 17]

"In the end, all of our explanations are reduced to finding what would have to be done to reproduce a given effect. This is all ours. It is narrow-minded. We have S (sense) and M (muscles)... Our world is confined within the combined whole of our perceptions and our actions. We have tried to connect this whole to a system of measurement, that is, of means to find... that is to say, digital formulas or recipes... A formula is nothing but a mathematical prescription". [VAL 33]

"Everywhere here, we feel the same fundamental process, the same style of thinking and creating. This is the mentality that targets the clarity everywhere and yet that, in doing this, recognizes the complication of life, never fully transparent, which targets the minutiae in the shaping of details and at the same time the simplicity of the great lines within the whole...". [CAR 61]

"The period of abundance is inseparable from an indefinite enlargement of the desired satisfactions and an inability to reduce the appetites of consumption, pure saturation of needs immediately accompanied by new demands". [LIP 06]

"Speed requires and creates the insensitivity to everything that could slow down, to the frictions, scraping, hesitations that make us feel that we are not alone in this world; slowing down is becoming capable of learning again, of getting to know, of becoming reacquainted with what holds us and makes us hold on, of thinking and imaging, and, in the same process, of creating relationships with others that are not seizures". [STE 13]

"Snow describes scientists in a somewhat simplistic way, as rational beings devoid of ideology whose process stripped of all emotional consideration is expressed in the anonymous and impersonal style of scientific magazines". [KAR 11]

"The event undoes the temporality required to invent another time from which another world, another view is introduced". [DEF 11]

"It is only when science has become a commercial product, a consumption good, that the distinction between science and public corresponds to a sharing between producers and consumers of science". [BEN 03]

"The entrepreneurs who most succeed never start with brilliant ideas; they discover them along the way". [SIM 11]

1.1. Introduction

As innovative as it may be, the technique which consists of creating objects through the successive superposition of matter is as old as the world, as the creation of pottery with columbine techniques before the invention of the pottery wheel or in the way that swallows build their nests, little by little. There, it is in fact the most natural form of constructing complex geometries [TRA 15]. It is only with computer science that additive manufacturing could be developed. The push was given by the car manufacturing industry, but numerous other applications also take shape.

While the imagination of industrial economies had installed material manufacturing in the world of factories, according to Ford's model, the effervescence surrounding 3D printers seems to open other horizons. There would be something behind this other than the clearly changing production methods, like that presented in Figure 1.1, representing the manufacturing of a wall with a rigid and light 3D structure within created through additive manufacturing [MOL 15]: through compression, the concrete and insulation are integrated into the 3D structure with a radical modification of practices. Other applications for very large objects to create houses are being developed (e.g. see company XtreeE in France) [MAD 15a]. According to Lipovetsky [LIP 06], "Ford's system spreading standardized products has given way to an economy of variety and reactivity in which not only the quality, but also the time, innovation, renewal of products have become criteria for competitiveness..."

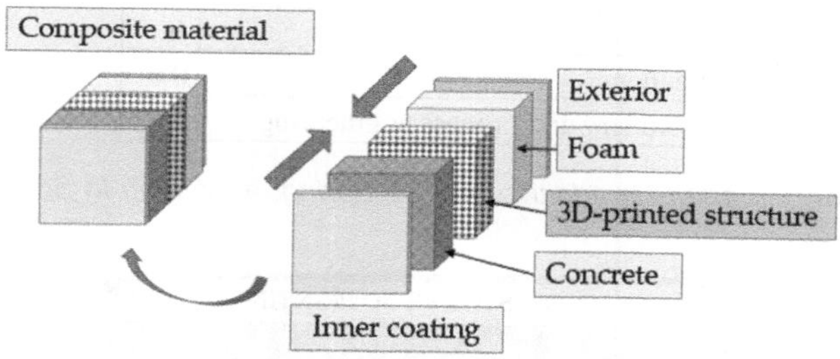

Figure 1.1. *New wall-manufacturing process using 3D network created through additive manufacturing*

Would not 3D technologies respond favorably to all of these comments? Would not they accentuate them? This statement could be found in clothing design with the possibility of specifically creating original personalized models [PAL 15]. However, other applicative niches exist which are presented in the following sections.

Figure 1.2 shows the current ecosystem of the application of 3D technology proposed by Rosenberg, Marion and Turcq [ROS 15]. Only some significant applications are dealt with in this chapter; they can also be found in Lipson and Kurman's work [LIP 14].

According to Sculpteo [SCU 16], the division of the applicative market is particularly vast, spanning every economic sector, as shown in Table 1.1.

	%
Consumer goods	26
High-tech products	15
Industrial goods	14
Electronics and electrical engineering	8
Services	7.5
University education	6
Metallic mechanics	5
Health sector	5
Aeronautics and space	3
Commerce	3
Automobile and land transport	2.5
Energy	2
Maritime domain	1
Food and cuisine	1
Weapons	0.6
Chemistry and process engineering	0.3

Table 1.1. *Current market for additive manufacturing, according to Sculpteo [SCU 16]*

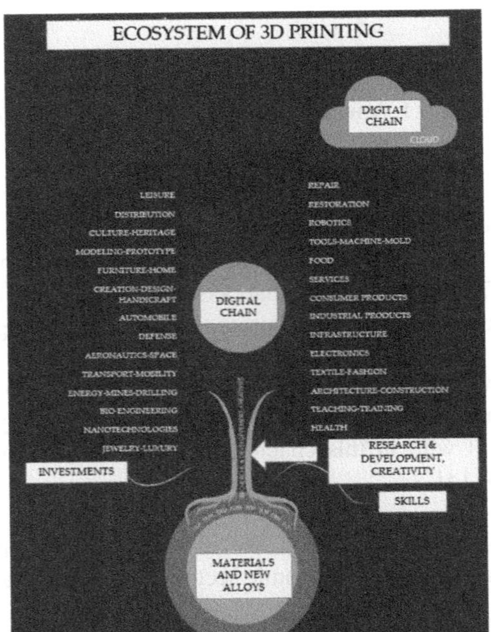

Figure 1.2. *Ecosystem of additive manufacturing (under creative commons)*

1.2. Maritime, military, aerial and spatial applications

The specificity, the functionality and at least the performance of the object are researched [MAH 15]. Thus, the US Department of the Navy intends to develop applications for additive manufacturing onboard warships for the purpose of autonomous repairs. However, this cutting-edge technology is still in its development stage relative to the repercussions that it could have for current manufacturing methods (DARPA [NEW 15, GAO 15a]). For example, in the henceforth classic applications, the creation of prototype parts (e.g. for wind tunnel tests) remains topical, as indicated in Figure 1.3 [AND 94]), (see also Wheeler [WHE 15]), or parts for rockets (see Figure 1.4 from 3ders.org [3D 14]).

Figure 1.3. *Part created through additive manufacturing for a study in a wind tunnel*

Thus, the American "United Launch Alliance" (ULA) uses 3D-printed components for its Atlas V space launchers. The component in question, the duct for the air-conditioning system (ACS), feeds sensitive electronic components within the rocket with nitrogen and is subject to extremely low temperatures. The duct used to be made up of 140 parts, but today it has no more than 16 thanks to additive manufacturing: a change that results in a reduction in costs of 1 million US dollars per year [BER 15]. For Coykendall *et al.* [COY 14], from Deloitte, concerning airplanes, the primary advantages of additive manufacturing are the following:

– production cost: possibly reaching 50% of the current cost;

– "time to market": possibly reaching 64%;

– weight reduction: possibly reaching 64%;

– "buy to fly" relation: 1 to 1.

With such great promises, it is thus reasonable to foresee technological developments of economic interest for additive manufacturing technologies and particularly those concerning metal.

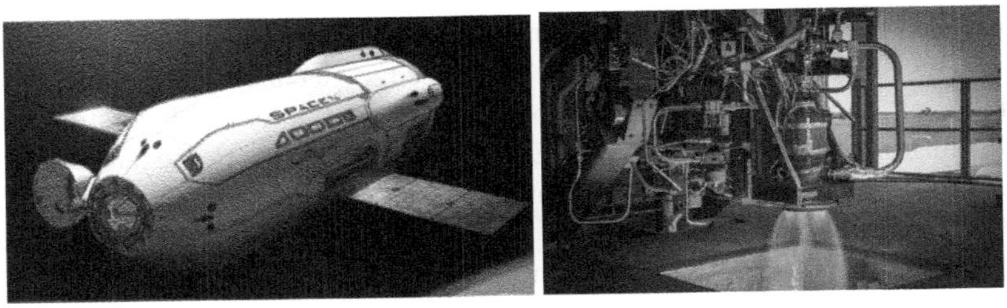

Figure 1.4. *Another niche example of very high added value: space (figure provided by 3ders.org, [3D 14])*

In 2016, General Electric (GE) in the United States developed a metal 3D machine with a diameter of 3.35 m to create massive airplane motor parts [HAL 16] in one stage, allowing a great reduction in weight. Likewise, the ESA (European Space Agency) has recently created an additive manufacturing unit in the United Kingdom in order to be able to create more robust, lighter and less expensive shuttle structures compatible with space (with a need to evaluate 3D technologies to ensure that they respond well to the demands and constraints of the hostile environment that is space). According to Newsroom [NEW 16], "the Agency has made the latest material available to the research team with a 'metal' 3D printer, an entire series of microscopes, an X-ray tomo-densinometric instrument, and ovens. So many devices that will help the researchers test the elasticity, durability, and micro-resistance of the materials."

Thanks to additive manufacturing, the inside of a titanium part can be designed with light structures by depositing material only where it is necessary. Zone Bourse [ZON 16] reminded us that the cost of putting 1 kg into orbit has risen to roughly $20,000 US; each gram saved obviously pushes the limits of space exploration.

Another representative example of the impact of 3D technologies is the project to put 3D machines in space to create tools and repair elements on-site [BRE 16], as presented in Figure 1.5.

Figure 1.5. *"Spatialized" additive manufacturing*

Furthermore, Turbomeca, an affiliate of the Safran group, has recently started mass-producing parts through 3D printing (selective laser melting or SLM). This involves motor fuel injectors designed and developed for helicopters in the near future. Turbomeca hopes to also use 3D printing to produce the swirler component for helicopter combustion chambers [CET 15].

According to Victor [VIC 16b], the military domain has seized additive manufacturing and is using it for the following applications:

– tanks: the Russian company Uralvagonzavod is said to be working on the development of titanium parts using additive manufacturing intended for armored vehicles other than prototypes;

– missiles from the American company Raytheon, leader in the production of guided missiles;

– 3D-printed foods personalized for the American army;

– 3D-printed drones capable of detecting and remotely detonating "anti-personnel" mines (USA);

– caring for those wounded in wars with bio-printing of skin;

– combat airplane elements (Israel, USA, UK, France, etc.);

– 3D printing to rescue aircraft carriers and warships; as with all confined spaces, it is of interest to create replacement or repair parts on-site from defective parts on demand (USA and DCNS-France according to Alex [ALE 16a]);

– military equipment for soldiers (relief and personalization in the USA);

– false "anti-personnel" mines printed for training bomb-disposal experts;

– and so on.

1.3. Conception: art and new domestic applicative niches

According to Notteau [NOT 15, COE 17], 3D printing will have a major disturbing effect on the conception and construction industry. As an example, a leading plumbing manufacturer has just introduced a sellable product using 3D printing technology; likewise, the US FDA (Food and Drug Administration) has just accepted the commercialization of an anti-convulsant medication to be created using this technology [MOL 15a]. An American laboratory, Aprecia Pharmaceuticals, therefore received permission to introduce a 3D-printed medication on the market. Despite all this, it is not possible to order this at the local Fab-Lab; the interest lies elsewhere: by using 3D printing, the manufacturer has managed to produce a particularly porous medication that dissolves in < 5 s, as opposed to 40 s for a classic medication, a sizeable innovation that for the moment is reserved to be used as an immediate treatment medication in case of epileptic fits. The company nevertheless specifies that it intends to manufacture other medications in this way [IND 15]. Today, there is a chance of printing 3D packaging containing medication or vitamins that spread throughout the body [MÉL 17c].

Independently of these aspects (which must not be neglected through the exploitation of original applicative niches), the process incrementally democratizes conception through additive manufacturing. Manufacturing is decentralized, which will result in overturning the design and construction industry and many other sectors, such as the food industry [COU 15, TWI 11]. Figure 1.6 presents an example of 3D food proposed by Alex [ALE 16d] (see also Koslow [KOS 15c], Notteau [NOT 16], Van der Linden [VAN 16], Mahon [MAH 16], Hall [HAL 16b], [HAL 16], Alex [ALE 17], Tucker [TUC 14]). In his Fab-Lab, Dupont [FAB 17] created a machine that enables chocolate to be printed, as illustrated in Figure 1.7. A new, more efficient business model developed for measuring design could also see the light of day (e.g. shoes and clothing as illustrated in Figure 1.8 (see also Mekkaoui [MEK 14]), clothes, in Figure 1.9 [ALE 15c] and jewelry, in Figure 1.10 ([GRA 15, PRO 17], see also Collectif Textile [COL 13] and Shapeways [SHA17]). With the possibilities of this technology, designing shoe soles and clothing specific to individuals becomes more easily accessible.

According to Mcor IRIS HD [MCO 17], the creation of "3D photography" becomes possible, as indicated in the left-hand side of Figure 1.11 (see also Tong [TON 11]). The creation of color parts is not as simple as we might imagine, even if an object can be printed using colored voxels (volume elements) with basic colors, due to problems with restoration quality [BRU 15].

These new forms of production would allow a reduction in the stocks that stem from mass production, all while opening a new world for conception and manufacturing. Moreover, by introducing determinism into the manufacturing of foams, for example, there is a possibility of manufacturing personalized shoe soles dedicated to a particular functionality (unlike foams whose random structure is linked overall to the energy provided (particularly from chemistry)) [ALE 15d, TUC 15, CLA 17]. However, we must then, for certain applications, reflect on Paquet's reflection [PAQ 05]: "Solutionism deals with problems like puzzles to which there is one corresponding solution, rather than as questions for which there could be one response?"

Figure 1.6. *Presentation of a plate of printed food [ALE 16d]. For a color version of the figure, see www.iste.co.uk/andre/printing1.zip*

Figure 1.7. *3D "Chocolate" printer (photograph conceived by the NYBI association at the Lorraine Fab Living Lab®, www.lf2l.fr (photo credit: Laurent Dupont, ERPI Laboratory, University of Lorraine, 2017)*

Modern Meadow in the United States proposes to print meat meant for human consumption. The system employs a biological printer, the NovoGen MMX Bioprinter, from Organovo. The so-called "bio-ink" is made up of muscular cells, adipocytes for storing fat and fibroblasts to ensure the coherence of the printed meat. The process relies on a substrate made up of agarose, a polymer derived from the gelating agent agar-agar. The goal is to produce 2 × 1 × 0.5 cm pieces of animal product, in other words, something to make ground meat. Phase II of the production process aims to produce larger pieces [BRI 16, USD 13, DOU 17]. A more "classic" application, but one that is lateralized in this regard and likely more "acceptable" in French society, which has some reservations regarding the quality of cuisine, involves creating artificial leather [MOD 16].

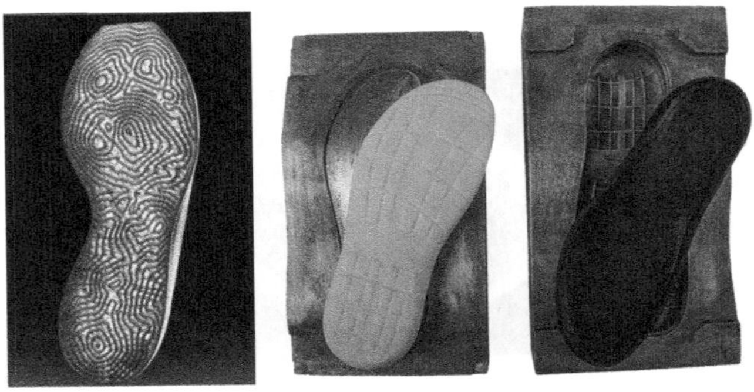

Figure 1.8. *Left: Shoe sole created through additive manufacturing [ALE 15d]; center and right: mold created through additive manufacturing by Prodways [PRO 17] - Copyright ©Prodways ©Hamilton de Oliveira*

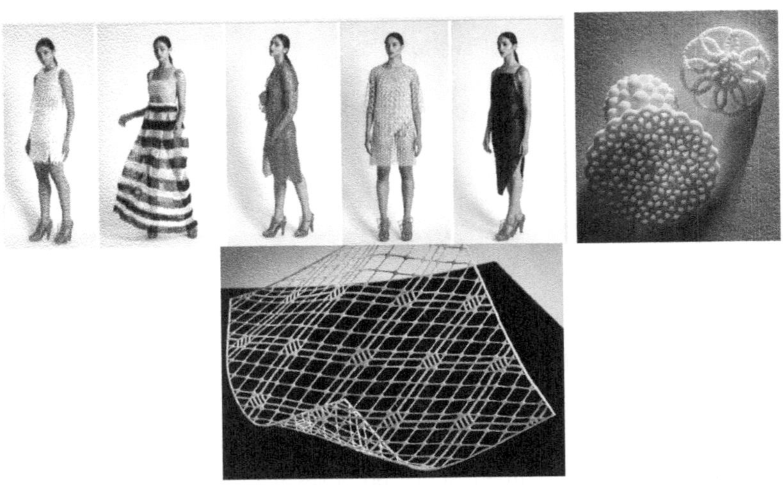

Figure 1.9. *3D clothing [ALE 15c], "haute couture" buttons and 3D lace [PRO 17] - Copyright ©Prodways ©Hamilton de Oliveira*

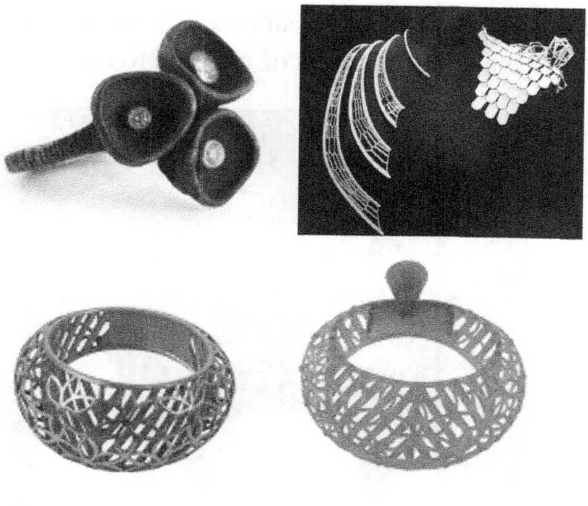

Figure 1.10. *Examples of jewelry (top left: Graham [GRA 15]; top right: Prodways [PRO 17]; bottom: rings; Prodways [PRO 17]) – Copyright ©Prodways ©Hamilton de Oliveira*

Figure 1.11. *3D "photographs" (agreement of Mcor IRIS HD, [MCO 17]; left and middle image generously provided by 3D Systems; right image corresponds to a photograph taken by Arte Journal [ART 17])*

1.3.1. *Art and additive manufacturing*

3D printing, in positioning itself as a technological revolution, necessarily inspires the manufacturing process. For some time now, works of art have been emerging in different areas: the reproduction of works of art (sculptures, paintings), the transformation of 2D drawings into 3D ones or even creations. Figures 1.12 and 13, extracted from Ars-Mathematica [ARS 16] and Marco Mahler [MAR 17], give some ideas about the interesting possibilities of this technology, where a virtual tool is used to achieve the transformation of material to create a real object of art.

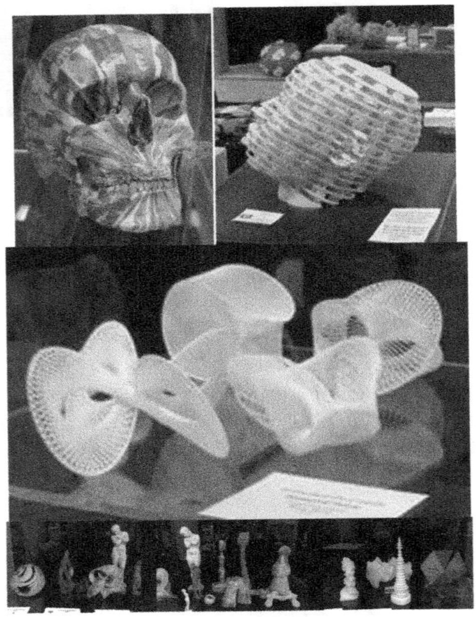

Figure 1.12. *Works by C. Lavigne (upper left and center), F. de Comité (upper right) and L. West, A. Werby and M. Neubauer (from bottom left to right) [ARS 16]. For a color version of the figure, see www.iste.co.uk/andre/printing1.zip*

Figure 1.13. *3D mobiles (reproduced with the authorization of Marco Mahler www.marcomahler.com)*

An increasing number of works focus on the light of day thanks to this new technique for working with material (Varin [VAR 16], J. Satterwhite cited by Cicetti [SAT 15]), as seen in Figure 1.14.

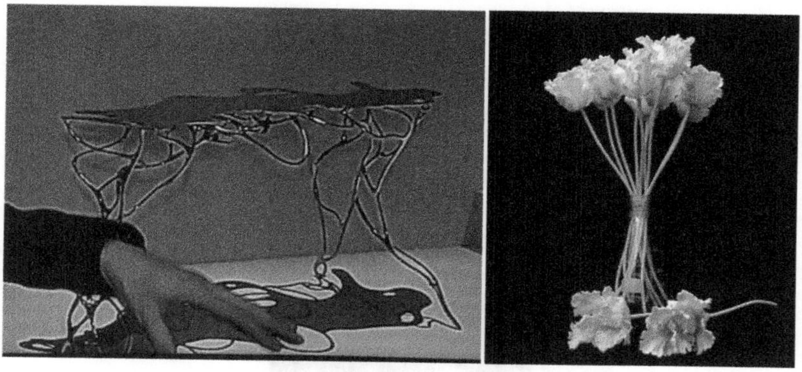

Figure 1.14. *Artistic table created through additive manufacturing (left: Arte Journal [ART 17]; right: Prodways) – Copyright ©Prodways ©Hamilton de Oliveira*

1.3.2. *Archaeology, museum restoration, reproduction*

The final effect on 3D copying corresponds to the goal sought by the museum that hoped to offer the public objects identifiable as facsimiles of glazed pipe terracotta objects [ROU 07] to allow the observer to notice different stages in the realization of ceramics (from the undecorated model to the prototype glaze imitating the final dream, going through the intermediate decorating phases). This final element responds to the museographic project, which then favors didactic presentation (see Figure 1.15). Other older applications had been Schmitt *et al.*'s target in 1994 as they used additive manufacturing to create small copies of the elements of Ancient Egyptian sculptures found in the Port of Alexandria to foresee optimal modes of connecting discovered pieces and their restorations. This application allows us to avoid contact with extremely degraded objects and these elements to then be kept intact so as to envision new forms of restoration. Other, more recent applications aim to create 3D prototypes to represent partially missing or damaged monuments [MÉL 17]. For Hazotte [HAZ 16], computer-assisted manufacturing processes involving the addition of material promoted the re-assembly of a fragmentary and lacunar archaeological pitcher.

A recent example of applying 3D technologies to art is the development of 3D structures from 2D paintings to make them appreciable for the blind (see Figure 1.16; see also Scalco [SCA 16]).

Figure 1.15. *Applications in restoration for museums; project to reconstruct a gate in Palmyra destroyed by ISIS [BEN 15]; project to reconstruct a bust of Nefertiti [BAL 16]; bottom, sculpture [PRO 17] - Copyright ©Prodways ©Hamilton de Oliveira*

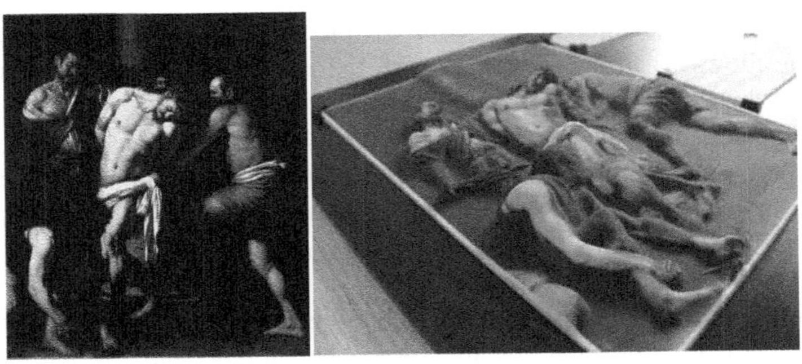

Figure 1.16. *Making paintings accessible to the blind (The Flagellation of Christ by Caravaggio) (figure provided by D. Scher info@3dprintingbusiness.directory)*

1.3.3. *Construction sector*

In what way could 3D printing revolutionize the building and construction sector? According to Morel [MOR 16], this technology's interest lies in the possibility of creating structures that are unimaginable with classical construction methods, with complex geometries and precision down to the millimeter scale. Construction will no longer be the same before and after large-scale 3D printing. In fact, 3D printing will allow this sector to be revolutionized by significantly reducing the costs and time of creation, as well as the manufacture of custom-made structures at the cost of standardized structures, which are of course less optimized for very specific applications (let us remember the leitmotiv of additive manufacturing, where it is often written that the complexity of the object is "free"...). Furthermore, the quality of construction is going to be heightened considerably with the appearance of new architectural designs (not yet invented), permitted by this technology. A new architectural movement should emerge from the exploration of joining architectural art, mechanical sciences and additive manufacturing.

Large-scale 3D printing will initially be used in certain stages of "traditional" construction, particularly for the pre-manufacturing of building operative systems (walls or flagstones) and the creation of molds (façade panels, posts, etc.). It will certainly be used for certain interior layouts with a strong added esthetic and technical value as well. The complete automation of construction will likely also take time, but the presentation in the coming months of concrete, large-scale creations should accelerate this transformation process. In any case, in a strictly economic sense, the tipping point between traditional construction and its costs (which have reached a lower limit after dozens of years of process optimization) and the costs associated with 3D printing could come without further delay. Figure 1.17 shows drawings from Morel's figures [MOR 16], illustrating the potential of 3D technologies for construction (see also Hall [HAL 16a]).

Figure 1.18 presents a "home" created by the Russian company Vocativ according to Alex [ALE 17b] (see also DDD News [DDD 15], Newsroom [NEW 16], Pierrakakis *et al.* [PIE 14], Benssoussan [BEN 16], and for a prospective view of a sea de-contamination project installation created with materials present in the water, Batiactu [BAT 16]). Figure 1.19, created in a laboratory, is a model of the Savoie in Pessac villa (in France).

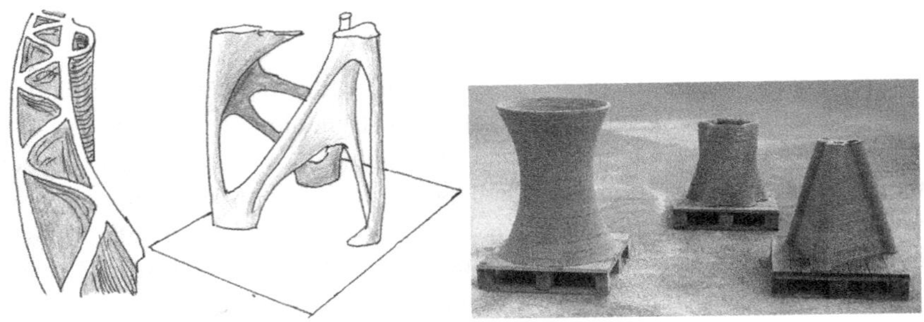

Figure 1.17. *Close-up drawing of new structures created for construction and construction elements generously provided by XtreeE*

Figure 1.18. *House being created through additive manufacturing [ALE 17b]*

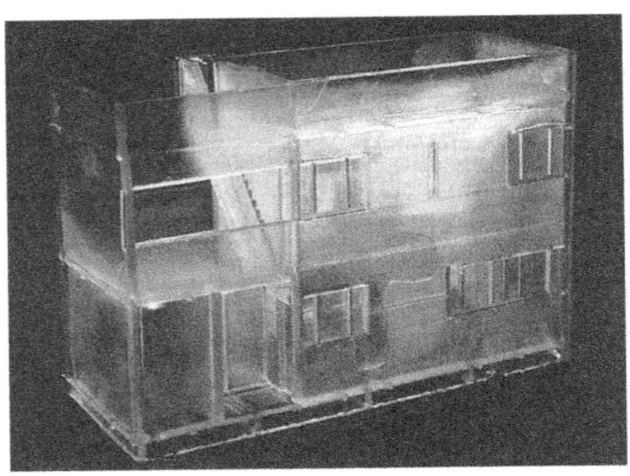

Figure 1.19. *Model of the Savoie de Pessac villa*

Researchers at the "Institut de Recherche en Communications et Cybernétique" in Nantes (France), in cooperation with engineers from the French company "Capacité", have created a non-standard 3D printer prototype, the INNOprint 3D, which is capable of manufacturing a large-dimension emergency home in less than a half hour. This mobile printer could be moved from one site to another and even play a role in the framework of emergency operations during certain natural disasters. Complementary research should be conducted to develop materials other than polyurethane, which has been used during the experimental phase [UNI 15]. UFUNK [UFU 17] reports the creation of homes in very modest times (24 h) at particularly attractive costs. In short, new projects are always ongoing.

1.4. Mechanical parts

3D printing processes, as indicated by Rosen [ROS 14], De Véricourt [DE 16] and Guo and Leu [GUO 13], must take into consideration couplings between the desired object, its possible creation and its use with different constraints (cost, performance, etc.). An example that respects this schema is presented in Figure 1.20 (see also ADDAERO [ADD 15]) and Figure 1.21, a drawing inspired by a figure by Chabaud (2015) illustrating a scientific application of 3D printing. The following illustrates the potentialities of metal 3D technologies (Figure 1.22, top left), 3D ceramics (Figure 1.22, top right) and a complex part created by the DED/CLAD (direct energy deposit) process of a part made of stainless steel (generously provided by BeAM).

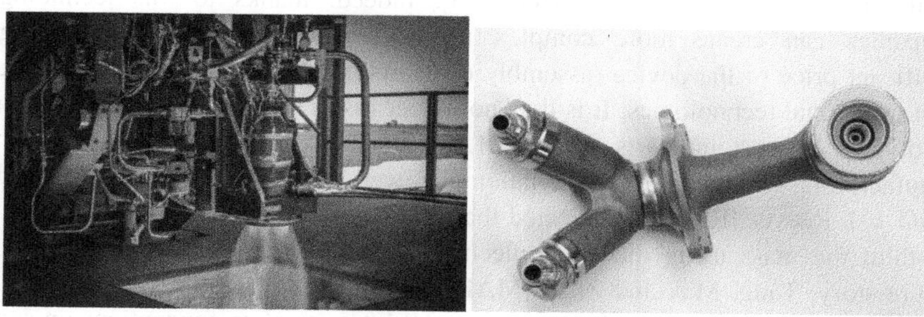

Figure 1.20. *Examples of additive manufacturing; left: part of a rocket motor (figure provided by 3ders.org [3D 14]); right: injector [GAO 15]*

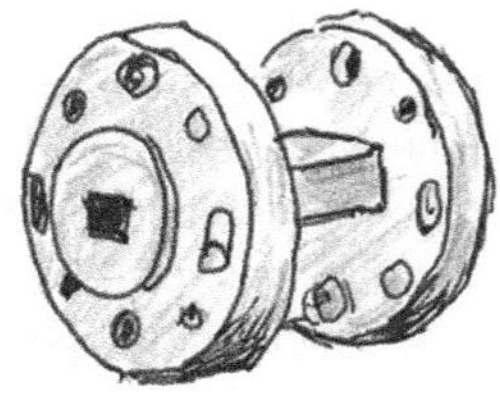

Figure 1.21. *Close-up drawing of the creation of a waveguide for research in electromagnetism*

Additive manufacturing technologies have long been reproached for the modest robustness of objects; this image is disappearing. Siemens, for example, has recently made a major breakthrough in the production of industrial parts that can be used in a nuclear power plant. Considering this sector's strict security and safety norms, "this achievement constitutes a veritable success and a major advance" [IND 17]. Nevertheless, it is worth noting that this (so far) only concerns components that are not used in the heart of these plants.

The Western industrial sector is no novice in the field of additive manufacturing and integrates itself well into the dynamics of competition between technologies. Indeed, additive manufacturing has long been used to create prototypes: car parts, airplane parts [JAC 16], tools and so on. However, leading industrialists have started creating finished parts, and some materials used by 3D printers have been certified by the responsible authorities [MOU 15]. Indeed, thanks to this technology, companies can create more complex parts than before for the increasingly significant price of the device (assembly of complex parts) than those with objects from traditional technologies. It is thus no longer just a matter of creating only test, prototypes or custom parts. Figure 1.23 illustrates the interest of additive manufacturing technologies for aviation (as well as so-called bionic methods) [CAO 17]. Reeves [REE 12] indicated that it is possible to create objects that have one-third the mass using the same mechanical strength for structural parts. In a feature story, Tang, Mak and Zhao [MAK 16], by taking lifecycle analysis as the objective function, presented the environmental interests of re-conceptions on some examples. Table 1.2, from Reeves [REE 12], shows all the interest of 3D technologies in terms of weight and consumption with the same amount of carbon waste. This view is confirmed by Schwartz, Director of Innovative Solutions at CENIT [CAO 17], who writes: "Among other things, we are aiming for a considerable reduction in time for the complete development of optimized 3D printing parts, as well as significant savings in terms of weight on airplanes, which will lead to a clear reduction in CO_2 emissions over its lifetime." These data are also significant enough to illustrate the potential interest of additive manufacturing (see also Ford and Despeisse [FOR 16]). They must still be corroborated, keeping in mind all the influential parameters of the production chain.

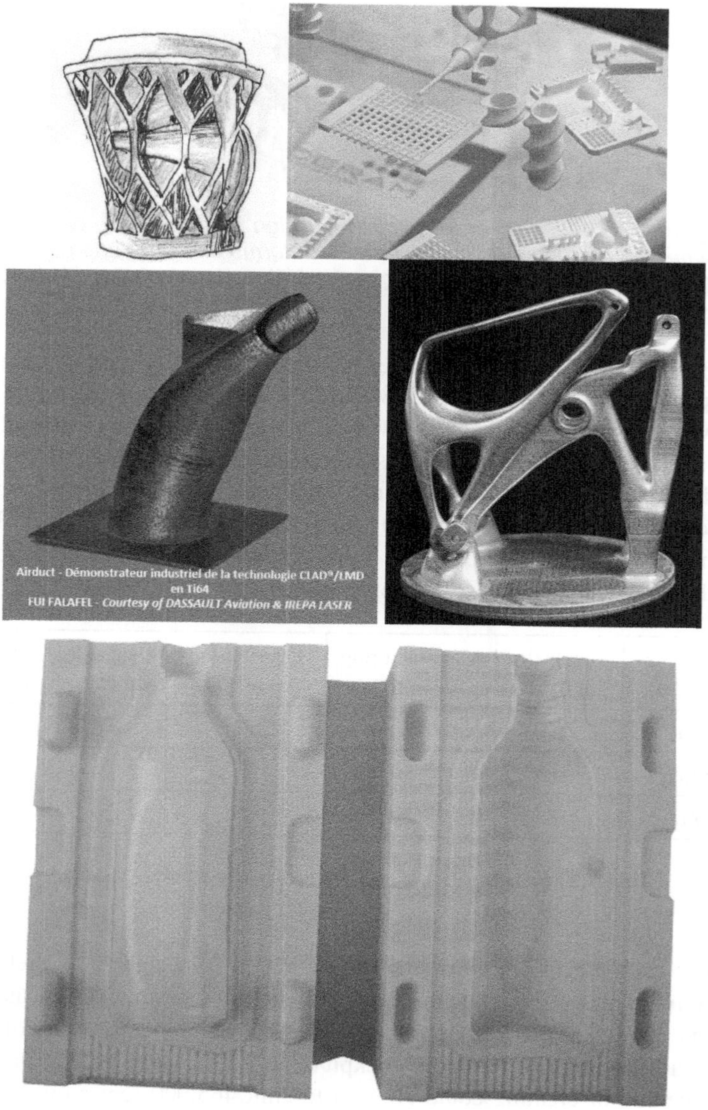

Figure 1.22. *Top: Drawing of a chemical reactor by AIF [AIF 16], ceramic parts (France 3, 2016; Mélanie [MÉL 17]; middle: industrial parts (BeAM left and 3A right); bottom: mold for bottles [PRO 17] - Copyright ©Prodways ©Hamilton de Oliveira*

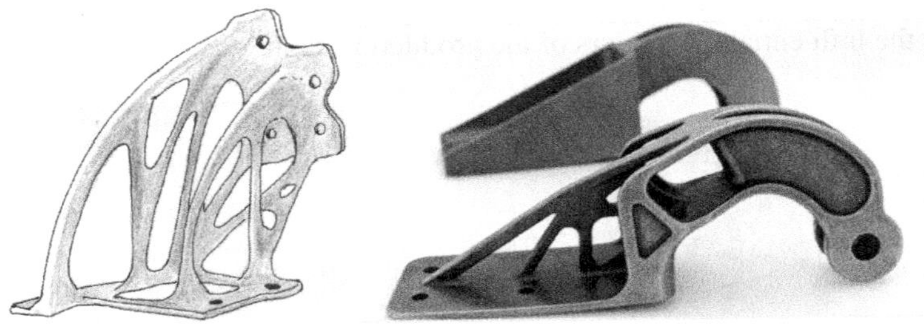

Figure 1.23. *Close-up drawing of a re-design for an airplane part (seat attachment), illustrating the improvements in terms of weight and comparison between traditional manufacturing and additive manufacturing [ALE 15e]*

Lavergne *et al.* [LAV 15] reminded us that, "today, the early phases have a major impact on the cost of products: 80% of the final cost is fixed, while only 5% of the development costs have been involved. It therefore becomes crucial to make methods and tools available to the designer which allow him/her to consider the specificities of [additive manufacturing] and which are integrated within the informational and human ecosystem used by companies: Product Lifecycle Management (PLM)" [GAR 05]. We must re-think conception because the 3D process allows it. (see Figure 1.27, for example.)

Manufacturing process	Matter (kg of CO_2)	Manufacturing (kg of CO_2)	Distribution (kg of CO_2)	Usage (kg of CO_2)	Lifecycle (kg of CO_2)
Classical	100	2	5	44	44
3D with mesh	16	5	1	16	16
3D optimized	18	7	2	20	20

Table 1.2. *Environmental advantage of the re-designed part*

Another example in air transport: the reduction of weight is a fuel economy factor. Combining a solid structure with a light micromesh, Airbus' new bionic partition is 45% (i.e. 30 kg) lighter than the current models. According to the creator, if this is applied to the entire cockpit of an A320 and to the current backlogs, this new method of creating could allow it to emit 465,000 fewer tons of CO_2 per year. The new bionic partition is manufactured on the basis of Scalmalloy, an aluminum–magnesium–scandium alloy finalized by APworks, a branch of Airbus specialized in additive manufacturing and cutting-edge materials. Specially designed for 3D printing, Scalmalloy offers excellent mechanical properties and a great extension capacity. This is the first time that it has been used on a large scale within an aeronautical component (see Figure 1.24 by Greg [GRE 15]).

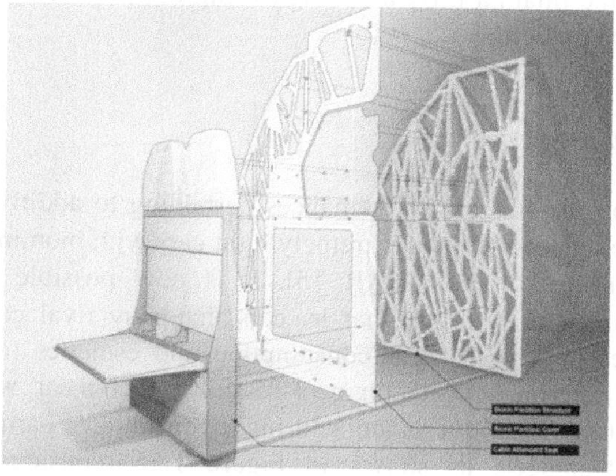

Figure 1.24. *Airbus' "bionic" partition (printed with the authorization of Grégori Pujol alias Greg)*

According to Atherton [ATH 16], Airbus has reinforced its advance in additive manufacturing by producing a 3D-printed drone. Named THOR (Test of High-Tech Objectives in Reality), the aircraft is made up of 55 parts created through 3D printing. "Only the undercarriage, the battery, the two electric motors, and the control commands were not created using additive manufacturing techniques. THOR is roughly 4 meters long and weighs approximately 25 kg. Its production cost does not exceed 25,000 euros, a particularly attractive price for this sort of solution. In addition to its low cost, the possibility of very quickly changing a damaged part after an accident constitutes the primary advantage of this device" [ATH 16].

The ORNL (Oak Ridge National Laboratory) currently holds the record for the largest industrial part produced with its "Big Aera Additive Manufacturing Machine". According to Angrand [ANG 16], "this part will be used in the scope of manufacturing the wings for the Boeing 777X. Its dimensions are 5.33 × 1.68 × 46 cm. Thirty hours (rather than 3 months) are necessary to produce this kind of part with additive manufacturing by using carbon fiber and composite materials made up of ABS thermoplastics, which allows the part's mass to be reduced to 748.4 kg."

Thus, according to De Véricourt [DE 16], 3D printing in two fields is developing into enhanced manufacturing and alternative manufacturing by integrating the concept of the factory of the future. For enhanced manufacturing, it is possible to create tools, patterns, fixations and so on. Alternative manufacturing uses 3D printing to create the final part. By avoiding (or limiting) breaks in production, new tools can be created on demand.

1.5. Land transport

The spatial and mechanical performances pertaining to additive manufacturing allow for the development of new, extremely light cars with biomimetic foundations [CAS 15]. According to Hart [HAR 15], it is now possible to create "real" automobiles through 3D printing, that is, cars that may rival cars created using classical processes in terms of gas consumption and esthetics (although still not quite comparable in terms of cost). Figure 1.25 shows such a car, which, with others created largely through additive manufacturing, demonstrates particularly attractive environmental advantages, as shown in Figure 1.26 (considerably lower CO_2 emissions than conventional vehicles). A complete analysis from conception to waste elimination (or recycling) could possibly validate or disprove this assertion. Figure 1.27 shows a partial drawing of a 3D electric motorcycle weighing 35 kg created through the selective sintering of aluminum alloy (with 60 μm thick layers) that can reach a speed of 80 km/h [JUL 16].

Figure 1.25. *Automobile created with 3D printing [ALE 17c]*

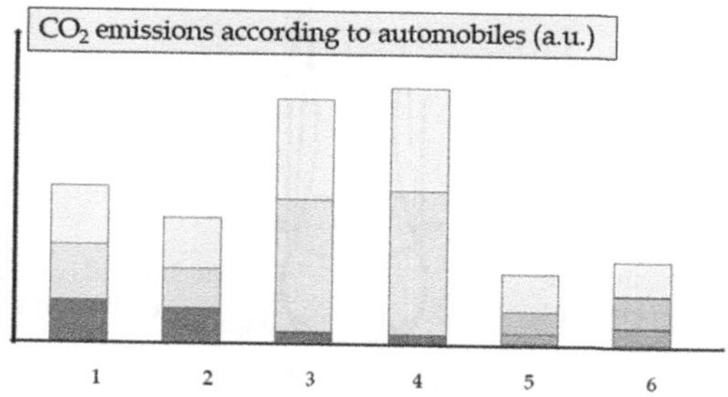

Figure 1.26. *Comparison of pollutant emissions from traditional vehicles and those created with 3D printing (a.u. for arbitrary units); 1: Gasoline-powered vehicle; 2: Hybrid vehicle; 3: Electric vehicle; 4: Electric vehicle (SUV); 5: Gas-powered vehicle created through additive manufacturing; 6: Gasoline-powered vehicle created through additive manufacturing; the lower part of the consumption indices represents the vehicle's functioning cost, median consumption and manufacturing height*

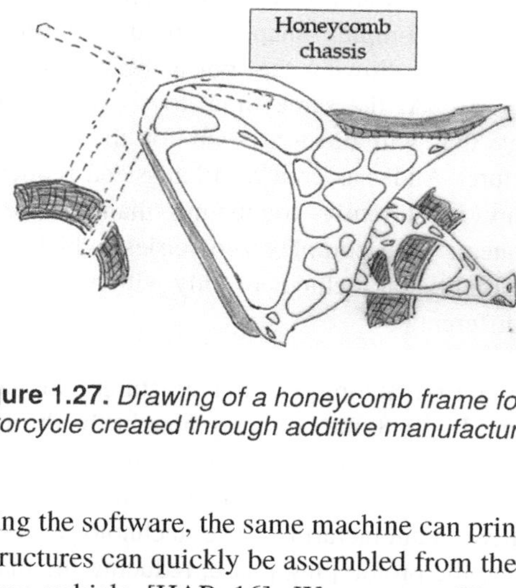

Figure 1.27. *Drawing of a honeycomb frame for a motorcycle created through additive manufacturing*

By simply changing the software, the same machine can print a sports car as well as a mini-van. The structures can quickly be assembled from their different elements so as to construct any vehicle [HAR 16]. We are possibly on the threshold of reinforced personalization of the shape of automobiles, knowing that a large part of the elements (motor, transmission, etc.) will continue to result from classical technologies. Some people think that this is the re-discovery of the individual initiative and financial gain for automobiles thanks to 3D technology, considered as energy-conscious relative to the traditional chain-production processes. For Giffi, Gangula and Illida [GIF 14], a car could be produced with a set of components created through additive manufacturing, as shown in Figure 1.28.

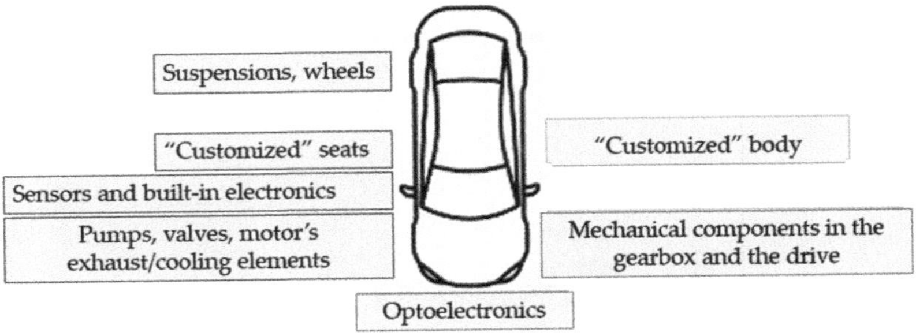

Figure 1.28. *Possible evolutions of automobile productions allowed by additive manufacturing*

Another example, returning to additive manufacturing: Opel in Germany has managed to reduce the production cost of its manufacturing tools by nearly 90%. 3D printing has also allowed technical teams at Opel to create tools in less than 24 h that are ready to be used on the assembly line. The company will henceforth be in a position to produce more complex shapes without the limitations imposed by traditional manufacturing by allowing the tool to be adapted to the user and to a specific vehicle [3D 15c]. At the same time, to show its know-how in metallic printing, a technology that will be used soon for their mass-produced cars, the German car manufacturer Audi has created a half-sized replica of the 1936 Silver Arrow. This company is currently conducting many studies to use it in the production of components with complex geometries, which would not have been possible according to traditional methods or only with a great difficulty [PRÉ 15b]. Ford [DIG 17] is no different.

Ford is one of the first automobile manufacturers to have had recourse to additive manufacturing. In 1988, the company bought the third 3D printer available on the market, using dozens of innovations, according to Alex [ALE 15]. From this initialization to the research stage, the company has printed more than 500,000 pieces through additive manufacturing. The technology allows the automobile company "to accelerate the prototyping phase, reducing the manufacturing of some parts from 6 weeks to only a few days, from motor components to simple rims to the body structure" [ALE 15]. Other examples have accumulated regarding this subject: the alliance between IBM and Local Motors to create a minibus through additive manufacturing [AUT 16]; diverse automobile activities [DEL 16a].

Divergent 3D is to make its expertise applied to automobiles useful through an agreement with PSA (a French car manufacturer) that will allow entire cars to be produced using 3D-printed metal structures with shapes and characteristics that were previously impossible. Even if the details of the partnership have not been published, according to Victor [VIC 16], Carlos Tavares from PSA has said, "We have the potential to optimize our industrial footprint, reduce the total weight of the vehicle, and the complexity of manufacturing, all while allowing ourselves nearly infinitive flexibility in terms of conception. This is a radical transformation for our activity." It is the same for Renault-Nissan (French–Japanese car manufacturer), which is trying to make its motors lighter thanks to additive manufacturing [PRO 17]. In short, the automobile sector has integrated additive manufacturing, even for F1 cars (see Mélanie [MÉL 17a], Clarke [CLA 17a]).

"Pushed to its limits of movement, it will also be possible to create one's own bicycle through additive manufacturing in the coming weeks (Soocurious [SOO 16]; see Figure 1.29 inspired by Alex [ALE 15a]), even one's own wheelchair…"

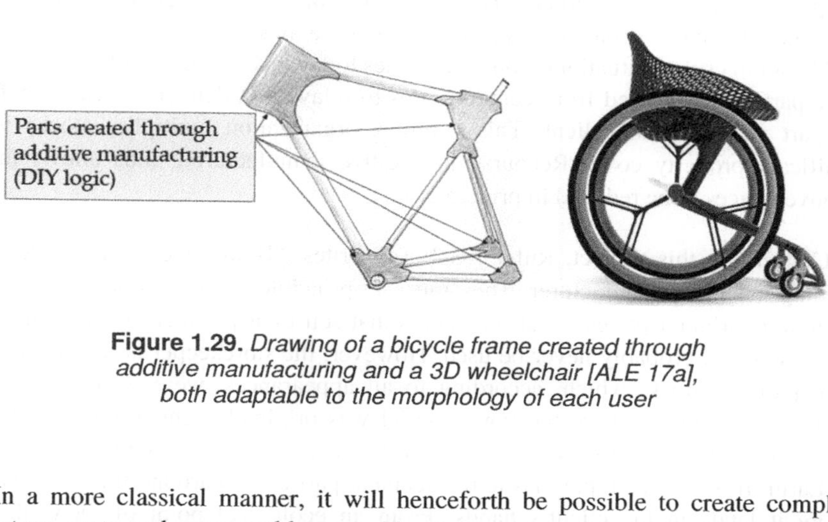

Figure 1.29. *Drawing of a bicycle frame created through additive manufacturing and a 3D wheelchair [ALE 17a], both adaptable to the morphology of each user*

In a more classical manner, it will henceforth be possible to create complex prototype parts and even usable parts like the rearview mirror element [PRO 17] presented in Figure 1.30.

Figure 1.30. *Rearview mirror element [PRO 17] - Copyright ©Prodways ©Hamilton de Oliveira*

1.6. The question of spare parts

Spare pieces participate in the good functioning of the equipment belonging to companies or private individuals. They must very often be delivered in a very short timeframe to avoid prolonging the cessation of a machine's functioning, which could lead to critical situations; some deliveries today take place in 24 h. Classically, spare parts are delivered from central stores to relays that deliver to dealers before the part arrives to the client. This complex organization multiplies stocks with significant property costs. Recourse to additive manufacturing thus allows these inconveniences to be reduced in principle.

Concerning this subject, Rufer [RUF 14] writes, "If we take the example of a washing machine manufacturer: They must, in principle, provide replacement parts (often at exorbitant prices) to all the stores that sell their products, even if most of these parts will probably never be used. However, the storekeeper is forced to have these parts in stock, which, according to all appearances, were manufactured in China, for example, where the raw material was originally sent. From China, the parts were then sent to the central warehouse that spread them to the four corners of the Earth. In short, a simple plastic replacement part can go around the world twice before it lands in the client's hands. From an ecological point of view, this is unacceptable. All the more so today as a collective conscience of consumers who largely integrate ecological impact into their purchasing act is developing today."

The market for digitized replacement parts is thus becoming a "profitable" economic market.

1.7. Toys for the young and the "not-so-young"

As an example, it is possible to create toys (Figure 1.31, Alex [ALE 14], see also Molitch-Hou [MOL 14], and Pauliat [PAU 14]).

Figure 1.31. *Toys created through additive manufacturing*

According to Cyberland [CYB 15], however, the Website Ghost Gunner (https://ghostgunner.net/) has been running a facility to produce automatic and semi-automatic weapons for $1,500 (shipping included!) since August 2015. It would also be possible to 3D-print one's own munitions and customize them as a function of the desired use. There are several accessible and open-source programs dedicated to 3D printing (see Benedict [BEN 16], Tampi [TAM 15]). As with weapons, printed munitions have no reference number and thus remain untraceable. The challenge is great (see Figure 1.32). The only technical solution would consist of sensitizing and mobilizing all of the builders of 3D printers to avoid abuses of their use. These could then equip their machines with a process capable of "algorithmically" recognizing the shapes of mechanical pieces involved in the composition of a weapon and automatically blocking the printing thereof. This example is great evidence of a security difficulty, given that this eventuality is just at the stage of pious wishes! According to Koslow [KOS 15b] pushing the technique to the limits of its application can move to other niches like the creation of sex toys [SEX 16], even some that could serve to provide sexual education to the blind [DUP 17]…

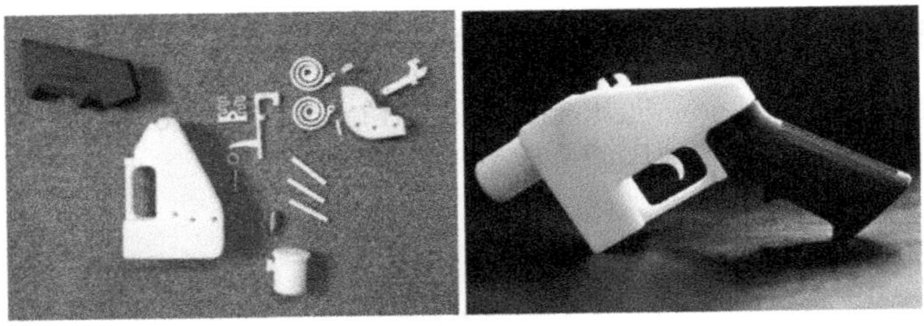

Figure 1.32. *Functioning gun with bullets created through additive manufacturing, made of polymer that is unrecognizable in an airport scanner [ALE 14a]*

1.8. "Traditional" medical applications

3D printing is a small revolution partially accepted for the medical and biomedical fields [BIG 11, REN 11, CHE 16]: prostheses, dental implants, molds for mouthpieces, recognition elements for the blind [ORT 14], bone repairs [BAG 11] and so on. There are very diverse uses and countless promises, but they have already posed the question of industrial property, as shown by Esmond and Phero [ESM 15]; the market is immense with estimated benefits in relation to the potential demand, which leads to this quest, while the biological feasibility stage of an emerging domain, which will be discussed in volume 3, that of bio-printing, essentially remains to be achieved. However, scientific and medical advances are taking shape, as seen in the work recently published by Johnson *et al.* [JOH 15], on nerve repairs and illustrated in Figure 1.33.

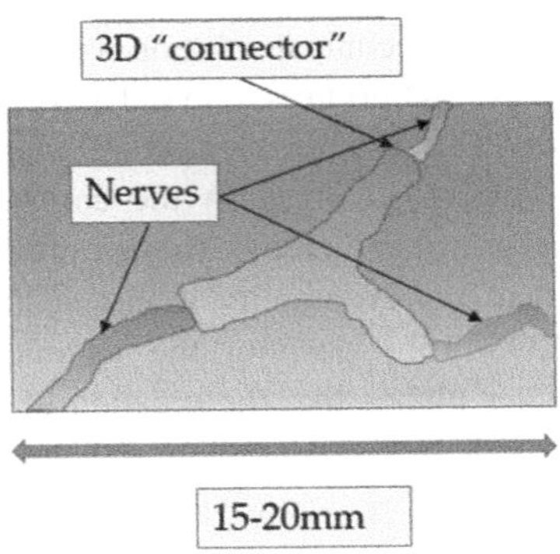

Figure 1.33. *Close-up drawing of using 3D technology to repair and regenerate nerves (rat)*

Some significant examples have also been collected for this part and are shown in Figures 1.34 and 1.35. Figure 1.34 presents a cranial bone repair: using the digitization of a skull and "purification" of the outline (see upper part), the object to be prepared as a future repair becomes possible [GIM 99]; a result from another "experiment" is presented in the lower part (images provided by 3DCeram [3DC17]. For more details about this field, the following references can be consulted: [GIA 09, SHE 15d, ERN 15, FOR 15, HAR 14], etc.). Likewise, ghosts that can be used in medicine (digital prints in legal medicine, artificial hands to calibrate medical instruments [ARO 13]) are being developed (Figure 1.35, e-Nable's prosthesis adapted to a young child, according to Alex [ALE 15b], and ear cartilage, according to Alex [ALE 16b]).

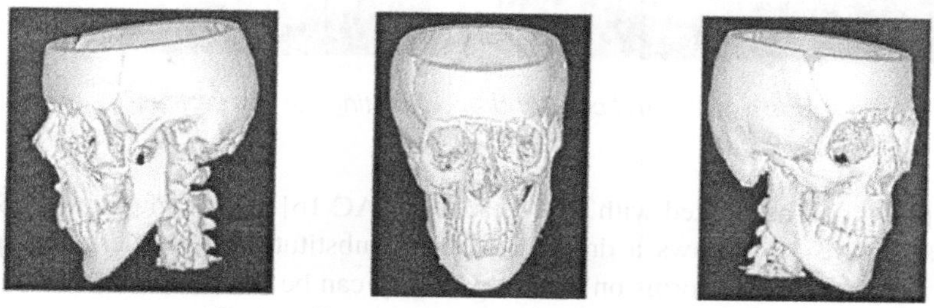

Data collection before the repair [GIM 99]

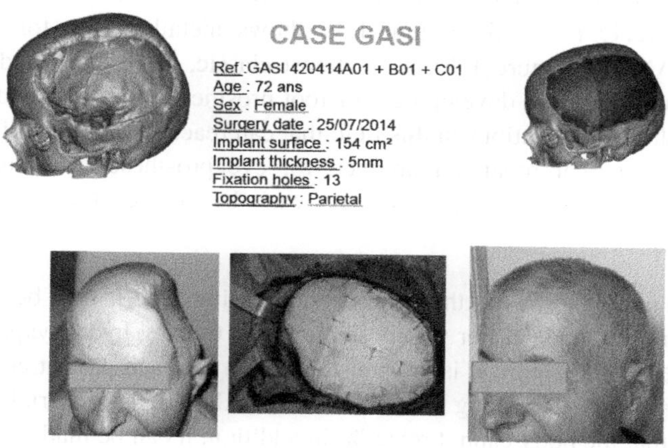

Figure 1.34. *Top: Porous ceramic repair part used during surgery (courtesy of 3DCeram); bottom: photographs before and after the operation*

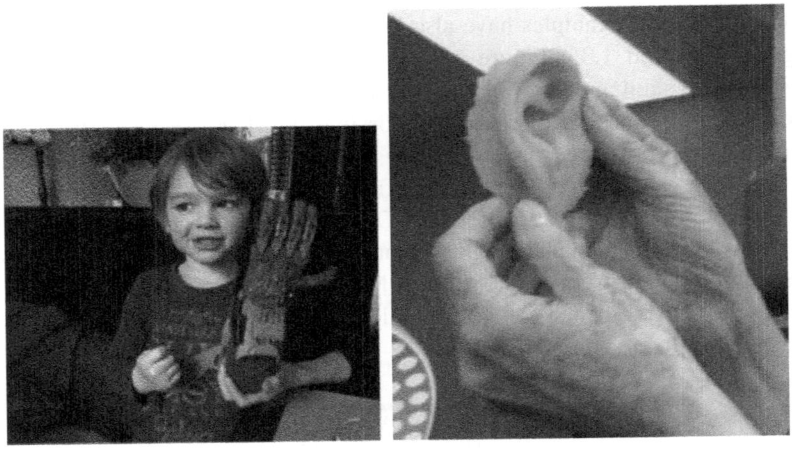

Figure 1.35. *Artificial hand and "printing" of ear cartilage*

Implants can be created with biodegradable [JAC 16] or biocompatible [CSI 15] objects. Figure 1.36 shows a drawing of bone substitutes. Hypoallergenic splints conforming to the ISO norms on biocompatibility can be created in less than 2 days [SOL 17]. Figures 1.37 and 1.38 show examples of applications in dentistry (see also [DIG 15, CAR 15, DEN 15, BRA 16, MÉL 17b]) in a quickly expanding market (SMARTECH, 2015). Figure 1.39 shows metallic parts for the kneecap [GAO 15, SAN 13]. Figure 1.40, now almost classic, represents hand prostheses. Other applications for the development of foot prostheses are also cited [NGU 13]. There are multiple applications in this field that can reach its full potential due to the personalized aspect of repairs; moreover, for eye prostheses, Jackson [JAC 17] showed that substantial improvements can be made thanks to additive manufacturing.

A printable bone allowing the bone or the missing part of it to be replaced and tissue regeneration made easier was possible in 2016 using hydroxyapatite (already present in the human bone). It is thus custom-made for each patient and is flexible, robust and synthetic. According to Manchuelle [MAN 16], the materials constituting it make it so that the body cannot reject it; in addition, it can be made in less than 24 h. These bones could serve as implants for not only the victims of very serious fractures but also the bones of children that could thus be strengthened and possibly continue to grow around the prosthesis (see Figure 1.36).

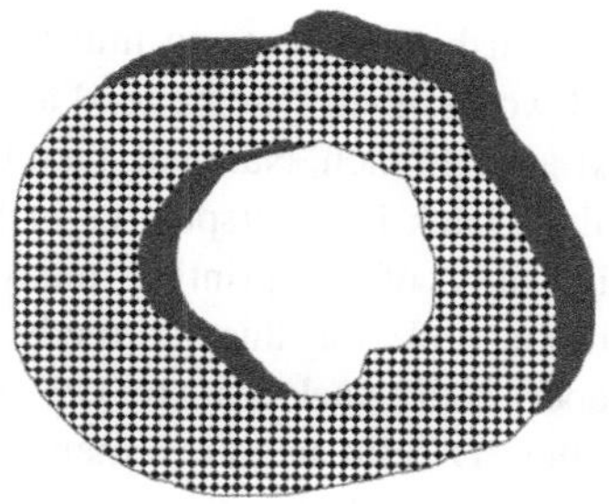

Figure 1.36. *Drawing of an artificial bone made of an elastic porous material [MAN 16, JAK 16]*

American researchers specializing in reparative surgery have used 3D printing to develop a synthetic bone made up of a malleable and resistant biomaterial that stimulates the natural bone regeneration of animals in laboratories. This is a promising study in this domain thanks to porous materials allowing cellular development [JAK 16]. This innovation paves the way for inexpensive implants and prostheses to treat a whole range of bone and dental injuries, not to mention plastic surgery. The researchers were thus able to successfully repair a spinal injury by promoting the "fusion" of the vertebrae in rats and a cranial malformation in a monkey. According to Jakus, the hole in this rhesus macaque's skull closed up in only 4 weeks without any signs of infection or side effects.

Unlike other existing synthetic bone grafts, this new material is both elastic and solid. It can also be easily cut as well as being capable of regenerating natural bone tissues without having to add substances to encourage bone growth (50% porosity). In these works, the "ink" used for the 3D printer is a mix of hydroxyapatite (90%), the primary mineral component of bone tissues and a binder to the biocompatible and bioresorbable polymer base (polycaprolactone or polylactoglycolic acid). In other words, once implanted, this synthetic bone quickly melts into the surrounding tissues, all while stimulating natural bone regeneration.

Among the recent scientific advances considered by Malaquin [MAL 15], let us mention that of the inner ear, skin substitutes, aortic valves, the use of 2-photon processes to create highly precise materials, the development of bionics and so on, the ability to create complex cellular assemblies respecting the morphology and anatomy of the tissues (Lee *et al.* [LEE 14] for the ear; Michael *et al.* [MIC 13] for the skin; Duan *et al.* [DUA 13] for the valves; Weiss *et al.* [WEI 09] for the materials; Watts [WAT 16] for heart valves; Mannor *et al.* [MAN 13] for bionic aspects). This results in a desire to create real biological tissues.

However, this domain obviously represents an immense scientific challenge with particularly stimulating initial works, even if some tend to confuse dreams and reality. Joëlle Amédée of INSERM (the French National Institute for Medical Research) [JOË 15] somewhat puts things back into perspective in Science & Santé [SCI 15], where she writes, "The difficulty of bio-printing today, even with the laser, is manufacturing functional tissue and, for this, correctly assembling the cells and layers." According to this paper, in the stacks of stem cells created by bio-printing, it is not yet possible to know if the 3D cellular arrangement corresponds to a viable and durable biological reality. As we see each day, there is still work to be done, presented and debated in Volume 3, to replace what nature makes, but the questions stem more from the biological science side than from the additive manufacturing technology.

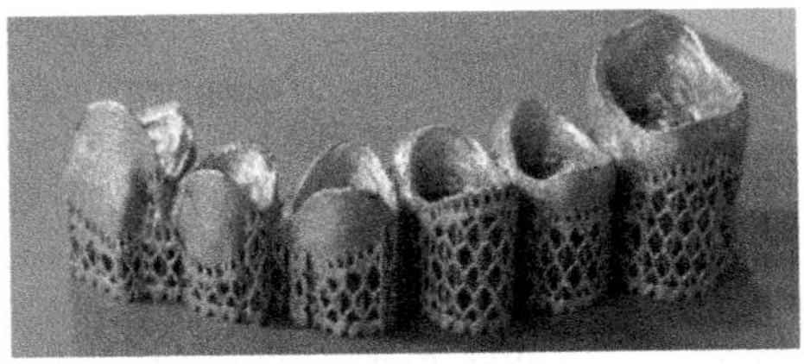

Figure 1.37. *3D dental prostheses (before completion) according to Alex [ALE 14b]*

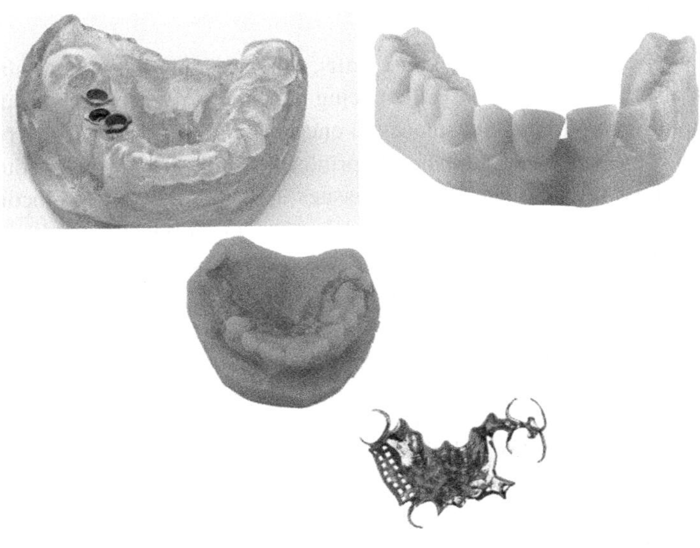

Figure 1.38. *Print for dental implants (top: Alex [ALE 16c], middle: Prodways [PRO 17], bottom: Prodways [PRO 17] - Copyright ©Prodways ©Hamilton de Oliveira*

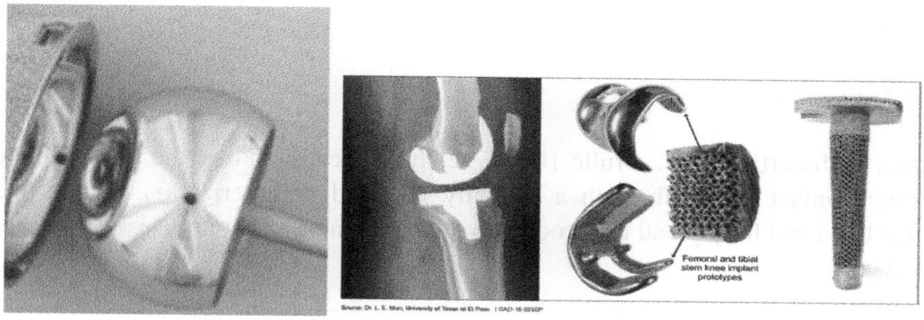

Figure 1.39. *Artificial kneecap [GAO 15]*

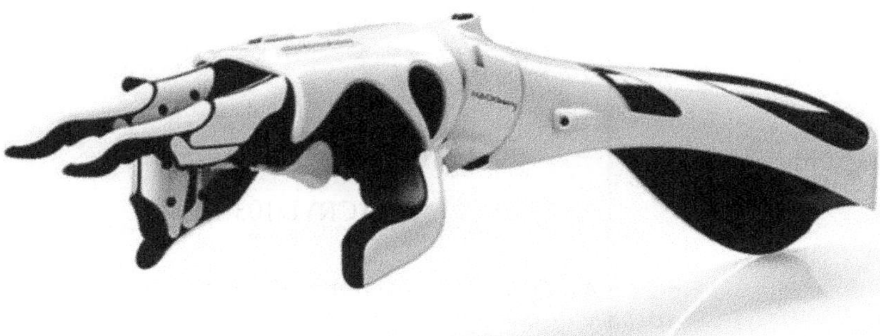

Figure 1.40: *Titanium hand prosthesis (with the permission of G. Kondo, Exiii [EXI 17])*

1.9. Animation

A movie was made with characters created through additive manufacturing, representing a long-term endeavor. "In fact, over 2 years, 2,500 parts were 3D-printed to produce the visual of the main character, as well as the trees that cover a darkened sun" [DAI 17].

1.10. Scientific applications

1.10.1. *Optics*

In his dissertation, Y. Brullé [BRU 92] had already shown the possibility of creating contact lenses through a spatially resolved polymerization process (see Figure 1.41) and had spread the process to the creation of multipurpose optical fibers [AND 91].

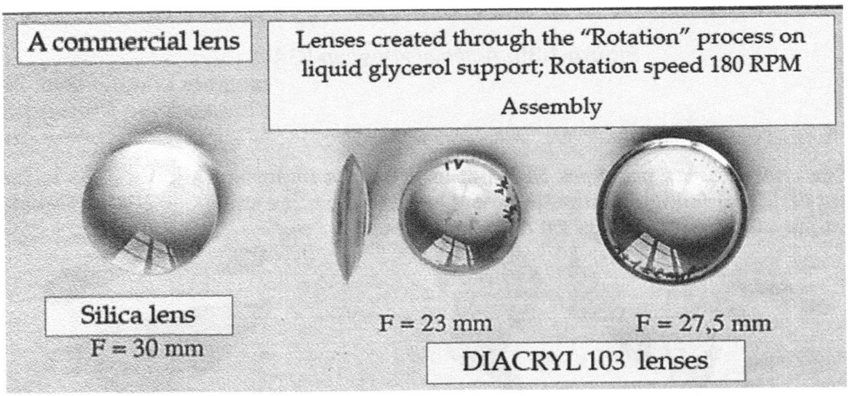

Figure 1.41. *Optical lenses created through spatially resolved photo-polymerization*

It was only recently that a Dutch company, LuXeXcel, developed a technology allowing 3D printing of the integrality of a pair of glasses, including the lenses, with no need for post-production work. This company is continuing to develop its technology and the commercialization of its service. Its technology is called PrintOptical, which allows functional contact lenses, among other things, to be 3D printed [ALL 16, CLA 17b].

Apparently, they use droplets of resin that spread over the part being built, which allows a good transparency to be obtained, as shown in Figure 1.42 from http://additivemanufacturing.com/2015/05/12/luxexcel-the-netherlands-proprietary-printoptical-technology-set-to-disrupt-markets/:

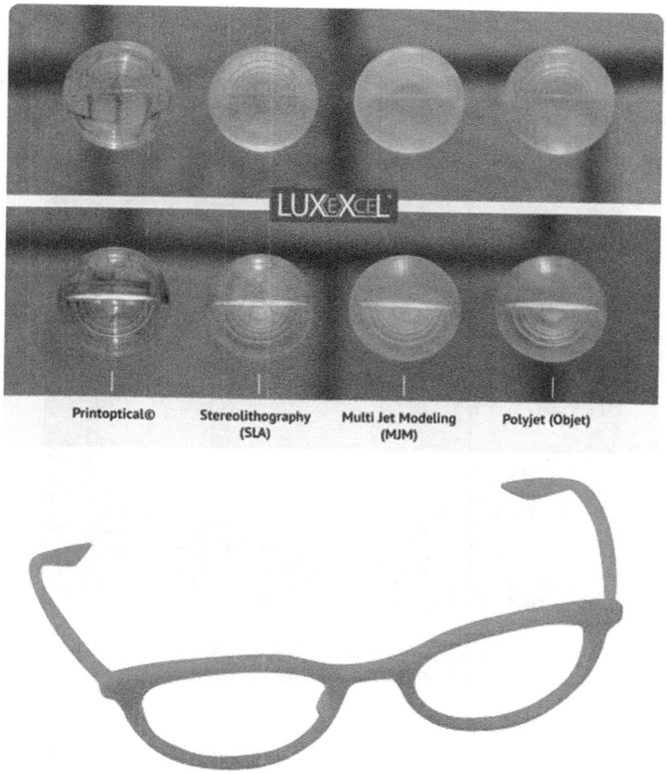

Figure 1.42. *Top: Optical lenses (from http://www.azom.com/article.aspx?ArticleID=11835); bottom: frames according to Prodways [PRO 17] - Copyright ©Prodways ©Hamilton de Oliveira; see also Lu-net.fr [LU 17]*

1.10.2. *Chemical and process engineering*

To realize the design for the internal elements of a reactor for the chemical transformation of matter, IFPEN [IFP 16] has adopted an innovative scientific process, still rare in the field of chemical engineering, which associates digital modeling tools simulating the flow of liquids (CFD, computational fluid dynamics) with experimental validation techniques using additive manufacturing. Once the reactor's geometry has been modeled, an inexpensive, quickly manufactured transparent resin prototype is printed using additive manufacturing to create hydrodynamic tests. Optimal internal geometries could be designed thanks to quick input–outputs between simulation, prototype manufacturing and experimentation. Finally, the final metal reactor was also manufactured through 3D printing.

1.10.3. *Complex structures*

In the 1990s, it was already possible to create complex parts, as illustrated by the fractal pyramid presented in Figure 1.43 [DIO 92/93].

Figure 1.43. *Example of an object that could not be created by traditional means: a fractal pyramid*

Currently, most 3D printers only allow a single material to be used for the creation of an object, yet the mechanical properties of this material may not totally suit the project. It is also possible for the project to require different rigidities according to the parts of the object, for example, with an exterior that is more resistant or less flexible.

To respond to these issues, researchers have studied how to put microstructures in place to respond to the expected physical specificities/constraints. Their solution is to create foam that is unpredictable whereas its mechanical properties are not. For this, they used homogenization, a mathematical tool that allowed them to translate the local properties that the final object must have into equations; this allowed them to define a homogeneous material "equivalent" to an unpredictable foam [CHE 16], as shown in Figure 1.44.

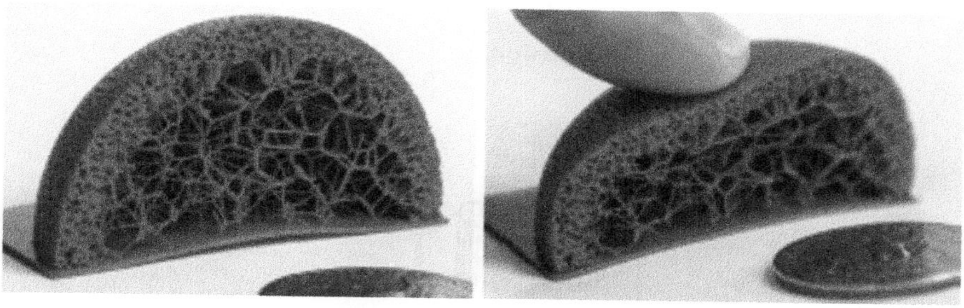

Figure 1.44. *Creation of a dome with a variable density (with the permission of S. Lefebvre)*

1.10.4. *Toward the infinitely small*

Without there being industrial applications, it seemed interesting to numerous researchers to push additive manufacturing technologies to their spatial limits. The following figures illustrate spectacular manufacturing possibilities (see, e.g. Stampfl [STA 08], André and Corbel [AND 94], Science Museum [SCI 16]). Among the first examples, we will cite the 1994 creation of micro-gears illustrated in Figure 1.45 from Bertsch *et al.* [BER 97] and "stemmed glasses" from Zissi *et al.* [ZIS 96]. Since this pioneering period, small metal (rather than polymer) parts have been created, as illustrated by the example in Figure 1.46.

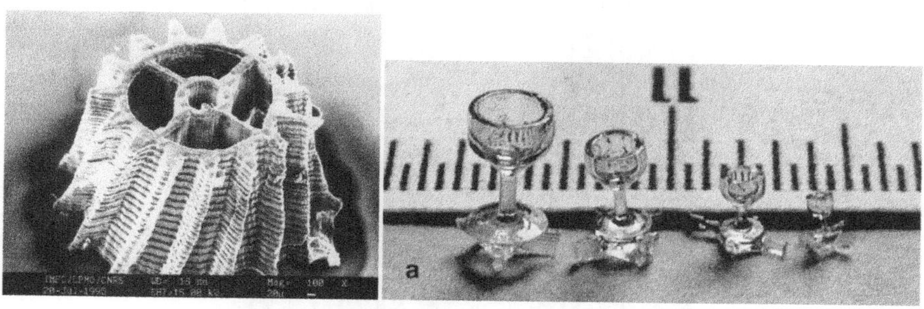

Figure 1.45. *Micro-gears created through stereolithography [BER 97] and "stemmed glasses" [ZIS 96]*

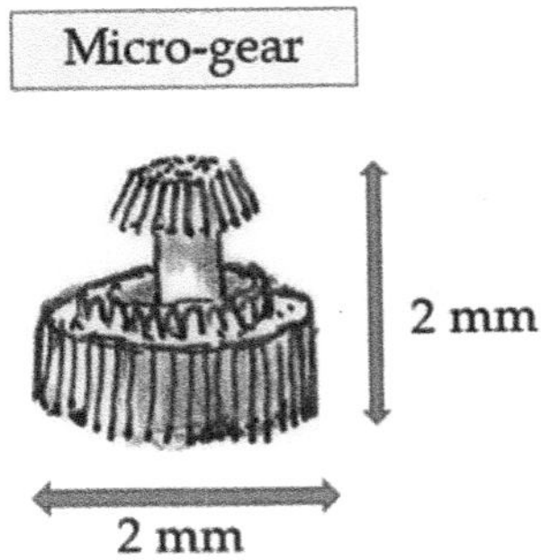

Figure 1.46. *Rough drawing of a metal mini-part (according to Designboom [DES 13])*

By using localization reinforced by light energy thanks to a "two-photon" photonic process, it is possible to make gains in spatial resolution and, as a result, to create objects in the immediate vicinity of the exit pupil of a microscope objective adapted to create very small objects, as brilliantly illustrated in Figures 1.47 and 1.48.

Figure 1.47. *Example of multi-photonic micro-stereolithography [PAR 09], presented with publisher's permission*

Figure 1.48. *Micronic objects (Nanoscribe [NAN 16] and its authorization, website: www.nanoscribe.com)*

1.11. Nanometric origami

DNA origami is a technique that consists of creating arbitrary shapes from bits of DNA. It was developed in 2006 by Paul Rothemund. The chemical properties of the molecules that make up DNA, the nucleic bases, are well understood by specialists. It is well known that pieces are connected to similar and other pieces by "Watson–Crick" pairing, and starting with this principle, it is possible to model two- or three-dimensional structures. Rothemund's technique uses a long chain of virus DNA, presenting a linear and rather simple structure with a known sequence. Then, using the appropriate software, precise locations where the DNA molecule must be bent to achieve the desired shape are found. Then, small "parts" are synthesized; these go on to precisely connect at a desired location to "pin down" the viral genome.

Only these chemically reactive "hooks" need to be synthesized; the basic supra-molecular material has a natural origin. These hooks, through the game of chemical pairings that they cause, attach themselves to the structure and deform it by attracting other molecules that also participate in the bending of the central DNA strand to give it a particular shape. It is this bending that gives it the original name "DNA origami". In Rothemund's publication [ROT 06], he presented images of variously shaped constructions that he had made. The most original are probably the map of the world, the snowflake and the smiley face. Of course, not only drawings can be reproduced with DNA; Hao Yan's team managed to create closed three-dimensional shapes ([HAN 11, SAW 11], and Figure 1.49 from [ROT 06]).

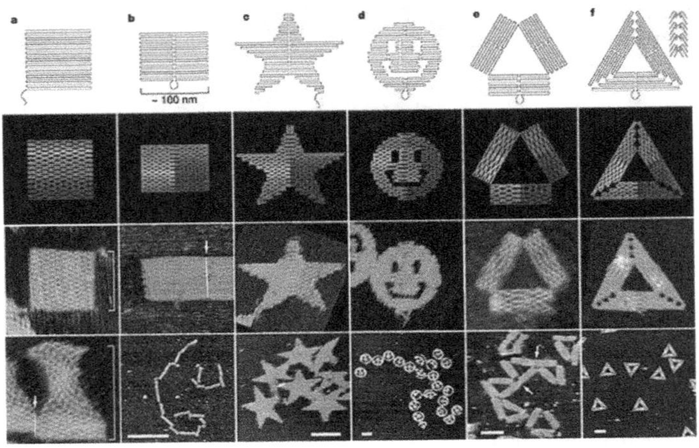

Figure 1.49. *DNA origami (with the authorization of Paul Rothemund)*

To do this, instead of a single strand of DNA, this team uses a much more complex scaffold made up of a series of concentric rings of DNA connected at strategic points. For this, the team uses DNA's pairing properties. Each of the bases, A, T, C and G, actually link up in a specific way: A with T and C with G to create a double-stranded molecule. The researchers' strategy thus starts with the manufacturing of 2D concentric rings with the double helix of DNA. The rings are then associated with one another at strategic points called "cross-over," where one of the strands of DNA is switched with another strand of the adjacent ring, structures that help support construction and stop the double helix from spreading. It is thus much easier, using this circular system, to create complete three-dimensional shapes. As such, there are many applications, from the envelope in which medicine could be placed before being directly placed in the cells to minuscule performing computer processes, to the calibration of super-resolved fluorescent microscopes and so on [WIK 15d].

According to Sussan [SUS 11], it will henceforth be possible to create this DNA origami using computer software, which presents interfaces that are rather analogous to those of comparably simpler software like "3D Studio" and "Blender," displaying 2D and 3D views side by side. Their operating procedure takes the specific characteristics of the DNA chains into consideration. First, the schema of the desired shape is defined. Next, the possible links between the different scaffolding points are defined. Then, the program will generate the list of chemical formulas for the chemical pairings necessary for the operation. Finally, overall synthesis is carried out in the real world. A problem linked to chemistry, it is not so easy to virtually describe this kind of origami. As always in biology, unforeseen interactions can prevent the molecular chains from creating the desired shape. This is where Rothemund's creativity can be found!

1.12. Conclusion

Through the presentation of a great (non-limiting) number of objects in this chapter, it is possible to highlight a multitude of ends, which impose the possession of adapted technologies, as the purposes of objects can be very diverse (shapes and functionalities). In accordance with Gausemeier *et al.* [GAU 11] and Pinkerton [PIN 16], it was possible to assemble the various functions of the applicative domains in Table 1.3.

This possibility, possibly arguable, highlights criteria that are broad enough for specialized methods of additive manufacturing to exist, even ones specific to an object's shape or even to a concept. The unitary aspect lies in the adjective "additive", which makes additive manufacturing an umbrella term based only on the appearance of a shared definition. In fact, the new domain could, from a scientific and/or technological point of view, be classified at least as much by ends, "systems of meaning", as by a field of issues or effective problems or a list of industrial results and so on. The existence of a certain fuzziness may be due to not only the absence of clarification but also the possibility of exploiting this unstabilized "magical" framework to act in an engaged way and to promote 3D printing as stemming from a considerable market likely to conceal a number of more "traditional" manufacturing techniques. In this case, dysfunctions can be expected, even ruptures in the dialogue between stakeholders of different forms of manufacturing, in this competition, and possible financial or governance failures. From this fact confirmed in cases where uncertainty reigns, users' knowledge of the diversity of representations is a prerequisite for positive exchange action and honesty between partisans of a given technological form. For these reasons, quickly brought up, the following chapter in this first part is dedicated not to technology but to the financial, cultural and assimilational impact of 3D technologies on society.

Domain/ Criteria	Space Army	Land transport	Foundry	Micro-techniques	Medical	Art	Home applications	Research
Desired material	+++	+++	+++	+++	+++	++	+	+
Spatial resolution	+++	+++	+++	+++	V	+	+	V
Production speed	+++	+++	+	+	+	+	+	+
Openness to other technologies	+	+++	++	++	++	+	+	V
Object size	< 1m	< 1–2 m	1 m	< 1 cm	10 cm	V	< 10 cm	V
Computer-aided design	C	C	C	C	MC	MC	S	V
Precision	+++	+++	+++	++	++	+	+	V
Material functionality	+++	+++	+++	++	+++	+	+	V
Operational cost	++	++	++	++	++	+	-	V
Material cost	++	++	++	++	+++	+	-	V
Multi-materials	V	V	++	++	+	V	-	V
Surface condition	++	++	++	+++	+++	+	+	V
Color	-	-	-	-	-	+	-	V
Biocompatibility	-	-	-	-	+++	-	-	V
Service life	++	+++	V	V	+++	V	V	V
Regulation	+++	+++	+++	+++	+++	+++	+++	V
Risk control	++	+++	+++	+++	+++	++	++	++
Standardization	++	+++	+++	+++	+++	+	=	V

Table 1.3. *Approximate relations between ends and manufacturing criteria imposing specific methods of additive manufacturing (+++: very high; ++: significant; +: modestly significant; =: indifferent; -: no importance; V: variable; C: complex; MC: moderately complex; S: simple)*

It is after having cleared some lines of force relative to additive manufacturing that it will be possible to present additive manufacturing technologies (and associated subtypes) by exploring, as much as possible, their advantages and their limitations, indeed calling for a rather incremental demand for innovation so that this earning venture can continue.

Thus, from the best possible knowledge of the diversity of representations, there is a chance to clarify the role/interests of the different actors implied in the development of additive manufacturing as a generic technology, to lend value to the diversity and possibility of reflecting more in-depth on the particular axes that should participate in this development (as well as to research other technological alternatives allowing the debate to be as balanced as possible). Through this sharing of significat lack of knowledge, it must be possible to state the uncertainties of scientific and technical knowledge, approximations of knowledge, abuses of interpretation and skill limitations to measure at least the breadth of these unresolved questions and the question marks. This approach, attempted by the author, should authorize the perspective with every (?) commercial ideology, overly re-assuring remark, untenable promise and abusive simplification that has not been scientifically proven. To do this, however, it is necessary to take a step back so as to have a framework that is partially "neutral" or serene enough to move forward.

The neutral aspect does not mean, however, a neutral decision for economic milieus, based on the search for a soft consensus. It will thus be the reader's job to judge if this goal has been achieved, at least partially. To start, it will be necessary to ask some questions, the first being how to know if the term "additive manufacturing", which is being used more and more as the process of manufacturing a part through the addition of matter, by piling up successive layers, as opposed to processes that subtract material, such as processing, still makes sense. Even if it is also the name given to three-dimensional printing technology, the samples presented in this chapter pose the question of re-defining the subject. In fact, it is possible to create objects by localizing the addition of matter, localizing it spatially (see optical lenses), forming matter in the image of DNA origami, localizing the energy necessary for the transformation of matter allowing the object to be manufactured or finally by combining several of these means.

NOTE.– Because it is often a matter of current industrial realizations concerning new applications of 3D printing, the bibliography for this chapter is largely from the Internet.

1.13. Bibliography

[3D 14] 3ders.org, "SpaceX unveils Dragon V2 spacecraft featuring 3D printed rocket engine", available at: http://www.3ders.org/articles/20140531-spacex-unveils-dragon-v2-spacecraft-featuring-3d-printed-rocket-engine.html, 2014.

[3D 15a] 3D NATIVES, "Des maisons de 200 m2 imprimées en 3D pour 3500 €", available at: http://www.3dnatives.com/habitations-impression-3d-shanghai/, 2015.

[3D 15b] 3D NATIVES, "Opel utilise l'impression 3D pour réduire ses coûts de production", available at: http://www.3dnatives.com/opel-impression-3d-production25112015/, 2015.

[ADD 15] ADDAERO, "Perspectives", available at: http://www.addaero-mfg.com/perspectives, 2015.

[AIF 16] AIF – Alliance pour l'industrie du Futur, "Construire ensemble une filière Fabrication Additive Française forte", available at: http://aif.diwi.org/, 2016.

[ALE 14a] ALEX, "Hasbro vous laisse imprimer en 3D ses jouets!", available at: http://www.3dnatives.com/hasbro-imprimer-3d-jouets/, 2014.

[ALE 14b] ALEX, "Cody Wilson, créateur de la première arme imprimée en 3D, se lance dans l'écriture", available at: http://www.3dnatives.com/cody-wilson-createur-premiere-arme-imprimee-3d-lance-lecriture/, 2014

[ALE 14c] ALEX, "L'impression 3D métal, pas si simple?", available at: http://www.3dnatives.com/impression-3d-metal-simple/, 2014

[ALE 15a] ALEX, "Gartner places 3D printing among the top 10 strategic technology trends for 2016", available at: http://www.3ders.org/articles/20151224-gartner-places-3d-printing-among-the-top-ten-strategic-technology-trends-for-2016.html, 2015.

[ALE 15b] ALEX, "ColorFabb imagine un vélo de route DIY grâce à l'impression 3D", available at: http://www.3dnatives.com/colorfabb-velo-impression-3d-26062015/, 2015.

[ALE 15c] ALEX, "e-NABLE, une communauté dédiée à la fabrication de prothèses imprimées en 3D", available at: http://www.3dnatives.com/enable-communaute-protheses-3d/, 2015.

[ALE 15d] ALEX, "Des imprimantes 3D pour une collection de prêt-à-porter", available at: http://www.3dnatives.com/imprimantes-3d-pret-porter-23072015/, 2015.

[ALE 15e] ALEX, "New Balance commercialisera la première paire de baskets imprimées en 3D", available at: http://www.3dnatives.com/new-balance-baskets-3d-20112015/, 2015.

[ALE 15f] ALEX, "La fabrication additive, source d'économie pour l'aéronautique!", available at: http://www.3dnatives.com/additive-economie-aeronautique-10062015/, 2015.

[ALE 16a] ALEX, "3d.FAB, un laboratoire français dédié à l'impression 3D du vivant", available at: http://www.3dnatives.com/3dfab-francais-impression-3d-vivant-19022016/, 2016.

[ALE 16b] ALEX, "TOP 5 des vidéos de la semaine: Visite des ateliers de Local Motors, la plus grande sculpture réalisée avec un stylo 3D…", available at: http://www.3dnatives.com/top-5-videos-04092016/, 2016.

[ALE 16c] ALEX, "Une première oreille fonctionnelle imprimée en 3D", available at: http://www.3dnatives.com/oreille-imprimee-3d-19042016/, 2016.

[ALE 16d] ALEX, "Le marché de l'impression 3D médicale atteindrait $280M en 2016", available at: http://www.3dnatives.com/impression-3d-medical-23112016/, 2016.

[ALE 16e] ALEX, "Food Ink, le restaurant éphémère de plats imprimés en 3D", available at: http://www.3dnatives.com/food-ink-restaurant-3d-12042016/, 2016.

[ALE 17a] ALEX, "TOP 5 des vidéos de la semaine: L'impression 3D en cuisine, interview de la startup française Syos…", available at: http://www.3dnatives.com/top-5-videos-22012017/, 2017.

[ALE 17b] ALEX, "TOP 12 des applications de l'impression 3D au service du handicap", available at: http://www.3dnatives.com/impression-3d-handicapes-11012017/, 2017.

[ALE 17c] ALEX, "TOP 5 des vidéos de la semaine: les prothèses d'Ambionics, l'impression 3D dans l'espace…", available at: http://www.3dnatives.com/video-impression-3d-05032017/, 2017.

[ALE 17d] ALEX, "TOP 12 des véhicules imprimés en 3D", available at: http://www.3dnatives.com/top-12-des-vehicules-imprimes-en-3d-02022017/, 2017.

[ALL 16] ALLARD F., "Les lunettes passent à l'impression 3D", available at: http://www.zesmallfactory.com/news/les-lunettes-passent-limpression-3d/, 2016.

[AND 91] ANDRÉ J.C., BRULLÉ Y., FAUGERAS P. *et al.*, Procédé de fabrication en continu de fibres et matériaux polymérisables, notamment de fibres optiques polymères, French patent no. 91 13 367 25, October 1991.

[AND 94] ANDRÉ J.C., CORBEL S., *Stéréo-photo-lithographie Laser*, Polytechnica, Paris, France, 1994.

[ANG 16a] ANGRAND A., "Le Boeing 777X aura la plus grande pièce en impression 3D", available at: http://www.air-cosmos.com/le-boeing-777x-aura-la-plus-grande-piece-en-impression-3d-82117, 2016.

[ANG 16b] ANGRAND A., "E teste un démonstrateur pour le futur ATP", available at: http://www.air-cosmos.com/ge-teste-un-demonstrateur-pour-le-futur-atp-84863, 2016.

[ARO 13] ARORA S.S., CAO K., JAIN A.K. *et al.*, 3D fingerprint phantoms, MSU technical reports, MSU-CSE-13-12, p. 6, 2013.

[ARS 16] ARS-MATHEMATICA, "Modèles numériques dans l'art et la nature", available at: http://ww.arsmathematica.org/IS2016-PDV/index.html, 2016.

[ART 17] ARTE JOURNAL, Reportage de D. Bornstein, N. Grimad et M. Maiffret, 19:45 Edition 27 March 2017.

[ATH 16] ATHERTON K.D., "Airbus 3D Printed This 13-Foot-Long Drone Named Thor", available at: http://www.popsci.com/airbus-3d-printed-this-13-foot-long-drone, 2016.

[AUT 16] AUTOPLUS, "IBM présente le minibus du futur imprimé en 3D!", available at: http://news.autoplus.fr/IBM-Voiture-autonome-Minibus-High-Tech-3D-1505778.html, 2016.

[BAG 11] BAGARIA V., DESHPANDE S., RASALKAR D.D. *et al.*, "Use of rapid prototyping and three-dimensional reconstruction modelling in the management of complex fractures", *European Journal of Radiology*, pp. 30–37, 2011.

[BAL 16] BALZER M., "Nefertiti hack", available at: https://plus.google.com/events/cmtahaof 4b8j6oo0qiaarfsivhg, 2016.

[BAT 16] BATIACTU, "Des villages imprimés en 3D pour dépolluer les océans", available at: http://www.batiactu.com/edito/villages-subaquatiques-imprimes-3d-depolluer-oceans-43181.php, 2016.

[BEN 15] BENEDICT, "3D printed replicas of ISIS-Destroyed Arch of Palmyra to be erected in London and NYC", available at: http://www.3ders.org/articles/20151228-3d-printed-replicas-of-isis-destroyed-arch-of-palmyra-to-be-erected-in-london-and-nyc.html, 2015.

[BEN 16a] BENEDICT, "West Virginia carpenter on target with (mostly) 3D printed semi-automatic gun", available at: http://www.3ders.org/articles/20160203-west-virginia-carpenter-on-target-with-3d-printed-semi-automatic-gun.html, 2016.

[BEN 03] BENSAUDE-VINCENT B., "La science contre l'opinion – histoire d'un divorce", *Les empêcheurs de penser en rond*, Paris, France, 2003.

[BEN 16b] BENSSOUSSAN H., "Interviewing XtreeE: 3D Printing Concrete to Push the limits of Construction", available at: https://www.sculpteo.com/blog/2016/12/07/interviewing-xtreee-3d-printing-concrete-to-push-the-limits-of-construction/?utm_medium=email&utm_campaign=NL%20US%2071216&utm_content=NL%20US%2071216+CID_d914f8c659 7202306458e9bfedf465f6&utm_source=Campaign%20Newsletter&utm_term=Learn%20 More, 2016.

[BER 97] BERTSCH A., ZISSI S., JÉZÉQUEL J.Y. *et al.*, "Microstereophotolithography using a liquid crystal display as dynamic mask-generator", *Microsystem Technologies*, vol. 3, pp. 42–47, 1997.

[BER 15] BERGOUNHOUX J., "Bourget 2015: L'impression 3D veut passer du prototypage à la production", available at: http://www.usine-digitale.fr/article/bourget-2015-l-impression-3d-veut-passer-du-prototypage-a-la-production.N336073, 2015.

[BER 16] BERTHOME K., PALACIN B., PUECH J. *et al.*, Les technologies additives pour les applications spatiales, Fabrication additive pour l'électronique, Workshop CNRS, Paris, France, 2016.

[BIG 11] BIGLINO G., SCHIEVANO S., TAYLOR A.M., "The use of rapid prototyping in clinical applications", available at: https://www.researchgate.net/profile/Andrew_Taylor6/publication/221916893_The_Use_of_Rapid_Prototyping_in_Clinical_Applications/links/0912f50ac1718c3374000000.pdf, pp. 20–40, 2011.

[BOU 06] BOURGINE P., LESNE A., "Morphogenèse. L'Origine des formes", Belin, Paris, France, 2006.

[BOU 11] BOURGINE P., LESNE A. (eds), *Morphogenesis: Origins of Patterns and Shapes*, Springer-Verlag, Berlin, Germany, 2011.

[BOU 17] BOUVERESSE J., *Le mythe moderne du progrès; décortiqué et démonté par le philosophe*, Agone, Marseille, France, 2017.

[BRA 16] BRAIAN M., JIMBO R., WENNERBERG A., "Production tolerance of additive manufactured polymeric objects for clinical applications", *Dental Materials*, vol. 32, pp. 853–861, 2016.

[BRI 16] BRISOU M., "Une start-up américaine veut imprimer de la viande", available at: http://www.industrie-techno.com/une-start-up-americaine-veut-imprimer-de-la-viande.13583, 2016.

[BRU 92] BRULLÉ Y., Procédés photoniques pour l'optique : lentilles et fibres, PhD Thesis, National Polytechnic Institut of Lorraine, Nancy, France, 1992.

[BRU 15] BRUNTON A., ARIKAN C.A., URBAN P., "Pushing the limits of 3D color printing: error diffusion with translucent materials", *ACM Transactions on Graphics*, vol. 35, Article 4, available at: http://doi.acm.org/10.1145/2832905, 2015.

[CAO 17] CAO.fr, "CENIT participe au projet de recherche européen BionicAircraft", available at: http://www.cao.fr/rubrique-toutes/actu/7311/CENIT-participe-au-projet-de-recherche-europeen-BionicAircraft.html, 2017.

[CAR 15] CARBON 3D, "Incredible new 3D printing technique looks like Sci.-Fi.", available at: http://www.popularmechanics.com/technology/a14586/carbon3d-3d-printer-resin/, 2015.

[CAS 15] CASEY T., "New "waterbone" light weight vehicle design could increase EV battery range", available at: http://cleantechnica.com/2015/04/27/new-waterbone-lightweight-vehicle-design-increase-ev-battery-range/, 2015.

[CET 15] CETIM, "Innover en mécanique", available at: http://www.cetim.fr/fr/Actualites/En-France/Zoom/Best-of-veille-d-avril-2015-un-bouquet-de-promesses, 2015.

[CHA 15] CHABAUD C., "Making an impact in electromagnetism", available at: http://www.sculpteo.com/blog/2015/11/18/sculpteo-success-story-making-an-impact-in-electromagnetism/?utm_medium=email&utm_campaign=Newsletter+181115+US&utm_content=Newsletter+181115+US+CID_6faa9b693a6a504f70d9a0e9fa4a4109&utm_source=Campaign%20Newsletter&utm_term=The%20Future%20of%203D%20Printing%20and%20Food, 2015.

[CHE 16a] CHEN X., XU L., WANG W. *et al.*, "Computer-aided design and manufacturing of surgical templates and their clinical applications: a review", *Expert Review of Medical Devices*, vol. 13, pp. 853–864, 2016. doi:10.1080/17434440.2016.1218758.

[CHE 16b] CHEN W., ZHANG X., XINY S. *et al.*, Synthesis of filigrees for digital fabrication and MARTÍNEZ J., DUMAS J., LEFEBVRE S., Procedural Voronoi Foams for Additive Manufacturing, SIGGRAPH '16 Technical Paper, Anaheim, CA 24–28 July 2016, available at: http://www.cnrs.fr/ins2i/spip.php?article2229, 2016.

[CHO 08] CHOMSKY N., HERMAN E., *La fabrication du consentement; de la propagande médiatique en démocratie*, Agane, Marseille, France, 2008.

[CIC 16] CICETTI R., "What is 3D printing art really?", available at: http://3dprintingindustry.com/2016/04/21/3d-printing-art-really/?utm_source=feedburner&utm_medium=feed&utm_campaign=Feed%3A+3dPrintingIndustry+%283D+Printing+Industry%29, 2016.

[CLA 17a] CLARKE C., "Adidas reveals plans for 3D printing 'Speedfactory'", available at: https://3dprintingindustry.com/news/adidas-reveals-plans-3d-printing-speedfactory-103519/, 2017.

[CLA 17b] CLARKE C., "Ferrari uses 3D printing to develop new engine for 2017 F1 season", available at: https://3dprintingindustry.com/news/ferrari-uses-3d-printing-develop-new-engine-2017-f1-season-104862/?goal=0_695d5c73dc-793b62ced2-64557777, 2017.

[CLA 17c] CLARKE, "Luxexcel meets ISO standards for 3D printed optical lenses", available at: https://3dprintingindustry.com/news/luxexcel-meets-iso-standards-3d-printed-optical-lenses-106931/?goal=0_695d5c73dc-db6f6977d0-64583257, 2017.

[COE 17] COE (Conseil d'Orientation pour l'Emploi), Automatisation, numérisation et employ, Les impacts sur le volume, la structure et la localisation de l'emploi, available at: http://www.coe.gouv.fr/IMG/pdf/COE_170110_Rapport_Automatisation_numerisation_et_emploi_Tome_1.pdf, 2017.

[COL 13] COLLECTIF TEXTILE, "Impression 3D", available at: http://collectiftextile.com/impression-3d/, 2013.

[COU 15a] COUNCIL A., PETCH M., *Future Food*, E. Lang, New York, USA, 2015.

[COU 15b] COUTELLEC L., *La science au pluriel*, Quae, Versailles, France, 2015.

[COY 14] COYKENDALL J., COTTELEER M., HOLDOWSKY J. *et al.*, "3D opportunity in aerospace and defense additive manufacturing takes flight", available at: https://dupress.deloitte.com/dup-us-en/focus/3d-opportunity/additive-manufacturing-3d-opportunity-in-aerospace.html, 2014.

[CSI 15] CSIRO, "Cancer patients receives 3D printed ribs in world-first surgery", available at: http://csironewsblog.com/2015/09/11/cancer-patient-receives-3d-printed-ribs-in-world-first-surgery/, 2015.

[CYB 15] CYBERLAND, "L'impossible contrôle des armes imprimées 3D", available at: http://cyberland.centerblog.net/265-impossible-controle-des-armes-imprimees-3d, 2015.

[DAI 17] DAILY GEEK SHOW, "Voici la toute première animation réalisée entièrement grâce à l'impression 3D", available at: http://dailygeekshow.com/chase-me-3d/?utm_source=newsletter&utm_medium=e-mail&utm_campaign=Newsletter_Journaliere_2017-01-10, 2017.

[DAR 15a] DARPA, "Boosting confidence in new manufacturing technology", available at: http://www.darpa.mil/NewsEvents/Releases/2015/05/29.aspx, 2015.

[DAR 15b] DARPA News, "Boosting confidence in new manufacturing technology", available at: http://www.doncio.navy.mil/chips/ArticleDetails.aspx?ID=6426, 2015.

[DDD 15] DDD NEWS, "L'impression 3D et l'architecture", available at: http://dddfactory.fr/ddd-news/, 2015.

[DE 12] DE GAULEJAC V., *La recherche malade du management*, Quae, Versailles, France, 2012.

[DE 16] DE VÉRICOURT M., "Impression 3D: les domaines qu'elle va (vraiment) révolutionner", available at: http://www.industrie-techno.com/impression-3d-les-domaines-qu-elle-va-vraiment-revolutionner.45255, 2016.

[DEF 11] DEFOURMANTELLE A., *Eloge du risque*, Payot, Paris, France, 2011.

[DEL 16] DELOITTE, "3D opportunity for the automotive industry", available at: http://dupress.com/articles/additive-manufacturing-3d-opportunity-in-automotive/, 2016.

[DEN 15] DENTCA, "CAD/CAM denture", available at: http://www.dentca.com/, 2015.

[DES 13] DESIGNBOOM, "nano-scale 3D printed metal parts with micro laser sintering", available at: http://www.designboom.com/technology/nano-3d-printed-parts-made-with-micro-laser-sintering-technology-12-18-2013/, 2013.

[DIG 15] DIGICAD, "Imprimantes 3D dédiées aux applications dentaires", available at: http://www.digicad.fr/imrpimante-3d-application-dentiste-prothesiste, 2015.

[DIG 17] DIGNAN L., "Ford: l'impression 3D de pièces d'un seul tenant", available at: http://www.zdnet.fr/actualites/ford-l-impression-3d-de-pieces-d-un-seul-tenant-39849396.htm, 2017.

[DIO 92/93] DIONNET V., HÉLIODORE F., LE MÉHAUTÉ A. *et al.*, Procédé de fabrication d'un objet fractal par stéréo-lithographie et objet fractal obtenu par un tel procédé, French patent no. 9206652, 2 June 1992 and European patent no. 93401389.7, 28 July 1993.

[DOU 08] DOURSAT R., "Organically grown architectures: creating decentralized, autonomous systems by embryomorphic engineering", in WÜRTZ R.P. (ed.), *Organic Computing*, Springer-Verlag, Berlin, Germany, 2008.

[DOU 13] DOURSAT R., SAYAMA H., MICHEL O., "A review of morphogenetic engineering", *Natural Computing*, vol. 12, pp. 517–535, 2013.

[DOU 17] DOUX J., "Innovation. Ils veulent créer un labo d'impression 3D culinaire", available at: http://www.ouest-france.fr/economie/entreprises/innovation-ils-veulent-creer-un-labo-d-impression-3d-culinaire-4722792, 2017.

[DUA 13] DUAN B., HOCKADAY L.A., KANG K.H. *et al.*, "3D printing of heterogeneous aortic valve conduits with alginate/gelatin hydrogels", *Journal of Biomedical Materials Research*, vol. 101A, pp. 1255–1264, 2013.

[DUP 17a] DUPÈRE K., "Quand l'impression 3D permet d'aider les aveugles dans leur éducation sexuelle", available at: http://mashable.france24.com/tech-business/20170127-impression-3d-sexualite-aveugles, 2017.

[DUP 17b] DUPONT L., "Lorraine Fab Living Lab", available at: https://laboratoire-erpi.wikispaces.com/Lorraine+Fab+Living+Lab?responseToken=021644860a63f191bab4570bf70bdd6e9, 2017.

[ERN 15] ERNOULT C., BOULETREAU P., MEYER C. *et al.*, "Reconstitution assistée par l'impression 3D en chirurgie maxillo-faciale", *Rev. Stomatologie, Chirurgie maxillo-faciale, Chirurgie orale*, vol. 116, pp. 95–102, 2015.

[ESM 15] ESMOND R.W., PHERO G.C., "The additive manufacturing revolution and the corresponding legal landscape", *Virtual and Physical Prototyping*, vol. 10, pp. 9–12, 2015.

[EXI 17] EXIII, "Exiii projects", available at: http://exiii.jp/projects/#hackberry, 2017.

[FER 15] FERRARI J., *Le principe*, Actes Sud, Paris, France, 2015.

[FEY 98] FEYNMAN R.P., *Vous y comprenez quelque chose Monsieur Feynman?*, O. Jacob, Paris, France, 1998.

[FOR 16] FORD S., DESPEISSE M., "Additive manufacturing and sustainability: an exploratory study of the advantages and challenges", *Journal of Cleaner Production*, vol. 137, pp. 1573–1587, 2016.

[FOR 15] FORMLABS.COM, "3D printing custom prosthetics", available at: http://formlabs.com/stories/3d-printed-prosthetics/, 2015.

[FRE 27/04] FREUD S., "L'avenir d'une illusion", PUF, Paris, France, 1927/2004.

[FRA 16] FRANCE 3, "Une entreprise de Limoges sur le marché chinois", available at: http://france3-regions.francetvinfo.fr/limousin/haute-vienne/limoges/entreprise-limoges-marche-chinois-1112697.html, 2016.

[GAO 15a] GAO, "3D printing: opportunities, challenges and policy; implications of additive manufacturing", available at: http://www.gao.gov/assets/680/670960.pdf, 2015.

[GAO 15b] GAO, "Defense additive manufacturing", available at: http://www.gao.gov/assets/680/673099.pdf, 2015.

[GAR 05] GARETTI M., TERZI S., BERTACCI N. *et al.*, "Organisational change and knowledge management in PLM implementation", *International Journal of Product Lifecycle Management*, vol. 1, pp. 43–51, 2005.

[GAU 11] GAUSEMEIER J., ECHTERHOFF N., KOKOSCHKA M. *et al.*, *Thinking ahead the Future of Additive Manufacturing – Analysis of Promising Industries*, Heinz Nixdorf Institute Ed., University of Paderborn, Paderborn, Germany, 2011.

[GHE 74] GHEORGHIU V., *La vingt-cinquième heure*, Plon, Paris, France, 1974.

[GIA 09] GIANNOTSIS J., DEDOUSSIS V., "Additive fabrication technologies applied to medicine and health care: a review", *Manufacturing Technology*, vol. 40, pp. 116–127, 2009.

[GIF 14] GIFFI C.A., GANGULA B., ILLINDA P., "3D opportunity in the automotive industry – Additive manufacturing hits the road", available at: https://dupress.deloitte.com/dup-us-en/focus/3d-opportunity/additive-manufacturing-3d-opportunity-in-automotive.html, 2014.

[GIM 99] GIMENEZ F., Prototypage rapide d'intérêt médical: application aux reconstructions osseuses, Thesis, INPL, Nancy, France, 1999.

[GOD 15] GODIN B., "Innovation: a conceptual history of an anonymous concept", available at: http://www.csiic.ca/PDF/WorkingPaper21.pdf, 2015.

[GRA 15] GRAHAM S., "Sarah Graham metalsmithing", available at: http://sarahgraham.com/product-category/jewelry/shop-by-categories/rings/, 2015.

[GRE 15] GREG, "Autodesk et Airbus dévoilent le futur de l'aéronautique avec la "cloison bionique", available at: http://www.journaldugeek.com/2015/12/02/autodesk-et-airbus-devoilent-le-futur-de-laeronautique-avec-la-cloison-bionique/, 2015.

[GUA 14] GUAL ORTÍ J., PUYUELO CAZORLA M., LLOVERAS MACIÀ J., "Three-dimensional tactile symbols produced by 3D printing: improving the process of memorizing a tactile map key", *British Journal of Visual Impairment*, vol. 32, pp. 263–278, 2014.

[GUÉ 15] GUÉRIN M., *La croyance de A à Z; un des plus grands mystères de la philosophie*, Encre Marine, Paris, France, 2015.

[GUO 13] GUO N., LEU M.C., "Additive manufacturing: technology, applications and research needs", *Frontiers of Mechanical Engineering*, vol. 8, pp. 215–243. doi: 10.1007/s11465-013-0248-8, 2013.

[HAL 16a] HALL N., "New 3D food printer coming soon", available at: http://3dprintingindustry.com/news/new-3d-food-printer-coming-soon-90710/, 2016.

[HAL 16b] HALL N., "GE builds the world's largest jet engine with 3D printing", available at: http://3dprintingindustry.com/2016/04/25/ge-builds-worlds-largest-jet-engine-3d-printing/?utm_source=feedburner&utm_medium=feed&utm_campaign=Feed%3A+3dPrintingIndustry+%283D+Printing+Industry%29, 2016.

[HAL 16c] HALL N., "Top 10 3D printed construction innovations", available at: http://3dprintingindustry.com/news/top-10-3d-printed-construction-innovations-83578/, 2016.

[HAL 16d] HALL N., "Try 3D printed food in a restaurant today!" available at: http://3dprintingindustry.com/news/pizza-legend-turns-3d-printer-84590/, 2016.

[HAN 11] HAN D., PAL S., NANGREAVE J. *et al.*, "DNA origami with complex curvatures in three-dimensional space", *Science*, vol. 332, pp. 342–346, 2011.

[HAR 16] HARRIS M., "Fini la vieille industrie automobile: imprimez votre voiture en 3D", available at: http://rue89.nouvelobs.com/2016/01/26/fini-vieille-industrie-automobile-imprimez-voiture-3d-262953, 2016.

[HAR 14] HARRISON B., "Bio-printing: the impossible can be possible", available at: http://www.3ders.org/articles/20140305-benjamin-harrison--tedx-talk-bio-printing-the-impossible-can-be-possible.html, 2014.

[HAR 15] HART W., "Faster than Ferrari and it was 3D printed", available at: http://3dprintingindustry.com/2015/11/13/faster-than-a-ferrari-and-it-was-3d-printed/?utm_source=3D+Printing+Industry+Update&utm_medium=email&utm_campaign=06418cce4f-RSS_EMAIL_CAMPAIGN&utm_term=0_695d5c73dc-06418cce4f-64557777, 2015.

[HAZ 16] HAZOTTE S., "L'impression 3 D au service du traitement de conservation-restauration d'une cruche en terre cuite gallo-romaine", *CeROArt*, vol. 5, available at: http://ceroart.revues.org/4865, 2016.

[IFP 16a] IFPEN, "Un réacteur chimique conçu par modélisation et fabriqué par impression 3D – Une première mondiale", available at: http://www.enviscope.com/energie/premiere-mondiale-un-reacteur-chimique-concu-par-modelisation-et-fabrique-par-impression-3d/48347, 2016.

[IFP 16b] IFPEN, "Un réacteur chimique conçu par modélisation et fabriqué par impression 3D – Une première mondiale", available at: http://www.euro-petrole.com/un-reacteur-chimique-concu-par-modelisation-et-fabrique-par-impression-3d-une-premiere-mondiale-n-f-13769, 2016.

[IND 15] INDUSTRIE ET TECHNOLOGIES, "Médicaments imprimés, famille robot, Google Science Fair: les innovations qui (re)donnent le sourire", available at: http://www.industrie-techno.com/medicaments-imprimes-famille-robot-google-science-fair-les-innovations-qui-re-donnent-le-sourire.39303, 2015.

[IND 17] INDUSTRIE-MAG, "Siemens franchit une nouvelle étape dans la fabrication additive", available at: http://www.industrie-mag.com/article12431.html, 2017.

[JAC 16a] JACKSON B., "Bone fractures fixed by 3D bio-printed silk", available at: https://3dprintingindustry.com/news/bone-fractures-fixed-3d-bio-printed-silk-97115/, 2016.

[JAC 16b] JACKSON B., "French aerospace manufacturers Safran expand into new frontiers", available at: https://3dprintingindustry.com/news/french-aerospace-manufacturers-safran-expand-into-new-frontiers-98529/, 2016.

[JAC 17] JACKSON B., "3D printing helps improve quality of life in reconstructive eye surgery", available at: https://3dprintingindustry.com/news/3d-printing-helps-improve-quality-life-reconstructive-eye-surgery-110044/?goal=0_695d5c73dc-0a072c18db-64557777, 2017.

[JAC 13] JACKSON T., *Prospérité sans croissance*, De Boeck, Brussels, Belgium, 2013.

[JAK 16] JAKUS A.E., RUTZ A.L., JORDAN S.W. *et al.*, "Hyperelastic "bone": a highly versatile, growth factor-free, osteoregenerative, scalable, and surgically friendly biomaterial", *Science Translational Medicine*, vol. 8, doi: 10.1126/scitranslmed.aaf7704, 2016.

[JOH 15] JOHNSON B.N., LANCASTER K.Z., ZHEN G. *et al.*, "3D printed anatomical nerve regeneration pathways", *Advanced Functional Materials*, doi:10.1002/adfm.201501760, 2015.

[JUL 16] JULIEN, "Light Rider: une belle moto électrique imprimée en 3D", available at: https://www.lesimprimantes3d.fr/une-belle-moto-electrique-imprimee-en-3d-20160524/, 2016.

[KAR 11] KARLHAUSEN L., *Les flux de la philosophie des sciences au XXème siècle*, L'Harmattan, Paris, France, 2011.

[KLE 11] KLEIN E., *Le small bang des nanotechnologies*, O. Jacob, Paris, France, 2011.

[KOS 15a] KOSLOW T., "3D printing brings closure to deep brain stimulation devices", available at: http://3dprintingindustry.com/2015/08/29/3d-printing-brings-closure-to-deep-brain-stimulation-microdevices/?utm_source=3D+Printing+Industry+Update&utm_medium=email&utm_campaign=ab7b6cdee2-RSS_EMAIL_CAMPAIGN&utm_term=0_695d5c73dc-ab7b6cdee2-64557777, 2015.

[KOS 15b] KOSLOW T., "3D printed micro-fish to explore the ocean of our bodies", available at: http://3dprintingindustry.com/2015/08/31/3d-printed-micro-fish-to-explore-the-oceans-of-our-bodies/?utm_source=3D+Printing+Industry+Update&utm_medium=email&utm_campaign=a215f7c7a5-RSS_EMAIL_CAMPAIGN&utm_term=0_695d5c73dc-a215f7c7a5-64557777, 2015.

[KOS 15c] KOSLOW T., "Sexshop3D bring safety to at-home sex toy 3D printing", available at: http://3dprintingindustry.com/2015/10/12/sexshop3d-brings-safety-to-at-home-sex-toy-3d-printing/, 2015.

[KOS 15d] KOSLOW T., "A fully 3D printed meal to satisfy every elder's appetite", available at: http://3dprintingindustry.com/2015/10/20/the-performace-concept-a-full-3d-printed-meal-to-satisfy-every-elderly-persons-appetite/?utm_source=3D+Printing+Industry+Update&utm_medium=email&utm_campaign=55e99ef15d-RSS_EMAIL_CAMPAIGN&utm_term=0_695d5c73dc-55e99ef15d-64557777, 2015.

[LAV 15] LAVERNE F., SEGONDS F., ANWER N. *et al.*, "Conception pour la fabrication additive: un état de l'art", available at: https://hal.archives-ouvertes.fr/hal-01148720/document, 2015.

[LEE 14] LEE J.S., HONG J.M., JUNG J.W. *et al.*, "3D bio-printing of composite tissue with complex shape applied to ear regeneration", *Biofabrication*, vol. 6, p. 024103, 2014.

[LIP 06] LIPOVETSKY G., *Le bonheur paradoxal; essai sur la société d'hyperconsommation*, Gallimard, Paris, France, 2006.

[LIP 14] LIPSON H., KURMAN M., *L'impression 3D; la prochaine révolution industrielle*, First Interactive, Paris, France, 2014.

[LU 17] LU-NET.FR, "Les Lunetiers 2.0", available at: http://lu-net.wixsite.com/lu-net, 2017.

[MC 15] MC LENNAN B., "The morphogenetic path to programmable matter", *IEEE Proceedings*, vol. 103, pp. 1226–1232, 2015.

[MAD 15] MADDYNESS, "Architecture: la startup XtreeE mise sur la démocratisation de l'impression 3D grande échelle", available at: http://www.maddyness.com/startup/2015/10/21/impression-3d-xtreee-architecture/, 2015.

[MAH 16] MAHON L., "3D printed candy fit Willy Wonka", available at: https://3dprintingindustry.com/news/3d-printed-candy-fit-willy-wonka-94265/#iLightbox[gallery94265]/0, 2016.

[MAL 15] MALAQUIN L., Impression tridimensionnelle de tissus cellulaires et d'organes, OMNT, Internal report to OMNT, Grenoble, France, p. 42, 2015.

[MAN 16] MANCHUELLE J., "Des chercheurs viennent d'élaborer une technique qui va révolutionner le soin des fractures", available at: http://dailygeekshow.com/os-synthetique-3d/?utm_source=newsletter&utm_medium=e-mail&utm_campaign=Newsletter_Journaliere_2016-10-25, 2016.

[MAN 13] MANNOR M.S., JIANG Z., JAMES T. *et al.*, "3D printed bionic ears", *NanoLett.*, vol. 13, pp. 2634–2639, 2013.

[MCO 17] MCOR IRIS HD, "True 3D color printer", available at: http://mcortechnologies.com/3d-printers/iris/, 2017.

[MEK 14] MEKKAOUI A., "Impression 3D: Les créations les plus insolites", available at: http://www.welovebuzz.com/impression-3d-les-creations-les-plus-insolites/, 2014.

[MÉL 17a] MÉLANIE, "Le Parthénon reconstitué grâce à l'impression 3D", available at: http://www.3dnatives.com/parthenon-impression-3d-16012017/, 2017.

[MÉL 17b] MÉLANIE, "TOP 12 des véhicules imprimés en 3D", available at: http://www.3dnatives.com/top-12-des-vehicules-imprimes-en-3d-02022017/, 2017.

[MÉL 17c] MÉLANIE, "Le projet français CIRCLE: l'impression 3D au service de la dentisterie", available at: http://www.3dnatives.com/circle-impression-3d-dentisterie-07032017/, 2017.

[MÉL 17d] MÉLANIE, "Multiply Labs et la fabrication de vitamines sur mesure grâce à l'impression 3D", available at: http://www.3dnatives.com/vitamines-imprimees-en-3d-01032107/ and https://www.multiplylabs.com/, 2017.

[MIC 13] MICHAEL S., SORG H., PECK C.T. *et al.*, "Tissue engineered skin substrates created by laser assisted bio-printing form skin-like structure in the dorsal skin fold chamber in mice", *PLOS ONE*, vol. 8, e57741, 2013.

[MOD 16] MODERN MEADOW, "Leather re-imagined", available at: http://www.modernmeadow.com/, 2016.

[MOL 14] MOLITCH-HOU M., "Trobok: a 3D printed toy brand from the future", available at: https://3dprintingindustry.com/news/trobok-3d-printed-toy-brand-future-30217/, 2014.

[MOL 15a] MOLITCH-HOU M., "Branch technology is 3D printing the future of construction on wall at a time", available at: http://3dprintingindustry.com/2015/07/28/branch-technology-is-3d-printing-the-future-of-construction-one-wall-at-a-time/, 2015.

[MOL 15b] MOLITCH-HOU M., "FDA approves the first 3D printed drug", available at: http://3dprintingindustry.com/2015/08/03/fda-approves-the-first-3d-printed-drug/?utm_source=3D+Printing+Industry+Update&utm_medium=email&utm_campaign=bda4f56508-RSS_EMAIL_CAMPAIGN&utm_term=0_695d5c73dc-bda4f56508-64557777, 2015.

[MOR 16] MOREL P., "La construction ne sera plus la même avant et après l'impression 3D à grande échelle", available at: http://www.batiweb.com/actualites/vie-des-societes/la-construction-ne-sera-plus-la-meme-avant-et-apres-limpression-3d-a-grande-echelle-philippe-morel-25-01-2016-27713.html, 2016.

[MOU 15] MOUSSET L., "L'impression 3D entre fiction et réalité", available at: http://information.tv5monde.com/info/l-impression-3d-entre-fiction-et-realite-48944, 2015.

[MOY 87] MOYER B., "History is a weapon", available at: http://www.historyisaweapon.com/defcon1/moyermap.html, 1987.

[MUR 07] MURATA S., KUROKAWA H., "Self-reconfigurable robots: shape changing cellular robots can exceed conventional robot flexibility", *IEEE Robotics & Automation Magazine*, vol. 14, pp. 71–78, 2007.

[NAN 16] NANOSCRIBE, "Sans titre", available at: https://3dprint.com/122337/nanoscribe-microscale-printing/, 2016.

[NEW 16a] NEWSROOM, "Amsterdam: Une cabane en bio-plastique imprimée en 3D au bord du canal", available at: https://humanoides.fr/amsterdam-cabane-impression-3d-cana/, 2016.

[NEW 16b] NEWSROOM, "L'Agence Spatiale Européenne se dote d'un laboratoire d'impression 3D", available at: https://humanoides.fr/esa-laboratoire-impression-3d/, 2016.

[NGU 13] NGUYEN T.L., ALLEN S.J., PHEE S.J., "Exploiting 3D printing technology to develop robotic running foot for footwear testing", *Virtual & Physical Prototyping*, vol. 8, pp. 259–269, 2013.

[NOT 14a] NOTTEAU Y., "Histoire de l'impression 3D", available at: http://www.additiverse.com/actualites/2014-09-29/histoire-de-l-impression-3d, 2014.

[NOT 14b] NOTTEAU Y., "Adieu les risques de fissure, NanoSteel a développé un acier dense spécialement conçu pour l'impression 3D", available at: http://www.additiverse.com/actualites/2014-10-01/adieu-les-risques-de-fissure-nanosteel-a-developpe-un-acier-dense-specialement-concu-pour-l-impression-3d, 2014.

[NOT 15] NOTTEAU Y., "Une nouvelle façon de customizer son intérieur", available at: http://www.additiverse.com/actualites/2015-06-09/une-nouvelle-facon-de-customiser-votre-interieur, 2015.

[NOT 16] NOTTEAU Y., "L'impression 3D repousse les limites de l'innovation culinaire", available at: http://www.additiverse.com/limpression-3d-repousse-les-limites-de-l-innovation-culinaire/, 2016.

[PAL 15] PALEG D., "How I 3D-printed a 5-piece fashion collection at home", available at: http://danitpeleg.com/3d-printing-fashion-process/, 2015.

[PAQ 05] PAQUET G., *The New Governance: A Baroque Approach*, University of Ottawa Press Ed., Ottawa, Canada, 2005.

[PAR 09] PARK S.H., YANG D.Y., LEE K.S., "Two-photon stereolithography for realizing ultraprecise three-dimensional nano/microdevices", *Laser & Photonics Reviews*, vol. 3, pp. 1–11, 2009.

[PAU 14] PAULIAT Y., "La fabrication additive, une nouvelle révolution industrielle?", available at: http://www.unilim.fr/pages_perso/thierry.cortier/0_Fabrication_documents/La fabrication additive ENSIL.pdf, 2014.

[PIE 14] PIERRAKAKIS K., KANDIAS M., GRITZALI C.D. *et al.*, "3D printing and its regulation dynamics: the world in front of a paradigm shift", pp. 1–20, available at: https://www.infosec.aueb.gr/Publications/ICIL-2014%203D%20Printing.pdf, 2014.

[PIN 16] PINKERTON A.J., "Lasers in additive manufacturing", *Optics & Laser Technology*, vol. 78A, pp. 25–32, 2016.

[PRE 15a] PREVÉRAUD J.F., "Un pont imprimé en 3D par des robots soudeurs", available at: http://www.industrie-techno.com/un-pont-imprime-en-3d-par-des-robots-soudeurs.38826, 2015.

[PRE 15b] PRÉVERAUD J.F., "Audi réalise une voiture en impression 3D", available at: http://www.industrie-techno.com/audi-realise-une-voiture-en-impression-3d.40893, 2015.

[PRO 17a] PRODWAYS, "Ensemble de photographies d'objets 3D réalisés par Prodways", Personal communication from C. Gorgé of Prodways Technologies (General Secretary), 2017.

[PRO 17b] PROTAIS M., "Renault Trucks allège ses moteurs grâce à l'impression 3D", available at: http://www.usinenouvelle.com/article/renault-trucks-allege-ses-moteurs-grace-a-l-impression-3d.N491389, 2017.

[REE 12] REEVES P., "Example of econolyst research – understanding the benefits of AM on CO_2", available at: http://www.econolyst.co.uk/resources/documents/files/Presentation_2012_AM_and_carbon_footprint.pdf, 2012.

[REN 11] RENGIER F., MEHNDIRATTA A., VON TENGG-KOBLIGK H. *et al.*, "3D printing based on imaging data: review of medical applications", *International Journal of Computer Assisted Radiology and Surgery*, vol. 5, pp. 335–341, 2011.

[ROQ 07] ROQUETTE M.L., *La créativité*, PUF, Paris, France, 2007.

[ROS 14] ROSEN D.W., "Research supporting principles for design for additive manufacturing", *Virtual & Physical Prototyping*, vol. 9, pp. 225–232, 2014.

[ROS 15] ROSENBERG J., MORAND P., TURCQ D., "L'impression 3D: porte d'entrée dans l'industrie du 21ème siècle", available at: http://www.adverbe.com/wp-content/uploads/2015/09/rapport-impression3d.pdf, 2015.

[ROT 06] ROTHEMUND P.W.K., "Folding DNA to create nanoscale shapes and patterns", *Nature*, vol. 440, pp. 297–302, 2006.

[ROU 07] ROUSSEAU V., ENGEN L., "L'emploi d'une imprimante 3D en conservation-restauration: une application spécifique pour le Musée communal de Huy", *CeROArt*, vol. 1, available at: https://ceroart.revues.org/362, 2007.

[ROU 03] ROUX E., "Approche marketing et sémiotique du luxe", in LIPOVETSKY G., ROUX E. (eds), "Le luxe éternel ; de l'âge du sacré au temps des marques", Gallimard, Paris, France, 2003.

[RUF 14] RUFER S., Quel est l'impact de l'impression 3d sur la supply chain?, Thesis, Université Paris-Dauphine, available at: http://www.memoireonline.com/04/15/9049/Quel-est-l-impact-de-l-impression-3D-sur-la-supply-chain-.html, 2014.

[SAN 13] SANCHEZ S., "A la découverte des matériaux d'impression 3D – Deuxième partie: Les métaux", available at: http://www.3dnatives.com/a-la-decouverte-des-materiaux-dimpression-3d-deuxieme-partie-les-metaux/, 2013.

[SAV 14] SAVAGE N., "Time for a change", *ACM Communications*, vol. 57, pp. 16–18, 2014.

[SAW 11a] SAWYER E., "Bio 2.0; dissecting the next revolution in biology", available at: http://www.nature.com/scitable/blog/bio2.0/dna_origami, 2011.

[SAW 11b] SAWYER E., "DNA origami", available at: http://www.nature.com/scitable/blog/bio2.0/dna_origami, 2011.

[SCA 16] SCALCO F., "3D printed art is coming of age as makersforart 3D print caravaggio's 'flagellation of christ'", available at: http://3dprintingbusiness.directory/3d-printed-art-coming-age-makersforart-3d-print-caravaggios-flagellation-christ/, 2016.

[SCI 15] SCIENCE ET SANTÉ, "Bio-impression; Quand la lumière crée le vivant", *Science et santé*, vol. 26, pp. 34–35, 2015.

[SCI 16] SCIENCE MUSEUM, "The future of industry", available at: http://www.google.fr/url?url=http://www.sciencemuseum.org.uk/~/media/documents/visit_us/antenna/3d%2520printing%2520the%2520future%2520of%2520industry.pdf%3Fla%3Den&rct=j&frm=1&q=&esrc=s&sa=U&ved=0ahUKEwjH0tWTo6HQAhXE1RoKHTuJAOg4FBAWCD0wBw&sig2=Mm84hpN85x8roPg60CE22A&usg=AFQjCNHO4m7kx4CSH0FYho97bDCvvXM4Sw, 2016.

[SCU 15] SCULPTEO, "3D printing and the tiny house movement in the US", available at: http://www.sculpteo.com/blog/2015/08/11/3d-printing-and-the-tiny-house-movement-in-the-us/?utm_medium=email&utm_campaign=Newsletter+12082015+US&utm_content=Newsletter+12082015+US+CID_66d4c457b681ed42b4dd366d94b6d4ce&utm_source=Campaign%20Newsletter, 2015.

[SCU 16a] SCULPTEO, "The state of 3D printing 2015 vs 2016", available at: http://www.sculpteo.com/blog/2016/02/17/the-state-of-3d-printing-2015-vs-2016/?utm_medium=email&utm_campaign=Newsletter%2021720l6%20US&utm_content=Newsletter%202172016%20US+CID_6038000b1c0963ea3c725ccb68a200a5&utm_source=Campaign%20Newsletter&utm_term=Predictions%20for%20the%20State%20of%203D%20Printing%20Report%202016, 2016.

[SCU 16b] SCULPTEO, "The state of 3D printing", available at: http://www.sculpteo.com/static/0.30.0-64/download/report/Sculpteo_State_of_3D_Printing.pdf, 2016.

[SER 11] SERVENT P., *Le complexe de l'autruche*, Perrin, Paris, France, 2011.

[SEX 16] SEXSHOP 3D, "Printable sex toys", available at: http://sexshop3d.com/store/, 2016.

[SHA 17] SHAPEWAYS, "The highest quality 3D printing materials in the industry", available at: https://www.shapeways.com/materials, 2017.

[SHE 15] SHER D., "Legal 3D Printing analyzes legal & ethical implications of medical 3DP&Bio-printing", available at: http://3dprintingindustry.com/2015/07/09/legal3dprinting-analyzes-legal-ethical-implications-medical-3dp-bio-printing/?utm_source=3D+Printing+Industry+Update&utm_medium=email&utm_campaign=0be10c61d6-RSS_EMAIL_CAMPAIGN&utm_term=0_695d5c73dc-0be10c61d6-64557777, 2015.

[SIM 03] SIMMEL G., quoted by LIPOVETSKY G., "Luxe éternel, luxe émotionnel" in LIPOVETSKY G., ROUX E. (eds), *Le luxe éternel; de l'âge du sacré au temps des marques*, Gallimard, Paris, France, 2003.

[SIM 11] SIMS P., *How Breakthrough Ideas Emerge from Small Discoveries*, Business Book, London, UK, 2011.

[SMA 15] SMARTECH, "3D printing in dentistry 2015: a ten year opportunity forecast and analysis", available at: http://smartechpublishing.com/reports/3d-printing-in-dentistry-2015-a-ten-year-opportunity-forecast-and-analysis, 2015.

[SOL 17] SOLANA A., "Comment une tablette combinée avec l'impression 3D peut aider à réparer des os brisés", available at: http://www.zdnet.fr/actualites/comment-une-tablette-combinee-avec-l-impression-3d-peut-aider-a-reparer-des-os-brises-39848200.htm, 2017.

[SOO 16] SOOCURIOUS, "Bientôt, vous pourrez rouler sur des vélos imprimés en 3D au design ultra futuriste", available at: http://soocurious.com/fr/velo-imprime-3d/?utm_source=newsletter&utm_medium=e-mail&utm_campaign=Newsletter_Journaliere_2016-02-20, 2016.

[STA 08] STAMPFL J., BAUDIS S., HELLER C. *et al.*, "Photopolymers with tunable mechanical properties processed by laser-based high-resolution stereolithography", *Journal of Micromechanics and Microengineering*, vol. 18, doi:10.1088/0960-1317/18/12/125014, 2008.

[STE 13] STENGERS I., *Un autre monde est possible! Manifeste pour un ralentissement des sciences*, Les empêcheurs de penser en rond, Paris, France, 2013.

[SUS 11] SUSSAN R., "L'origami d'ADN assisté par ordinateur: vers une nanotechnologie de garage", available at: http://www.Internetactu.net/2011/05/10/lorigami-adn-assiste-par-ordinateur-vers-une-nanotechnologie-de-garage/, 2011.

[TAM 15] TAMPI T., "One gun & three ways to make it – A ghost story", available at: http://3dprintingindustry.com/2015/06/07/one-gun-three-ways-to-make-it-a-ghost-story/, 2015.

[TAN 16] TANG Y., MAK K., ZHAO Y.F., "A framework to reduce product environmental impact through design optimization for additive manufacturing", *Journal of Cleaner Production*, vol. 137, pp. 1560–1572, 2016.

[TEC 14] TECHAMAJIG, "13 amazing things you can make with a 3D printer", available at: https://www.youtube.com/watch?v=0wWG_3MeyHk, 2014.

[THO 83] THOM R., *Paraboles et catastrophes*, Flammarion, Paris, France, 1983.

[TIN 94] TINLAND F., "Le site de la technologie; éclairages théoriques et enjeux pratiques", in TINLAND F. (ed.), *Ordre biologique; ordre technologique*, Champ Vallon, Paris, France, 1994.

[TON 11] TONG T., "Capturing Color and Texture of an Object in 3D Scanning", available at: http://go.lmi3d.com/capturing-color-and-texture-of-an-object-in-3d-scanning, 2011.

[TÖP 35] TÖPFFER R., quoted by BOUVERESSE J. (2017), *Le mythe moderne du progrès; décortiqué et démonté par le philosophe*, Agone, Marseille, France, 1835.

[TRA 15] TRANCHARD S., ROJAS V., "L'avenir en fabrication 3D", available at: http://www.iso.org/iso/fr/home/news_index/news_archive/news.htm?refid=Ref1956, 2015.

[TUC 15] TUCKER E., "New balance partners with nervous systems to design personalized 3D printer trainer soles", available at: http://www.dezeen.com/2015/12/06/new-balance-nervous-system-3d-printed-personalised-soles-trainers-footwear/, 2015.

[TWI 11] TWILEY N., "3D Candy Printing: An Interview with Designer Marcelo Coelho", available at: https://www.good.is/articles/3d-candy-printing-an-interview-with-designer-marcelo-coelho, 2011.

[UFU 17] UFUNK, "La première maison imprimée en 3D, en 24h et pour moins de 11.000$", available at: http://www.ufunk.net/tech/first-3d-printed-house/, 2017.

[UNI 15] UNIVERSITÉ DE NANTES, "Une imprimante 3D pour construire un habitat d'urgence en 20 minutes", available at: https://www.univ-nantes.fr/1435591545031/0/fiche_actualite/&RH=ENS, 2015.

[USD 13] USDA, "Modern Meadow submitted to CRIS", available at: http://www.reeis.usda.gov/web/crisprojectpages/0228895-engineered-comestible-meat.html, 2013.

[VAL 33] VALÉRY P., *L'idée fixe*, Gallimard, Paris, France, 1933.

[VAN 16] VAN DER LINDEN, "3D printing makes food unique", available at: https://www.tno.nl/en/focus-area/healthy-living/food-nutrition/food-innovations/3d-printing-makes-food-unique/, 2016.

[VAR 16] VARIN P., "L'impression 3D et l'art, ou comment stimuler la créativité", available at: http://www.fabulous.com.co/blog/art/limpression-3d-et-lart-ou-comment-stimuler-la-creativite/, 2016.

[VIC 16a] VICTOR, "Peugeot signe un partenariat avec Divergent 3D sur l'impression 3D métal", available at: http://www.3dnatives.com/peugeot-metal-20092016/, 2016.

[VIC 16b] VICTOR, "Le premier bâtiment européen imprimé en 3D", available at: http://www.3dnatives.com/batiment-impression-3d-27102016/, 2016.

[VIC 16c] VICTOR, "TOP 10 des applications de l'impression 3D dans le domaine militaire", available at: http://www.3dnatives.com/top-10-armee-impression-3d-05102016/, 2016.

[WAT 16] WATTS L., "Denver University researchers use 3D bioprinter to create artificial body parts", available at: http://www.thedenverchannel.com/money/science-and-tech/denver-university-researchers-use-3d-bioprinter-to-create-artificial-body-parts, 2016.

[WEI 09] WEISS T., HILDEBRAND G., SCHADE R. *et al.*, "Two-photon polymerization for microfabrication of 3D scaffolds for tissue engineering application", *Engineering in Life Sciences*, vol. 9, pp. 384–390, 2009.

[WHE 15] WHEELER A., "Popular science bring the SR-72 spyplane to life with 3D printing", available at: http://3dprintingindustry.com/2015/06/09/popular-science-brings-the-sr-72-spyplane-to-life-with-3d-printing/, 2015.

[WIK 15] WIKIPÉDIA, "Photolithographie", Wikipédia, available at: http://fr.wikipedia.org/wiki/Photolithographie, 2015.

[ZEE 13] ZEESE K., FLOWERS M.,"Major social transformation is a lot closer than you may realize; how do we finish the job?", available at: http://www.alternet.org/activism/major-social-transformation-lot-closer-you-may-realize-how-do-we-finish-job, 2013.

[ZIS 96] ZISSI S., BERTSCH A., JÉZÉQUEL J.Y. *et al.*, "Stereolithography and microtechniques", *Microsystem Technologies*, vol. 2, pp. 97–102, 1996.

[ZON 16] ZONE BOURSE, "Atos: et Materialise créent un composant révolutionnaire pour structures spatiales en impression 3D métallique", available at: http://www.zonebourse.com/ATOS-4612/actualite/Atos-et-Materialise-creent-un-composant-revolutionnaire-pour-structures-spatiales-en-impression-3D-23151815/, 2016.

2

Integration of Additive Manufacturing Technologies into Society

"Scientifically organizing humanity is the final word of modern science, it is its audacious, but legitimate pretention". [REN 14]

"The behavior of the project based on iteration. Unlike the classic process where a project's phases are chained together, one after the other (framing, conception, realization, etc.) in a linear way, 'Design Thinking' proposes a rather experimental method based on comings-and-goings and a more contorted progression. One of the great originalities of this method is the use of prototypes as tools". [MAR 17]

"Society is in any case divided between leaders and followers, obeyers and commanders, the governed and the governing". [MIC 09]

"The emerging potential for using 3D printing is illuminating some of the inefficiencies in mass production: the stockpiling of components and parts, the large amount of working capital required for such stockpiling, profligate waste of materials, and of course the expense of employing large numbers of people… It continues and accelerates a longer-term trend towards de-dematerialization of manufacturing goods – a trend that has already kept the total tonnage of global goods constant over the past half century, even as their value has increased more than three-fold". (Al Gore, The Future [GOR 13a], quoted by Taylor [TAY 14])

"Little by little [...], we are getting so used to living with an electronic wave that it seems to disappear, until a crisis breaks out: a lawsuit, a scandal, a survey. Then, caught off guard, we look back and realize that we were the instruments of our own surveillance. But most of the time, we behave as if this were simply invisible". [TUR 15]

"From transitive, technology today is becoming reflexive, Man's action on himself and no longer a simple instrumental meditation on Man's action on things. This is why the old philosophical cliché of the existential, ethical, and political neutrality of technology should be questioned". [KUN 05]

"Technology as a meditation between Man and Nature". [HOT 94]

"Work that has become immaterial is also virtually omnipresent and makes us the "[computer] terminal Men" of a connected society in which our time crumbles between our private and professional lives, which is no longer separated by any border". [ETT 92]

"Thus, innovation has become a sort of general horizon and conditions of possibilities. A self-fulfilling prophecy in the field of technoscientific knowledge and related cultures, it is more and more often orienting and justifying all of the institutional policies for transforming and standardizing the dominant regime of legitimous knowledge production". [RUA 17]

2.1. Introduction

At present, additive manufacturing processes represent an economic market with a value of several billion euros/year. Their applications concern numerous technological domains: mechanics, transport, medical applications, Materia Medica, cosmetics, computer science, tires, paintings, inks, textiles, clothing, art and so on. 3D printing is a breakthrough technology with a dedicated ecosystem presented in this chapter. The industries concerned are health, space, aviation, pharmacy, food processing, automobile, transport, chemistry, cosmetics, defense, electronics, clothing and so on.

It is nevertheless difficult today to distinguish things stemming from the possible from things that are simply science fiction or promises. According to Pouzot, Bourguet and Jolly [POU 14], this technology is not yet mature and still has a very long way to go. This field, which is becoming more and more significant (more than a double-digit growth rate for a few years now), is the object of different forms of essentially technological work with a socioeconomic interest and some aspects connected to normalization and regulation and with little effect on the mastery of risks, toxicology and so on, likely in an attempt at specific regulation.

Emerging technologies like 3D printing are developing, as indicated above, with the following five principal concerns present and oriented toward more prospective aspects:

– showing that additive manufacturing is part of an industrial process allowing the "time-to-market" to be reduced;

– finding new openings for the technology: from children's games (Mattel®), to the printing of organs to the construction of bridges or houses (see Volumes 2 and 3);

– innovating to maintain dynamics and for durability: incremental approach to new processes (see Volume 2) and particularly materials and multi-materials; break-away approach revisiting the concept (e.g. 4D printing; see Volume 3);

– performing concept tests and communicating: the initial demonstrators are developing in light structures with a modest investment but high creativity;

– manufacturing finished products in the least possible number of steps.

At present, 3D printing is considered one of the digital technologies likely to profoundly change methods of production and, as a result, the current economic models ([AND 17, ING 15, GEB 14, WEL 15, BLA 14, MER 15, TEC 15, MOR 14, ACA 15, BER 14, BAR 15, CCI 15, HAU 14, HOS 14, GIB 15, SCU 15a SCU 15b, HLA 14, PAN 13, KAZ 14, RAT 12], etc.). We are dealing with a new boundary-object, as demonstrated by Star and Griesemer [STA 08], that has been recently integrated into the public memory, whether we are speaking of 3D printing, 3D photocopying or additive manufacturing ("Sufficient plastic to adapt to local needs and the constraints of various groups that use it, all while being sufficiently robust to maintain a common identity from one site to another"). Moreover, the US Government Accountability Office [GAO 15] places additive manufacturing in the collective imagination as a technology that can/must be among the foundations of sustainable development, in the sense that it is possible to avoid material losses.

Access to new forms of freedom, in the creation of objects with very diverse ends and the possibility of creating otherwise unfeasible parts, probably constitutes highly attractive major stakes to be considered in the "all-round" development of 3D technology. It is well known today that different technologies are dependent on one another, certainly to different degrees. Gille [GIL 78] reminded us, however, that there must be, among them, a minimum coherence for them to be employed. This is indeed the case of 3D technologies, which are, in particular, a material accomplishment of the development of computer science.

After creating a balance sheet of the application fields and economic markets allowing the economic activities of additive manufacturing to be located, it seems significant in this chapter to examine how society "is impregnated" with the contributions of technology and with what assimilation and inclusion bring into the collective imagination. In this framework, limitations with various origins as well as new desirable aspects appear, which constitutes an unstable discussion forum.

2.2. Markets and application domains of 3D printing

2.2.1. *Markets*

Additive manufacturing processes, governed by action, the real component of a "ubiquitous revolution", represent, in most developed states, a factor of technological progress, a real form of technological utopianism often with a top-10 position among emerging technologies. Here is for the discourse of the powers, supported more or less by many companies and to a lesser extent by scientists, who are isolated and at best followers of this development.

For many years, the authority of experts has limited the democratic operation by denying the right to decide (even to judge); there are several examples (Tignes dam in the 1950s, French nuclear choice, etc.). The "ignorant" citizen has been led to accept the evolution imposed from above as compensation for certain "advantages" stemming from the technicist society. However, recently, in the image of GMOs, whose advantages are far from being felt by opinion, the propositions of those in power concerning certain technologies like GMOs and/or nanotechnologies have been strongly limited, as they are the subject of controversy. The case of information and communication technologies (ICT), whose importance no longer needs to be shown, is apparently different because it is more "desirable"; this is also the case for one element contributing to ICT, that is, additive manufacturing.

According to Raynal [RAY 16], additive manufacturing technologies are developing in a very significant way. According to the forecasts of the IDC cabinet that he quotes, the domain's world market could be as high as US $35.4 billion by 2020, compared to the figure predicted for 2016 (i.e. US $15.9 billion). Between

2015 and 2020, the market could then progress at an average annual growth rate of approximately 24% (about 50% for machines and materials). In fact, different surveys agree on the double-digit evolution foundations (but with considerable differences). Table 2.1 represents some results concerning both the market and the growth of additive manufacturing.

Market (billion €/year)	Growth (%)	Year	Reference
0.556 (4D)	43	2025	[MAR 15]
1.36 (mechanics)		2023	[CET 15], quoted by [MAD 17]
1.5 (bio-printing)		2021	[BIO 16]
	17.2	2016–2020	[BIO 17]
10	~20	2020	[OEC 17]
21		2020	[CLA 15, CAN 16]
19.5		2020	[KOH 16]
17	25		[LEC 17]
3.5 (metal)	33	2020	[OBS 17]
27		2019	[ALE 16]
0.063 (4D)	43	2019	[MAR 15]
12		2018	[MAD 15, MOU 15, CON 15, KOH 16]
3		2016	[HEI 13]
4	43.5	2016	[CAN 16]
6	20–25	2016	[FRE 13]
15	24	2016	[RAY 16]
11	25–30	2016	[CAN 16]
	50	2016	[GHI 16]
	50	2016	[VIC 16]
4.5		2015	[KOH 16]
300 (potential)			[MAN 13]
230–550 (potential)			[ADD 16]

5	26	2015	[ALE 16]
	60 (companies)	2015	[ALE 15b]
	75 (machines)	2015	[XER 16]
5.2	25	2015	[LEC 17]
4.3	100 (machines)	2015	[PAR 16]
5	38	2014	[PIP 17] (average values of their proposition)
3	20	2013	[JUT 14]
0.34 (mechanics)		2013	[CET 15], quoted by [MAD 17]
1		2008	[JUT 14]
100 (bio-printing potential)	36	2016–2023	[VIC 16]

Table 2.1. *Present and estimated markets for additive manufacturing*

Table 2.1 not only shows the difficulties in mastering the additive manufacturing market but also the decreasing cost of 3D printing machines [DEH 16]! In the case of 3D printing, the criteria shown in Figure 2.1 [CHE 12] are taken into account: the temporal effects concerning incremental innovations are reduced, the quality of productions is improved and the price of machines and the cost of use are reduced.

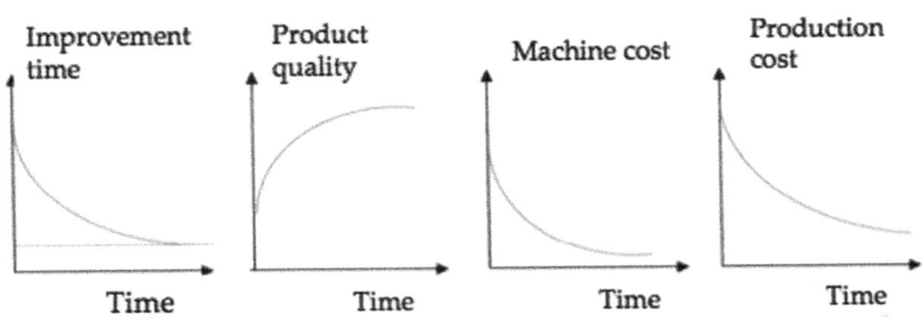

Figure 2.1. *Temporal effects on the improvement of processes*

In addition to R&D aspects that are not targeted by the survey by Sculpteo [SCU 16a], taking one of the examples presented in their report, the authors provide (with great precision) an indication of the "great" economic masses on certain industrial and societal pillars linked to 3D printing in Figure 2.2.

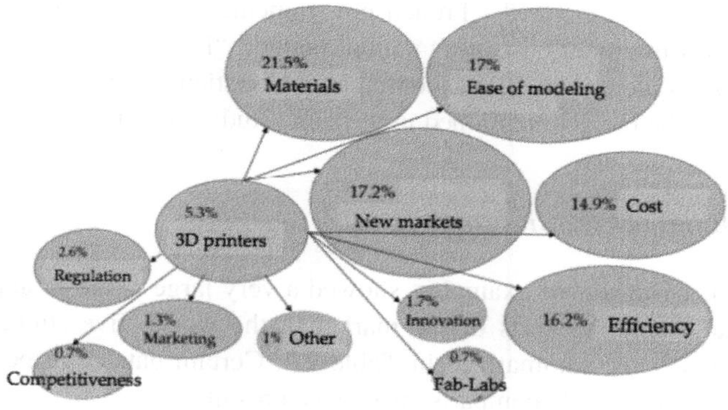

Figure 2.2. *Division for the "high-tech" rubric of 3D activities*

Moreover, new manufacturing technologies are seeing the light of day with high manufacturing costs at the start of production and a decrease with the maturity of these techniques. For example, Baumers *et al.* [BAU 16] compared two technologies, again implying the difficulty in quantitatively judging the evolutions of additive manufacturing. Nevertheless, obviously, black and white literature is developing, the number of exhibitions dedicated to 3D is exploding; in short, the world of additive manufacturing is moving and integrating itself into the economic and social world.

NOTE.– French national support for additive manufacturing:

According to 3D Natives [3D 16a], following the initial aid concerning industrial robots in 2015, the new financing law voted on by the French government envisions a reduced tax system for SMEs interested in purchasing 3D printers with a tax mechanism foreseeing a more rapid amortization for these SMEs. This motivation should serve businesses interested in making up for the technological delay in this sector in France. "In terms of additive manufacturing equipment, the United States represents 40% of all the industrial machines installed, compared to 29.3% in Europe and 26.2% in Asia and the Pacific. In Europe, France, with only 3%, is behind in comparison to Germany, which boasts 9%," explains Socialist Party congressman Jean Grellier from Deux-Sèvres concerning the amendment quoted by 3D Natives. As has been shown, 3D printing constitutes veritable stakes for the future of certain industries (biomedical, automobile, space, aeronautics, etc.).

The tax measure will be applied to SMEs with a limit of 200,000 € of aid over a 3-year period for all equipment acquired or created between October 1, 2015 and

December 31, 2017. As for the French government, it estimates that this measure will represent a marginal cost in its annual budget. "In 2017, it could be around a million euros, as well as in 2018", commented Christian Eckert, Secretary of State, responsible for the budget mentioned in the same study quoted by 3D Natives.

2.2.2. *Principal application niches*

Chapter 1, by means of examples, showed a very large range of applications. In an attempt to frame the parts of the market in the world, five studies have been conducted, which are summarized in Table 2.2. Certain data have been integrated and fused to draw possible comparisons between results.

Domain / %	[GOR 16]	[LUB 16]	[BER 14]	[AGA 12]	[SCU 16a, SCU 16b]
Aerospatial	20	34	12	12	8
Automobile	7	15	17	19	8
Army	0	0	5	6	5
Metallurgy + machines + chemistry + electronics	0	10	19	11	27
Medicine + dental	25	12	14	15	5
Retail products	0	0	9	21	29
Energy	0	15	0	0	2
Tertiary sector	20	0	9	0	10
Building and architecture	0	0	4	3	0
Academic sector	10	0	6	8	6
Luxury, textiles, food	0	9	0	0	1
Miscellaneous	18	6	5	5	0

Table 2.2. *Attempt to frame the application markets of additive marketing*

Figure 2.3 shows the unweighted average of Table 2.2, allowing an approximate estimation of the large domains where additive manufacturing has been implemented. It is indeed in the space and aeronautic domains that additive fabrication is spreading, because it is possible to create lighter parts and/or parts more quickly (see Chapter 1). The other remarkable domain is that of medical applications that, with 3D printing, display new possibilities for implants and various repairs, of which there are also abundant examples in Chapter 1. In every case, there is a possibility to create the desired and, in the final example, personalized object on site (an obvious example of autonomized confined spaces).

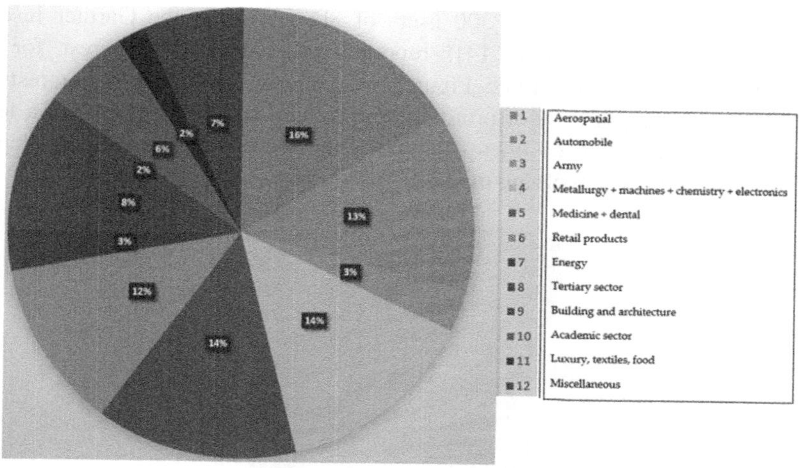

Figure 2.3. *Average of the large application markets. For a color version of the figure, see www.iste.co.uk/andre/printing1.zip*

Furthermore, the OECD [OEC 92] has already explained, "the renewal of the most current technologies is no longer done quickly and at a low price unless these technologies are directly linked to scientific disciplines that have achieved the predictive stage, but whose development is still being pursued" (see also [CRO 12]). Such is the case of 3D technologies developed by American laboratories since 1985 (and which could have come out of French laboratories in the same time window). Moreover, numerous arguments, some of which will be mentioned later, give them a good image and real attraction: offering support to help produce meaning with much hope and desire. In addition, additive manufacturing obviously has well-defined specificities that make it a technology of the future [BOS 15, GEB 14,

OLL 15, REE 08a, REE 08b, REE 12a, REE 13, LIN 12, BAU 12, MCK 13]. Let us recall that, among others, this rather recent technology (approximately 30 years old) allows for the creation of:

– complex geometries;

– integrated assemblies;

– “mass-produced” personalized products;

– “redesign” engineering enabled by additive manufacturing technologies.

In a recent survey involving 300 users of 3D machines, the Gartner Institute (quoted by Grunewald, [GRU 14]) reported the centers of interest for this technology among manufacturers: Figure 2.4 indicates the fields of interests for them, proving the importance of prototyping, product development and, to a lesser degree, innovation. It will be interesting to follow the evolution of this perception over time with a quickly evolving market (~40%/year).

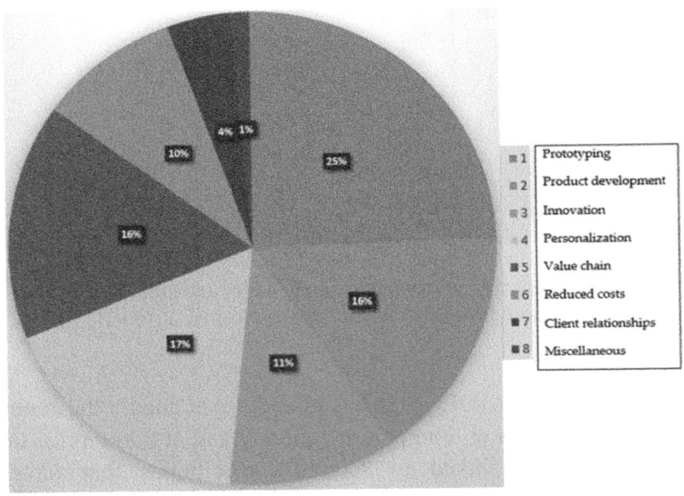

Figure 2.4. *Centers of interest of additive manufacturing for manufacturers. For a color version of the figure, see www.iste.co.uk/andre/printing1.zip*

To summarize, the new economic model involving 3D printing could overturn entire facets of industry for various reasons:

– political: in 2013, Barack Obama, the 44th President of the United States, stated that the mastery of this technology was strategic and that it would allow the re-industrialization of the United States; his successor has made no statement on this topic to date;

– technological: single-stage manufacturing of very complex parts, object personalization and reduced consumption of prime materials (traditional processing can incur an 85% material loss to create the final piece, while it seems possible to say that this loss could be zero or nearly zero with additive manufacturing);

– re-conception: manufacturing a complex assembly into a single part is translated by a reduced weight for the same mechanical stability. This is important in a domain where the energy to be provided once or sequentially is a very significant critical parameter: satellites, avionics, automobile, and so on (domain 1 in Figure 2.3);

– modest maintenance: no further need to store replacement parts as they can be created on-site using additive manufacturing; only manufacturing materials would need to be stored;

– re-localized production: near users (from the 3D machine in space to the washing machine repairman) in micro-factories (e.g. Local Motors in Géneau and Caulier [GÉN 16]), in the homes of "makers", in Fab-Labs or in equipment providers' shops (3D and production files cached on site);

– logistic simplification: only the materials used for additive manufacturing must be transported;

– personalization: in the cases of prosthetics, art, jewelry, clothing, shoes and so on which use a standardized software base that can be adapted to the client's demand;

– reduced transport costs: the extension of 3D printing and the re-localization of certain production as close as possible to consumers could reduce the degree of air and sea traffic;

– legal problems: a legal battle begins, with an industrial property right that could be abused (and that will be discussed later). The democratization of 3D printers could not only allow millions of individuals but also dishonest competitors to copy or adapt all or part of a business's retail products, even works of art, at a modest cost.

– however, at the same time, for Obaton [OBA 16], the oracle has not yet spoken: "Thus, it has been confirmed, for example, that concerning the powder used, it degrades over time and during the reuse of unconsumed material. This has a direct impact on the properties of the part created. It is therefore important to precisely measure these elements and to understand how the properties of the powder used evolve and impact the creation of the parts being produced." The robustness of 3D processes is therefore an important subject to be perfectly mastered in order to maintain their developmental dynamics.

2.3. Growth dynamics

Should we not consider this attractive and sometimes exciting potential (and the possible "poaching" that is part of it) in this reflection by Taguieff [TAG 01], who writes, "Exploiting to produce, producing to exploit, all in anticipation of unlimited profit: the spirit of capitalism has been introduced at the very heart of democracy, dictating its axiological presumptions and its normative orientations"... Certainly we are still far from considering additive manufacturing technologies to be a new omnipresent form of "opium for the masses," but this technological intrusion as a "3D" printer, a home tool today, like all mobile computers in our society, can have numerous effects on the society and its behaviors. Indeed, this promising technological innovation may be an instrument of power and of a certain domination (in particular of those who perfectly master the operation of the 3D machine), not anticipated in July 1984, because it allows methods of production to be renewed, even reinvented, our ways of life to be created and then modified in some way.

In continuing with the past, the swift transition that we are undergoing is given over more and more to the technological determinant that provides elements of technical progress and a quantification of everything or nearly everything. Yet, at present, the absence of effective piloting of the technical choices is seen, the difficulty for a politician to impose the conditions of a lasting dialogue on the technological options, to anticipate risks for the population and environment, and the "sour" polarization of certain stakeholders who know how to make themselves heard. It is more and more common on the market (encounter between consumers' solvent needs and producers' offer capacities) to guide the economic change and help society evolve (which some, with a certain cynical tone, call "acceptance"). Politicians find themselves in an uncomfortable situation of adapting public policies to the more and more globalized socioeconomic reality, essentially targeting employment and GDP (and the absence of overly loud revolts or demonstrations on the citizens' part).

Entrepreneurial globalization is upsetting the economy, cohabitation, health and so on, thanks to ICT and the associated technologies. All of the digital technologies allow for the combination of technoscientific advances among themselves to such a degree that the public may think that all the limitations of interdisciplinarity have been exceeded and that bringing the disciplines together is an accessible “art”. According to scientists, it is indeed a matter of moving away from the desire to know in order to obtain a result whose public can seize it, of going from an individual quest to that of a collective approach with its own non-scientific limitations [GRA 01].

According to Pouzot, Bourguet and Jolly [POU 14], the development of additive manufacturing processes is naturally not inconsequential for the economy. These authors believe that the logistics sector will be affected first, because, due to the need for specific materials with a high added value, the market of prime materials should become a market with a great potential for development (the materials aspect is generally not considered enough in most studies and surveys). The second market emerging from 3D printing could be a “cultural transformation” of the final base consumer becoming a production actor, directly or indirectly via Fab-Labs. Thus, by 2025, 5–10% of products could be impacted by 3D printing. New markets will therefore develop and particularly those of CAD (computer-aided design) files. Products will have a dual existence, the digital file and the physical product, both able to be fully dissociated spatially.

It is then necessary to search for what developed states desire for their citizens, but to examine the goals of large industrial groups and their financial supporters. Faced with this statement, several options present themselves to us today: continuing to count on the technosciences to maintain (*a minima*) a certain well-being, even improving human performances (promises mentioned in Volume 3…), continuing to consume, accepting individualized control, practicing vexing fundamentalism or thinking about human progress as a work on oneself (humanist, even spiritual option) with a change in the social context (with a possible failure). In a hypothesis of apparent continuity, the modes of cooperation between man and machines, the Internet of Things (IoT), robotics, additive manufacturing and so on will explore more and more forms of “Internet” convergence and subsequently deeply modify the citizens’ culture.

When we speak of additive manufacturing, we cannot help but reflect on the subject of convergence [BEC 07], which underlines the growing interconnection between informatics (I), processes (E: Engineering), materials (C: Chemistry), a true stake in studying and mastering the beginnings of hard science complexity. Convergence or not, since 1984, the number of publications has been constantly growing, as indicated in Figure 2.5.

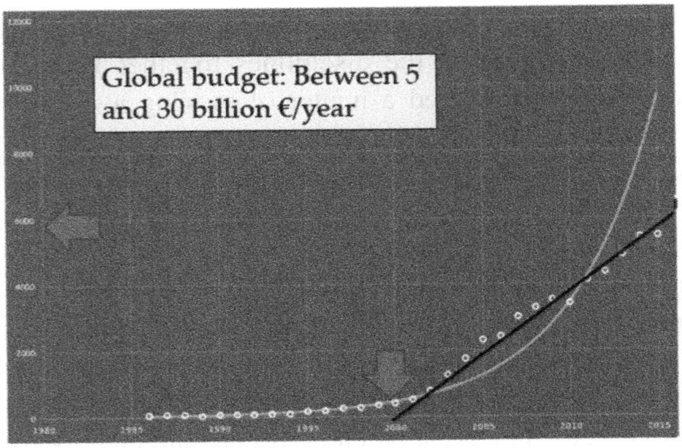

Figure 2.5. *Evolution of additive manufacturing publications over time from the BU-Univ-Lorraine*

One of the strengths of additive manufacturing devices, beyond the considerable field of personal initiatives permitted by this technology, rests in the fact that what was costly about traditional processes (e.g. processing) becomes practically free. According to Anderson [AND 12], the arguments are as follows:

– "variety is free": thanks to computer science, it is possible to easily create different unique parts or to become integrated into mass production;

– "complexity is free": it is the set of 2D transfers that define a part's creation time in a sweeping laser process; it is reduced even more if irradiation is performed through a mask, even less if the placement of layers can be reduced and so on;

– "flexibility is free": computer science's great ability to modify one or multiple parameters without the machine itself being modified.

However, for computer science and its other ICT applications, it seems that there is a critical transformation point in organizations corresponding to all the levels capable of re-employing human resources and acquiring future digital talents; a certain trend toward different losses of impulse with different origins can be felt. The different talents to be studied will be those of people who integrate technologies and knowledge about these technologies, with the difficulty of managing disjointed disciplinary knowledge and the ability to organize it, to develop it and make it live. [SUC 16]. Nevertheless, according to the technologies considered, the dynamics are variable as a function of parameters that would have to be looked into further. Thus, Figure 2.6 considers all of the publications corresponding to the key words "additive manufacturing process" and Figure 2.7, the unique criterion "stereolithography". A comparison of the two types of results is interesting in that the "historic" original domain of additive manufacturing undergoes a distinctly linear evolution, while other processes emerge and try to overtake the initial technology. If a linear approximation is accepted for recent years, the "starting" time would be the year 2000, while for only stereolithography, it would be 7–8 years earlier. In any case, the stabilization pillar has not been attained.

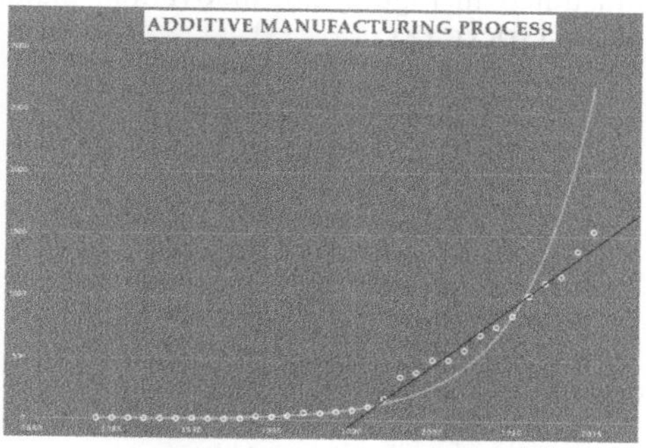

Figure 2.6. *Evolution of publications on additive manufacturing processes over time from the Univ-Lorraine Library*

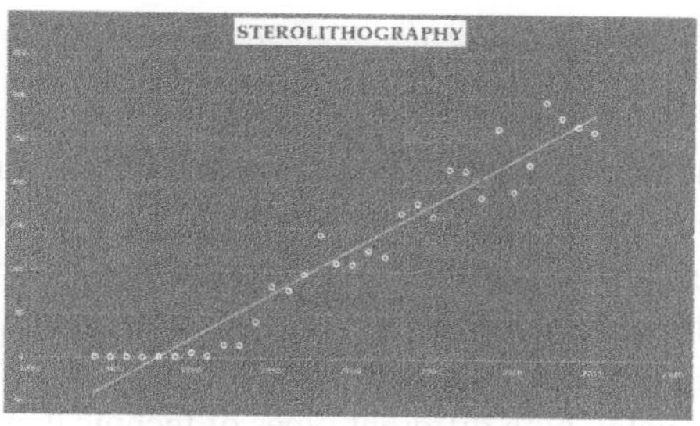

Figure 2.7. *Evolution of publications on stereolithography over time from the Univ-Lorraine Library*

Access to this information is simple, but can be considered displaced relative to the current economic market and its evolution. It is actually difficult to precisely know these data, to base things on the number of machines (whose costs vary from a few hundred euros to millions of euros) and so on. 3D printing nevertheless presents a unique innovation potential for many. Applied to industrial production, it should allow the "time-to-market" to be reduced over time and the personalization of products to develop, as was illustrated in Chapter 1. If the technology is not new, the democratization of additive manufacturing makes it a true breakthrough. The additive manufacturing revolution is comforted by analysts: In 2015, the Gartner Institute had foreseen 200,000 3D printers for a global revenue of US $1.6 billion. In the next three years, the volume of 3D printers should explode to reach 3 million units in 2018, that is, a revenue of US $13.4 billion [MAD 15, MOU 15, CON 15]. For 3D printing, as illustrated in Figure 2.8, innovation, mature technologies and creativity are interconnected [POT 14].

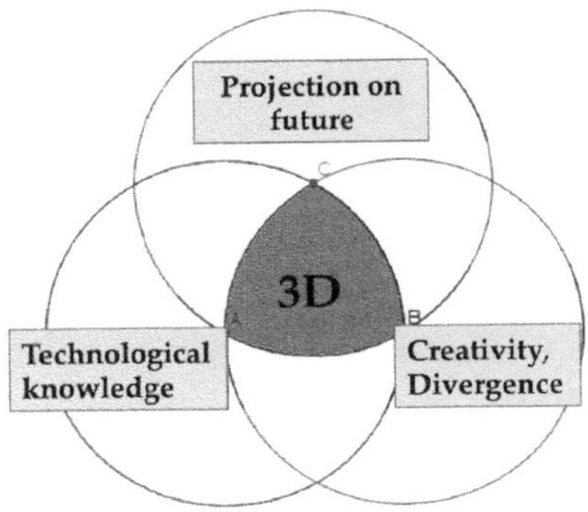

Figure 2.8. *Specific position of 3D printing*

3D printing is sometimes presented as an essential vector for the "third industrial revolution". Its impact will be significant, beyond a shred of doubt; however, some recurring incorrect ideas on the subject must be done away with. This is particularly the case concerning the freedom of design and the functionality of objects. Even if additive manufacturing processes allow new possibilities to be explored, this brings with it limitations associated with different types of technical, educational, economic and/or cultural difficulties that will be mentioned in this chapter.

2.4. Studies on the dynamic of growth

2.4.1. *Convergence*

According to Wikipedia (https://en.wikipedia.org/wiki/Technological_convergence), the term "technological convergence" aims to integrate separately developed components/concepts into one technical system (that will then be called "convergent"). The expanded fertilization of convergent technologies could be a source of innovation (see, for example, the GPS) [LAU 05]. For AFTU [AFT 06], it is indeed an interdisciplinary approach. The expression "technological convergence" refers to a voluntary encounter of innovations in the domains of microelectronics, computer science, energy–matter interactions, digital control, materials and so on. Technology is advancing to the borders of scientific knowledge, under the pressure of powerful, sometimes speculative, commercial or political interests [GUI 11, JEO 15]. This statement is confirmed by Cornu [COR 08], who writes: "Does technology, beyond its undeniable progress, help support the dream to make applications acceptable by the greater masses or does it rather give some fans of science-fiction legitimacy to influence research programs? Whatever the case may be, a 'convergence of interests' has allowed people to gather around this inspiring program".

Figure 2.9 (already presented in Chapter 1) illustrates the principles associated with the concept of convergence/divergence and allows some barriers that will be discussed later to be illustrated/explained (see [JEO 15, PAR 99]).

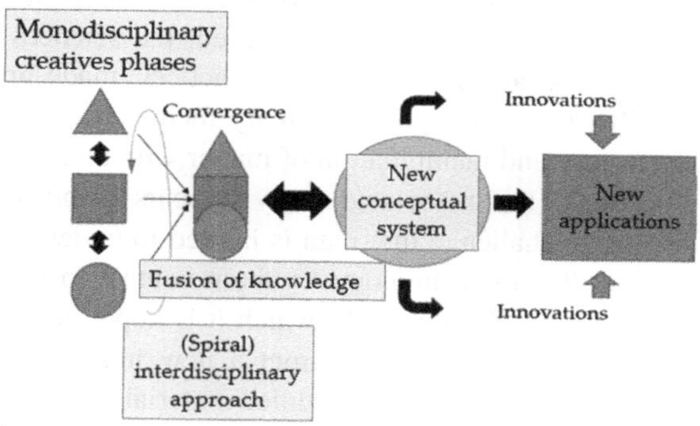

Figure 2.9. *Principles of convergence/divergence*

The basic idea of convergence is to extract disjointed scientific and/or technical knowledge from paths allowing for applications through the fusion of disciplinary knowledge, which leads to coupling between scientific, economic and social aspects, as shown for a successful innovation in Figure 2.10 [BUT 16]. There is thus a need to make production systems not only with technological interest available but also with certain attractiveness/social desirability.

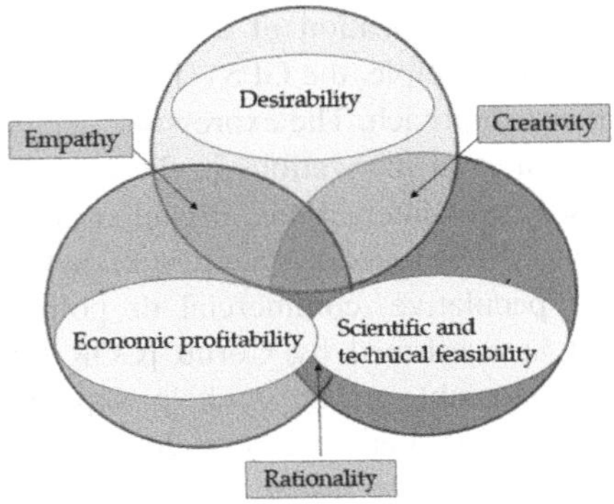

Figure 2.10. *Reminders on the founding principles of a successful innovation*

Discussions held concerning convergence applied to additive manufacturing stem from a symbolic dimension and remain dependent on mythic thoughts about the mastery of matter. To seize all the major stakes of new technologies, it is essential to know the imaginary motivations. Indeed, the sciences and techniques (what we call technologies) are, like all human practices, made up of images and ambivalent desires [CHI 09]. Through the primary themes evoked in the symbolic "3D printing" (the mastery and manipulation of matter with an aim to create objects with a desired function), the associated story appears a bit like the modern expression of a demiurgic challenge that man is invited to undertake, becoming the actor in a new religion, that of technology "without contact to matter" (where the manual tool of yesteryear has disappeared), which it is hoped to personally master. Additive manufacturing thus tends to construct a new imaginary of "à la carte" manufacturing with interchangeable and scheduled materials.

But how does this research come to join an interdisciplinary perspective? In several texts on interdisciplinarity, the idea according to which it is the complexity of certain objects of study [BOI 04, DEF 04] that dictates the interdisciplinary practice is recurrent. Indeed, the nature of those objects is conceived as a dynamic system, "irreducible to a single dimension" [KLE 04]. The accumulation of knowledge and the growth of views, the interactions between phenomena, undeniably characterize objects of research: complexity has become an intrinsic value of knowledge and research in general [BÜH 06]. However, in using the machine, this complexity must be hidden.

In these knowledge-fusing activities, it is best to try to limit the recourse to operations that increase the risk of losing mastery of the theorization process as much as possible, yet the social communication at the heart of interactions necessarily draws on common sense. It is difficult to "purify" the rational dimension of the project – caring on a cognitive level – of its largely implicit affective and social components, and, more fundamentally, of the symbolic dimension of every human process. Sociabilities naturally play an eminent role in the exploration of crossed reflexive conditions; there is a need to know oneself, to have common references and to overcome strong shared implicits [LE 12]. In fact, the disciplines concerned are rather close to one another, and the pressure on innovation is strong enough for the creative dynamics of new knowledge to be pursued. However, as will be discussed at the end of Volume 1, the further we are from proofs of concept, the more difficult it is to satisfactorily and comprehensively deal with all of the concerns from a scientific standpoint. The loss of dynamics illustrated by the results on the evolution of the number of scientific publications from Figure 2.1 (going from exponential growth to "simply" linear growth) is possibly associated with this limit, the difficulty pertinently blending scientific disciplines.

2.4.2. *"Attractiveness" of additive manufacturing technologies*

2.4.2.1. *General framework*

The liberal economy has developed a mass industry with very advanced standardization; it gives way, for various reasons, to an "economy of variety and reactivity" [LIP 06]. Within this framework, what novelty does the virtual bring to the conception and creation of objects? At first, a creator has, at least, one intention expressing a desire. To further his/her conception, he/she makes sketches or models; yet, he/she has largely lost his/her skills as an artisan or modeler (like most of us), which limits his/her creative action. With additive manufacturing (which keeps the tool in its place), he/she can plan the creation of the shape and/or functionality of the object he/she desires without constraints. "After a battle against the technical difficulties that it involves, he/she can contemplate the finished work and deliver it to the client. Painters, sculptors, fashion designers, and architects have always drawn sketches to specify their project and discuss with their client; they have therefore always made the 'virtual' and it can therefore be said, remaining strictly logical, that the computer provides nothing new to them" [VOL 15].

However, its contribution becomes apparent if the practical conditions of creation and dialogue with oneself or the client are considered; it does not lose the commercial logic with respect to constraints and hence standardization that develops around additive manufacturing [ANS 17]. A sketch is nothing but a suggestive drawing, a model must be enriched by the imagination; in the virtual space located behind the computer screen, on the contrary, a 3D object can be rotated, placed on a landscape to see the scale, be shaded, colored in and so on. This is what physically allows additive manufacturing. In this context, but in another framework, Michel de Certeau has been exploring the techniques and tricks that lead to the behaviors of certain "makers" in the face of accessible content since 1994; these are likely to re-route the propositions of content producers for their own operatory efficiency (for seniors, let us recall Minitel's twist with its "pink" (sexual) applications) (see also [MÉA 09]).

If the domain has disappeared with its "knowledge" and experience as a virtuous craftsman, additive manufacturing is a means of leaving the virtual spaces occupied by numerous users of information and communication technologies to return to the real, to the palpable and to physical experimentation. Since the late 1990s, all information has been digitized, that is, transformed into series or tables of numerical data, which has led to their nearly complete dematerialization and to their being processed by all sorts of algorithms. Digitization has thus, according to Berry [BER 15], brought about mental inversions: re-routed perceptions concerning elementary actions of everyday life.

The links between virtual relation and narcissistic relation are notable with computer science and, in a rather general way, ICT, a form of relation facilitating introspective attitudes and thus going in the direction of a lack of will to meet with the other subject, disconnection, which, itself, seems encouraged by certain forces in our current society, possibly revisiting the notion of work, even that of associated rights [GOM 16]. From this standpoint, 3D printing offers an original path of material resourcing and may break away from a certain imprisonment connected to the unrestrained use of computer science. For example, it allows people who are not skilled with their hands to prove their ability to create objects so long as they possess the necessary know-how using a 3D machine and are open to others physically or virtually thanks to the possibility of a large-scale collaborative work.

Questioning an "economy of the multitude", to the detriment of an "economy of proximity" protecting work is misleading. Since the works of Nicolas Colin and Henri Verdier [COL 15], the economity of the multitude refers to the digital economy close to "devouring the world", according to Marc Andreesen's expression. It seems unfounded, however, to counteract it using the argument that work in the digital era distances the worker from the consumer even more than in the industrial era. It is at least necessary to describe what this economy reconfigured by digital technology is experiencing. It is indeed undeniable that the companies of the new economy are seeking alliances with "the multitude", that is, all individuals [MAI 16]. Yet, actually, additive manufacturing processes enable this merging of digital technology with the physical domain and of the global with the local. It is therefore a very interesting opportunity for the public, as long as the price of the basic machines become accessible, sometimes the price of a classic printer. And at the same time, the fun aspect allows a space to be created that is generally favorable, without there being a need to produce "informed" knowledge [NEU 86].

Finally, at present, 3D printing allows personalization, the possibility to produce objects for a single-person market without paying money for the prohibitive price of an industrial prototype. Thanks to 3D printers, within a very short time (sometimes a little more the normal), it is possible to create customized objects which are often more creative than those found on the commercial market. Personalizing an object costs nearly nothing: it suffices to change the instruction code and modify some parameters. Printing a complex model, as has been pointed out, costs no more than printing a simple model. It is the consumer who can henceforth have the final word as long as he/she works to return to forms of reality which now seldom exist in his/her daily practices.

Even if an astronaut at the International Space Station used a 3D printer to create a socket wrench not long ago, and even if in this perspective, there was a reference to a near future where the digital code could replace the need to send specialized tools into orbit, this technology would essentially lead to a plethora of small creative companies (start-ups). Table 2.3, inspired by Taylor [TAY 14], allows for exploration of the impact of additive manufacturing technologies on society's activities; it would notably affect modes of consumption, creativity in manufacturing production, services and certain forms of productivity. Currently, there is oscillation between effective technical progress that is useful for the society and passing fads (as happened a few decades ago with cell phones). In 1968, Galbraith wrote [GAL 68]: "most goods serve to satisfy needs that the individual discovers, not due to the secret discomfort that comes along with privation, but as a result of a mental awareness of their possession". Regardless of their origin, the facts speak for themselves. The attractiveness of technology is spreading with growing success thanks to 3D manufacturing processes.

General theme	Specific theme	Impact (from 1 for "weak" to 3 for "strong")
Social framework	Quality of life	1
	Effects on consumption	**3**
	Nature of work	2
Evolution of companies	Evolution of companies	**3**
Governance; openings	New products and services	**3**
	Organizational rearrangement in companies	1
	Organizational rearrangement concerning users	2
Economic implications	Economic growth	**3**
	Competitiveness	2
	Effects on employment	2
	Regulation	1

Table 2.3. *Influence of 3D printing, emerging technology, on the socioeconomy*

But even so, more and more machines are being bought privately (just as we have color printers hooked up to our computers) by "DIYers" (DIY, Do It Yourself), independent creators, teachers or educational centers. Furthermore, for the past 5–6 years, "open-source" kits have been available at costs below 1,000 €, basic machines for less than 500 €, which, in principle, allows a mass market to be achieved (in comparison with the price of an iPhone when they were put on the market). Some of these machines even allow other machines to be produced, thereby exploiting a self-replication principle [CAH 14].

These low-cost units therefore lie at the heart of a new kind of self-producer that can then take care of conception, artist, engineer, producer and consumer functions; in short, practically the craftsman of yesteryear, moving from blue collar to white collar, capable of creating a tool adapted to a specific need, yet without claiming to be a specialist [BER 14]. Additive manufacturing machines only acquire their meaning of a "derived need" [BAU 72] in the social framework in which they are integrated. These facts thus lead us to think of a great attractiveness of additive manufacturing technologies vis-à-vis the public. But does a 3D printer user need to be "educated"?

2.4.2.2. *Acceptance of the technology*

With acceptance, we here mean the potential attraction of this novelty for citizens. Social acceptance is today considered to be a necessary element to carry out projects. More and more, it constitutes a concern for manufacturers, investors, citizens, business communities and so on [ARC 16, TCH 17]. In the field of additive manufacturing exploration, knowledge of the risks (regardless of their origins and consequences on man and the environment) has not been stabilized. "The existence of unknown and uncontrolled risks, implicit in the 'media chorus' of the argument of an 'absence of zero risks' is not part of reality, but inherent of a deficit not only of knowledge, but also of coherent concepts [and of perception] in relation to the technological ambitions pursued" [POU 06]. In a situation of uncertainty, several criteria on the performance and impact of the research program have already been thought up by Chevassus-au-Louis [CHE 07]. They have been recently completed by Porter and Youtie in 2009, by Kim and Moon in 2013 [KIM 13] and then by Kim in 2015 [KIM 15]. On the basis of these works, mapping (see Figure 2.11) has been proposed to some thesis students at the LRGP and student engineers at the ENSIC (which is home to the research unit) on the 12 following criteria (already presented in the introduction of this first volume):

– Frame of attractiveness of a technology:

- attractiveness/self-image;

- ease of use;

- cognitive aspects to be implemented (aid to the intellectual activity, games, etc.);

- cost;

- efficiency (price/quality relation);

- positive perception of the technology.

– "Coherence" of the risk:

- risk observability (short term);

- analogy with other risks;

- non-reducibility of the risk (possibility or impossibility of taking preventive measures);

- severity (gravity potential; population concerned);

- plausibility (potential knowledge of effects);

- irreversibility of the effects on man (e.g. cancer).

By attributing a value to each of these considerations (from 1 to 5), it is thus possible to position a "center of gravity" for the possible risk/attractiveness of a technology on this dodecagon, thereby attempting to understand questions that may affect society (and the innovator). For this analysis, which could lead to appropriate measures, there is a possibility to implement adapted deliberation processes (which, however, stem from the subject of this work) internally (possibly with external support).

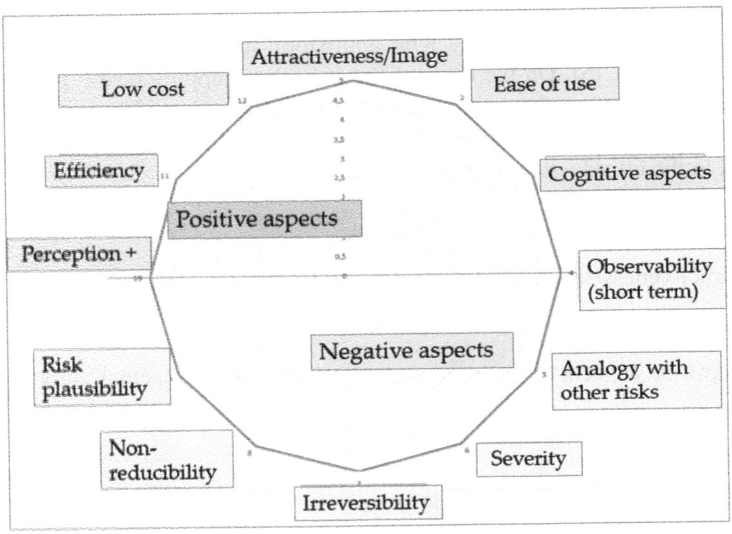

Figure 2.11. *Mapping the 12 criteria connected to the attractiveness of a technology and the associated risks*

After having defined a restricted group of students from ENSIC and LRGP – CNRS – Nancy (who, in their first year after the preparatory classes at the "Grandes Écoles", have only a general scientific education) represented by the NBIC convergences (nanotechnologies, biology, information, cognition) and ICT (information and communication technologies), it was possible to perform the following mappings, as presented in Figure 2.12.

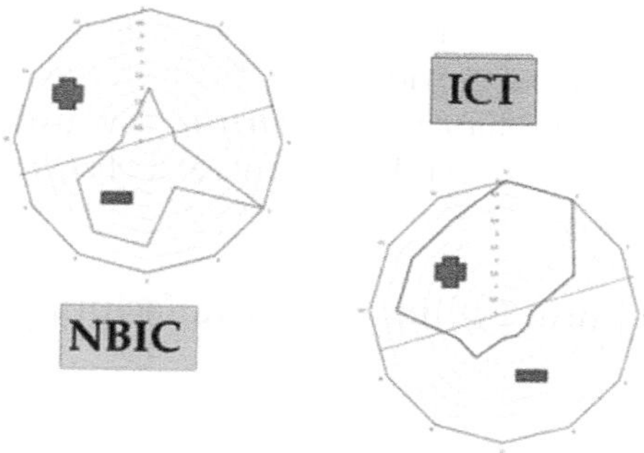

Figure 2.12. *Differences in perception of NBIC and ICT technologies [AND 16]*

While the image of the asbestos plays a major role in the results concerning NBIC convergences, those connected to ICT (which could be affined) are qualitatively a good illustration of the differences between two technologies, the first that is not the subject of very large interest (although there are not yet commercial NBIC products [AND 17]), while the second indeed falls under the principles of general attractiveness. However, in this latter domain, is anything known about the risks? Nordmann [NOR 04] thus comments on the ethical debate rasied by NBIC convergence: "Taken in isolation, these various technologies have already brought about controversies and anxiety. Their convergence represents a major challenge not only for the world of research, but also immediately for policy makers and European societies".

Is the appreciation of a possible risk connected to ICT linked to an insensitive approach (there is no knowledge of deaths connected to its use, unlike an "accepted" risk like automobile transport), or strongly rejected like that associated with asbestos exposure, the fear of the effects of GMOs, carbon nanotubes and so on or, on the contrary, to a certain desirability? According to Verdo [VER 16], it is important to take into account mental invariants, those cognitive biases that stop us from forming a proper idea of the truly incurred risk. "Our brain is made such that it has a natural tendency to overestimate (by a factor of 10–15) the risks presenting both a very low probability of occurring and catastrophic consequences if it occurs. This fact was established by the neuroeconomics pioneer […] Drazen Prelec. Yet, many technicoscientific risks belong to this category." This is indeed what is observed for the "nano" domain with, for its NBIC component, a true overestimation (as there are no products integrating all of the NBIC facets on the market, to the author's knowledge) and, to the contrary, likely an underestimation on the side of informatics.

By applying the same procedure to 3D printing with the same group, results are obtained that are close to those provided by ICT (which may seem rather reasonable insofar as additive manufacturing is a domain that stems largely from this convergence with ICT and the fact that the potential chemical risk is hidden). The results are presented in Figure 2.13.

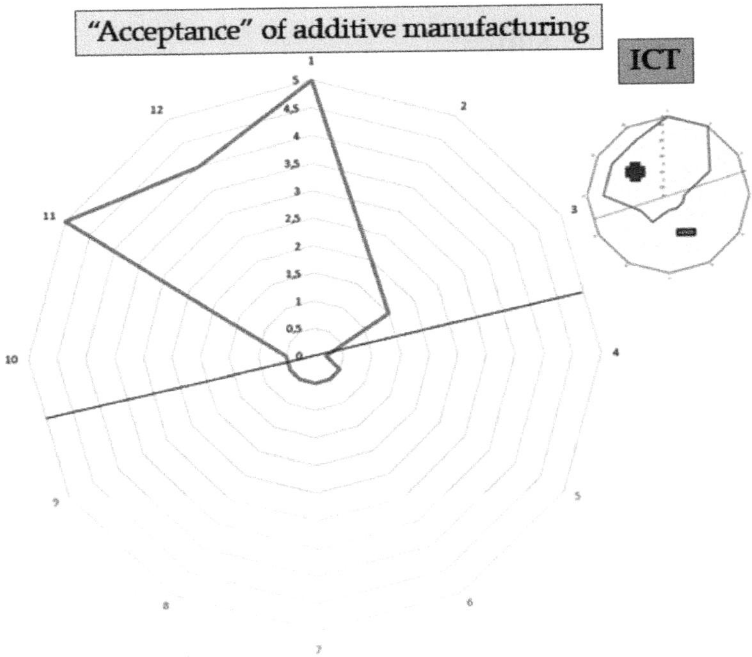

Figure 2.13. *Attractiveness of additive manufacturing for an educated but untrained public (with the same mapping for ICT as a reminder)*

This figure illustrates how obvious attractiveness can be obtained (in the image of the evolution of the 3D market and its democratization over time). But looking a little closer, the cognitive aspects that were perceived as attractive for ICT are much less so, as with efficiency aspects. Indeed, the students ask themselves the question of programming objects and the ease of their creation. According to Gomez [GOM 16], two desires for freedom are frontally opposed to one another in citizens of the modern world: being free through work and/or via fun aspects or those of consumption. Between (often immaterial) production and use, there is, thanks to the perceived performances of additive manufacturing, visibly a possibility to merge the two standpoints, the social utility and the desired. Current limitations to the development of 3D technologies are nevertheless identified by the students (before and beyond the survey, during restitution):

– the current slow timeliness of 3D printing;

– the part-time need for human intervention to create an object;

– the relationship between machine and object size, associated with the quality of the final objects;

– the need for strong technicity.

Maybe this situation has led to not only the existence of interfaces which are represented at present by the "maker" movement and Fab-Labs but also a still modest democratization of this technology.

On the contrary, a recent study published by the "World Economic Forum" [WOR 17] conducted with manufacturers on the risks associated with new technologies and summarized in Figure 2.14 shows that 3D printing is considered rather attractive (more attraction than untoward consequences) relative to other technologies like geo-engineering (e.g. shale gas).

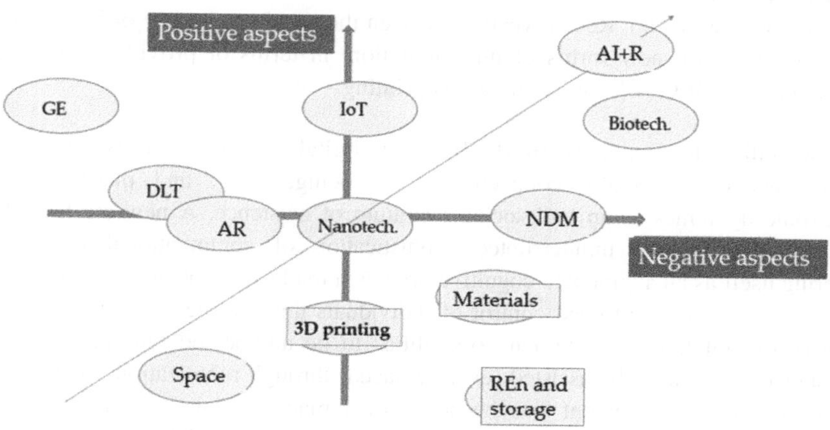

Figure 2.14. *Consequences of the technological advances associated with the revolution 4.0 on man and the environment (GE: geo-engineering; IoT: Internet of Things; AI+R: artificial intelligence and robotics; DLT: distributed ledger technology (e.g. Bitcoins); AR: augmented reality: Nanotech.: nanotechnologies; NDM: new digital methodologies; Materials: new and nanomaterials; REn: renewable energy)*

What quickly seems to be understood in these two very different forms of survey is that additive manufacturing can fulfill two missions of speed and flexibility by allowing the direct passage from a digital 3D file to the physical object (gained time and mastery of the process by the designer by short-circumventing the traditional intermediaries). By responding to questions, it seems that the quality of the objects from 3D printers can sometimes be deceiving, particularly for low-cost machines (less than a few hundred euros). The printers capable of printing metal offer better performance, but remain very expensive and rather slow, still with a limited choice of materials, yet this is improving (see also [SAA 15]).

At the same time, in an attempt to understand this decrease in the rise of revenue, number of publications, machine sales (even if the timeliness remains rather good, even very good), it is good to look at how the revenue of the domain's large medium-sized enterprises (MSE) is behaving vis-à-vis the stock market. 3D activity is probably too "oversold", too strongly manipulated by different media and different stakeholders, actually leading to unkept promises. Because they are not the authors of the action, citizens never have access to the full meaning of this action, and a lack of belief can limit the range of innovations, including those concerning additive manufacturing, even if technical progress has upturned the situation of informatics: each user can have a device allowing him/her to process data. He/she finds himself/herself faced with a display tool, while the data are processed again where the computer capacity is found (on the contrary, it is more and more difficult to know where this is). Recent years have seen the rapid emergence of breakthrough technologies with new forms of intermediation, in terms of providing services as consumption [OEC 16], for example, 3D printing.

According to Rapoport [RAP 16], the digital revolution aims for integral interconnection, "virtually connecting every being, thing, and place, enrolling electronic dynamics as an indissociable stratum of existence. A new era henceforth characterized by the uninterrupted sophistication of computational intelligence, offering itself as an augmented cognitive organism made available to all." At the same time, it allows for reinforced control of individuals through the collection of a great deal of quantitative data, human possibilities to be augmented (augmented reality, exoskeletons), man to be assisted (even replaced) through robotization, in short, help to be provided to the extent that the power of human decisions is marginalized and overtaken by this incessant development (see also [SAD 13, SER 16]).

Associated to this uncontexted reality are more questionable elements which allow us to cite Alexandre [ALE 15]: "We are already tracked by our behaviors, our cosmetic or eating habits, our surfing on the net and other social networks. The 'prosthetization' of the body, already presented when deficient sight, hearing, knees, or hips must be healed, adds itself to connected objects, intrusive nanoparticles, digital chips included, made commonplace through their mediatization if not by the reality of their usage…"

To return to the subject of this first volume, it seemed important to better understand why a recent, emerging history, of patents falling into the public domain in great numbers, is being slowed in its acceleration. Indeed, many official reports [ALE 15] have placed 3D technologies among the top 10 technologies of the future (see e.g. [PWC 16, NFI 16, DAS 16, ALE 15b]), but currently, however, estimates indicate that it may only represent approximately 5–6 billion €/year for the world (very imprecise estimation, but for all domains of 3D printing). It is simply one order of magnitude, which can be compared to the single (more precise) debt of France which exceeds approximately 2 trillion € and that of the machine tool; for the planet, it would be 80 billion € (according to the survey conducted by PIPAME in [PIP 17]). There is thus much progress left to be made.

Thus, before dealing with the current markets and those projected for the near future, it seemed significant to see how small shortcomings, often minor (or considered as such), can limit the promotion of a technology that always makes us dream, with the possible vision, for some supporters of 3D printing, of never arriving at the hegemonic position that they advocate.

2.4.2.3. *On object design for the public, the stage of digital design*

As previously mentioned, long reserved for an audience of professionals or educated DIYers, 3D printing is hereafter open to the public with ever more accessible machines (with machine costs of a few hundred euros). The British cabinet Context [CON 15], in a study concerning the 3D printing market, indicated a 24% progression in the first trimester of 2015 in model sales with costs below US $5000. This could be in part due to the growing number of actors on the market. It is worth mentioning that the offer proposed to individuals is only progressing by simplifying the use of matter, by proposing more and more multi-materials, software with more practical uses, Wi-Fi links and so on.

This desire to reach the market covers all of the elements that are involved in the object creation process [VER 15, BOU 14, DIR 14, MAD 17]: In a general way, the creation of a 3D-printed object requires a three-dimensional digital model created using CAD (computer-aided design). These models exist in the form of digital files saved in standard formats that will be briefly presented and discussed here. Once opened in print preparation software, they are analyzed and transformed into instructions sent to the specific printer (stage, so-called slicing) in the additive manufacturing process. The localized addition and progression of matter/energy through successive layers gradually builds the physical object. Once this process has taken place, postproduction operations may be necessary depending on the materials used and the final quality desired [AND 94, COT 14]. These different stages are represented in Figure 2.15, which illustrates rather well the comments made by the group of students on the apparent complexity of the comprehensive process of manufacturing an object for a layperson.

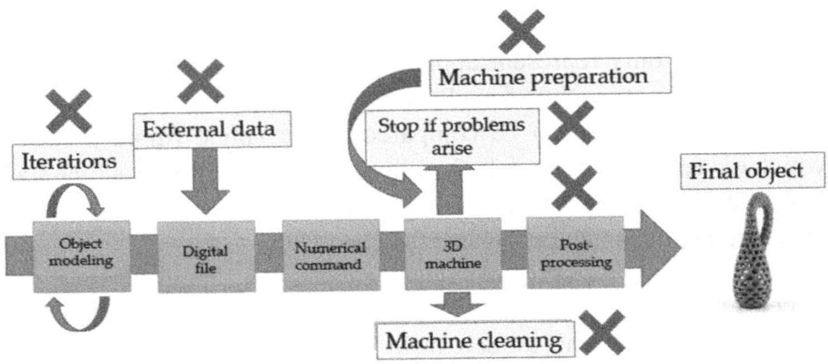

Figure 2.15. *Stages of manufacturing an object through additive manufacturing (red crosses indicate human intervention in the process)*

As can be seen with reference to this figure, it is not the same as that which an average user does with a classic printer due to the increased number of human interventions. In the “modeling” component, the producer designs the object (unless external data are available, with the risk of rather “incompetent” use of additive manufacturing lacking creativity), bringing it to life through iterations. Furthermore, in this stage, he/she can plan to create complementary elements for the operational creation of the object (this addition depends on the process) with the possibility of planning to work with various materials. The first difficulty for a novice user is thus created through a need to master the computer elements that he/she is faced with.

Then comes the question of which software to use, which to start with? Whether this is for designing or modifying a 3D file oneself, it is best to ask oneself what the most “appropriate” 3D software is. According to Alex [ALE 15], it is difficult to make this choice among the numerous programs in existence; he reminds us that the Belgian company “i.Materialise” published a list of top 25 most popular design programs in the 3D printing universe, with scores between 80 and 11. While this classification in no way intended to hierarchize 3D software, this information provides a lead “for those who wish to attempt 3D modeling, for whoever speaks of popular software speaks of a large and active community to respond to his questions”.

Laly [LAL 14] provided a practical presentation of the questions that a novice user wishing to create an object may ask himself. He foresaw several stages before arriving at the desired part.

– obtaining/creating the digital file: The first step is, of course, to see if the part to be created exists in a database (generally open source) and to import it; otherwise, it is necessary to virtually "create" the device to be produced using the appropriate software, unless it is possible to scan a similar part that could be "reworked" using appropriate software; we must remember, however, that these "educated DIY" operations, as interesting and educational as they may be, can be time consuming;

– performing the "slicing": most additive manufacturing software is capable of dividing the object to be constructed into "slices" with automatically determined thicknesses;

– choosing the manufacturing process: in general, owners of 3D printers have only one, but Fab-Labs are developing; thus, at present, it is possible to access different types of machines to create parts using very different types of materials. In this decision-making stage, it is possible to ask for the assistance of Fab-Lab members;

– Laly's view [LAL 14]: "It can be difficult to obtain a perfect part the first time. Once printed, it is quite simple to notice a design error... You will only lose a few days and euros...". Success is thus not immediate, and additive manufacturing, therefore, is not exactly a tool like others. However, as with informatics that has been strongly democratized [what a joy to see the end of punch cards, as the author did!], progress can be made in ergonomics and the ease of using software. Let us believe Laly [LAL 14] when he writes: "Do the test and you will see that these tools and services are truly starting to be within our grasp. You will have understood that it is quick, useful, and simple, and soon, there will no longer be a single person with an excuse not to dive in!".

2.4.2.3.1. Software to deal with a digital "object"

– *Modelers*:

- before printing an object, except with rare exceptions, it is generally necessary to possess a 3D file that can be created using 3D modeling software. For design, according to Alex [ALE 15a], 3D drawing software is divided into three large principal groups:

- volume modelers: adapted to simple shapes (cylindrical, cubic, rectangular, spherical, etc.), it is possible to "manufacture" the part through the addition, subtraction or assembly of shapes (this is slightly more complicated than the use of drawing software). The following are the existing and classical mentioned programs: Solid Edge, CATIA, Autodesk (Inventor) and Solidworks;

- surface modelers: the surface of the object is mathematically defined; the primary users would be design and 3D art specialists, with programs like Rhinocéros and Catia from Dassault Systèmes;

- parametric modelers: the object is defined by customizable equations; OpenSCAD and Rhinocéros (via the Python functionality) are thus used by a number of design engineers and architects.

Alex [ALE 15a] reminds us that, regardless of the modeler, it is possible to introduce other parameters like materials, colors, textures and luminosity. "The choice of modeler will be a function of the object's nature, its applications (sculpture, mechanical part, architectural part, etc.), but also of the budget and skills of the user in this domain."

– *File recovery*:

- One simple method consists of recovering a file from a sharing platform, often a free one. There are a number of sites, some of which are proposed in Alex [ALE 15a].

– *Scanning an object*:

- There are numerous methods of defining an object for additive manufacturing (see [3D 16] for example), ranging from old mechanical sensors from the 19th Century to 3D digitization methods with or without lasers, allowing us to obtain digital data through triangulation methods. Possible shadow effects, however, sometimes require corrections to be made to the digitization (which sometimes appears upon the part being manufactured). Figure 2.16 corresponds to C. Lavigne's demonstration of the possibility of creating a digitization of a person with the appropriate optical system during the science festival in Verdun (France) in 2016 (see also [ANI 17]). Using this knowledge, an anamorphic bust could potentially be produced, as shown by an old cutout from the French newspaper Est Républicain (05/09/1990) in Figure 2.17 taken from the research laboratory.

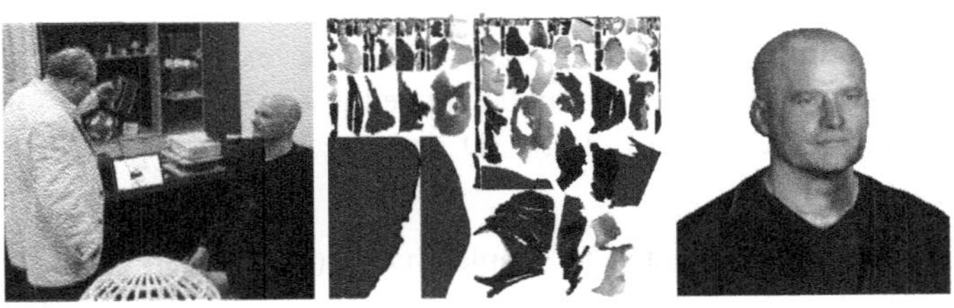

Figure 2.16. *Digitization of a person's head (Patrick Visentin, 3D artist from Quebec). For a color version of the figure, see www.iste.co.uk/andre/printing1.zip*

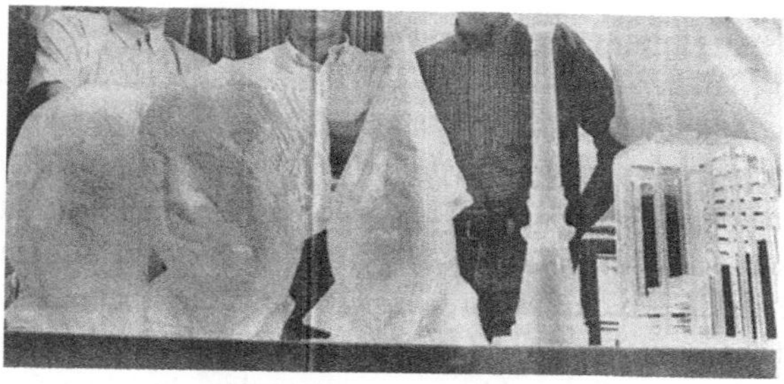

Figure 2.17. *Figures digitized using laser sheets, deformed by anamorphosis for creation using laser stereolithography in 1990. For a color version of the figure, see www.iste.co.uk/andre/printing1.zip*

– *Preparation software*:

- STL (Standard Tessellation Language) files for 3D printing are used to store sets of links whose basic principles were developed by 3D Systems [ALE 15a]. The STL file describes an object according to its external surface. "This surface is necessarily closed and defined through a series of triangles (or facets). Each triangle is defined by Cartesian coordinates (x, y, z) in a direct trihedron of its normal vector (n) oriented towards the outside and its three vertices ordered in the trigonometric sense. The coordinates of the vertices must be positive. Each triangle must share 2 vertices with each of the adjacent triangles. In other words, the vertex of a triangle must not be on the edge of another triangle. There is no information on scale. The unit of length is arbitrary. It is recommended to classify the points by increasing z to make it easier for certain programs to read the file" [SCU 16].

The digitization of contours for a given dimension in fact passes again from an unbroken model, as shown in Figure 2.18, to a discrete model (approximation of an unbroken contour through a polygon). The precision, obviously, is much better when the number of points is higher.

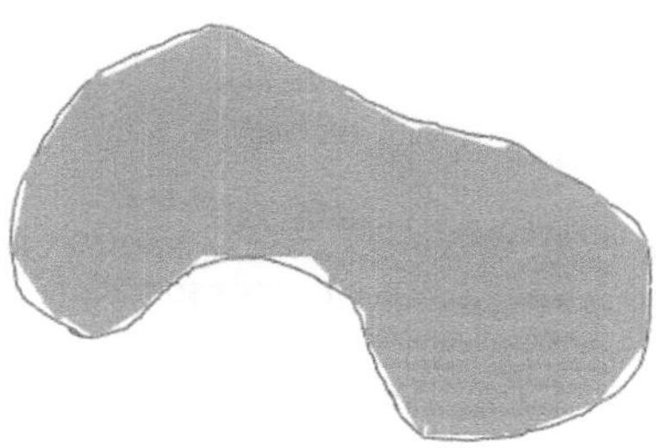

Figure 2.18. *Contour approximation through a polygon*

"Tessellation" constitutes the phase used in which the geometry of the object is transformed into triangles in order to save them. This phase can change the origin linking if the resolution is too high, as shown in Figure 2.18, taken from the same source. For the computer engineer, modeling consists of connecting the neighboring contours with triangles. This operation is generally rather simple, but in some cases, it may be necessary to develop specific algorithms. The first example corresponds in part to Figure 2.19, in which the surface is reconstituted using two contours with reconstitution of the associated surface with the help of triangles (see Figure 2.19). The second example is represented by a more complex situation: for a surface, there is a contour at one level and two contours at the next level; contours are created at an intermediate level, enabling the reconstruction of the surface. This operation, presented in Figure 2.20, can be undertaken step by step; the surface is then the union of different created surfaces.

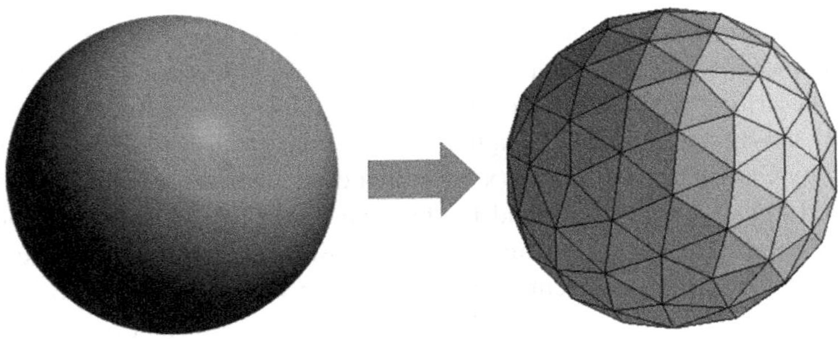

Figure 2.19. *Effect of the quality of triangulation on the shape of the object in the computer memory*

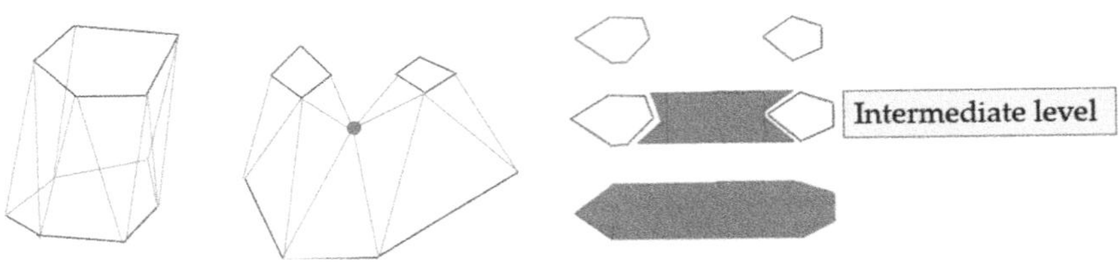

Figure 2.20. *Reconstruction of an object using triangles (left) and contour association (right)*

The "STL" system, developed in 1987, is simple and robust and has become a point of reference in this domain; it can be (or is) integrated into numerous CAD software packages like Autodesk Inventor, Alias, Blender and Solid Work [ALE 15a, REC 16]. Alternative software to STL files is being developed. According to Recrosio [REC 16], the most interesting is the 3MF file from Microsoft, which aims to improve the software (new functionalities) by making it easier to use through the help of computer engineers and machine producers. We must not forget that there is a specific norm for this software (in France, see ISO [ISO 16])

– *Adaptation of the object to a functionality*:

- According to Tancogne-Dejean, Spierings and Mohr [TAN 16], it may be advantageous to check if the part to be created cannot be made lighter while maintaining its mechanical properties (in the case of space and air transport in particular, see Chapter 1). According to these authors, the weight can be reduced by more than 30% without the part's functionality being modified (this is often associated with the mechanical strength). Software adaptations are thus necessary to attain this goal. This topological optimization technique consists of eliminating volumes of matter where the efforts are weak enough, with the help of optimization software [ROD 16]. Among the most known cited by this author are DesignSpace from the publisher Ansys, Tosca from Dassault Systèmes, Within Labs from Autodesk or even Inspire from SolidThinking.

– *Toward the creation of the object*:

- Several elements can or must be added when the digital stage is not at all over; first, it is generally necessary to foresee a manufacturing grid called a "raft" (underneath the part) or "brim" (around the piece). According to Alex [ALE 15a], this support is meant to prevent the object coming loose from the platform (a phenomenon called "warping"); layer-by-layer deposition is carried out using this base to stop the piece from falling apart. This presence of a complementary element introduces a physical (generally manual or using tools and/or portable machines) or chemical (if the base is created using material other than the part with the possibility of dissolution in an appropriate solvent) separation at the end of manufacturing.

Furthermore, if the part to be created has a complex shape like the object presented in Figure 2.21 or if the part being constructed is not supported at the z_0 height, the final object will never be true to the digital projection (see e.g. [MIR 16]).

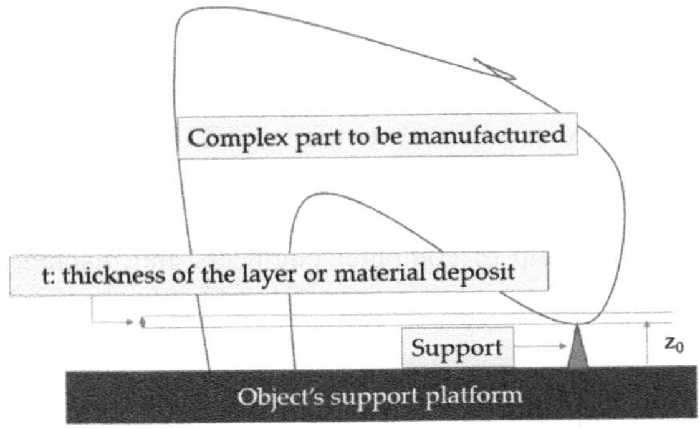

Figure 2.21. *Need for supports for complex parts (depending on the process used)*

Under these particular conditions, it is necessary to add complementary additions to the digitization of the object which enable its feasibility. At the end of manufacturing, as for the base, there will be a need to eliminate the scaffolding elements by physical or chemical means. This principle being defined, if the part is hollow, it will probably be difficult to expect to eliminate internal supports by mechanical means, which will lead to working with special "multi-material" machines or ones that do not require supports [AND 16].

2.4.2.3.2. Digital object – process relation

A 3D object can be manufactured through the addition of material transformed by different means as follows:

– contour manufacturing with the risk of having a fragile structure (but obviously rapid production as only the surface is processed and not the volume);

– contour manufacturing with internal reinforcement to promote mechanical strength; in the two cases, post-processing is often necessary (particularly with stereolithography);

– surface manufacturing: the materials brought or transformed during additive manufacturing undergo temporarily evolving modifications from geometric perspectives or superficial tension (for certain powders); under these conditions, resting time may be necessary to avoid deformation of the object being constructed; if this is the case, it is necessary to define a transformation located on the surface such that these temporal effects are minimized; in any case, the manufacturing time is greater than that of the first two cases;

– an alternative to this aspect has been exploited in the laboratory [AND 94] using the principle of polymerization induced by light (which has a certain volume reduction between the monomer and the polymer) in resolved spatial polymerization; this is defined in Figure 2.22. First, the disjointed elementary bricks are created, not adhering to the object being constructed (arrangements are thus made for the depth of polymerization to be less than the thickness of the layer). Once these voxels have relaxed, a second passage of light allows the voxels to be linked to one another and to the support. Under these conditions, the reduction only applies to the not-yet polymerized volumes, which, in general, is translated by deformation linked to the much more modest reduction. There are other "tricks" to limit reductions (such as the programmed presence of unpolymerized zones to reduce the internal tension). Let us note that there are numerous computer sophistications to limit this problem (see [JIN 17] for example). Figure 2.23 shows some examples of methods of voxel production displacement that play a role in the deformation of created parts (singeing patches). This problem, linked to the aspects of volume anisotropy, is one of the most difficult problems to properly deal with.

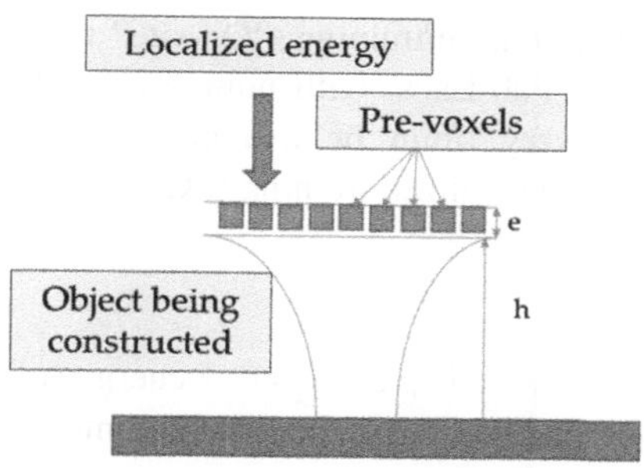

Figure 2.22. *Two-step layer transformation*

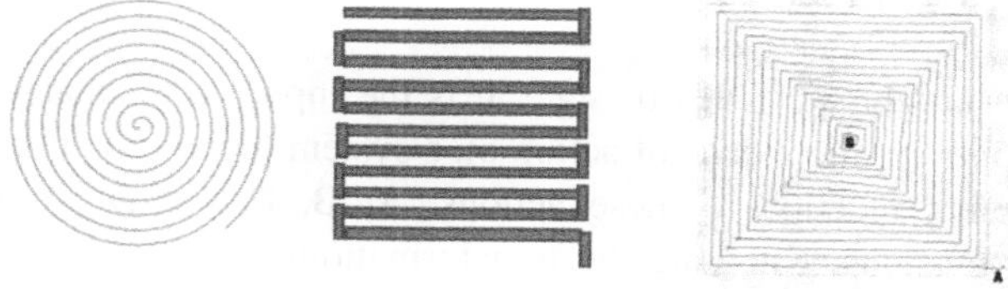

Figure 2.23. *Examples of possible displacements that introduce different memories/deformations according to the space considered (interactions between displacements)*

Upon reading these comments, it seems that the software component enabling the realization of rather complex parts with mechanical, spatial and other demands is not immediately accessible to "the average citizen" without preparation, which allows real competence to be achieved step by step through the incremental increase of knowledge. In fact, with the help of a command relationship model, attempts are always being made to reduce things to a few monovariable (one input and one output) problems that automation can process. In other words, the following question is being answered: is it possible to render a (supposedly simple) system's transfer matrix so that an output is controlled by one and only one entry? This is what business does out of need (with some knowledgeable DIYers, like the one mentioned in Figure 2.23); however, this is often a limitation for the amateur who can find immediate attraction in commercial systems, followed by disappointment as soon as he/she tries to output standard memory objects (generally provided with the machine to encourage the initiation of the education process). These comments bring some contributional elements to a certain disenchantment with 3D technologies among the public (even if the number of home machines is always increasing with a double-digit progression). A fundamental element of this development is the existence of open labs for initiation/training. "On-site" training can take place in a simple and friendly way in Fab-Labs, with numerous partners convinced of their "mission" and who may very often be available. Moreover, to maintain this dynamic, Sculpteo proposes specialized training [SCU 17].

NOTE.– Small reminders of automation:

For a specific layer, the local displacement of energy, of deposited matter or of the two entities, the law of linear control, the most common, is generally represented by a relation of the form:

$$dx/dt = A.x(t) + B.u(t)$$

$$y(t) = C.x(t) + D.u(t)$$

where x represents the variables of state, u is the input vector and y is the output vector. The inputs are the means of action on a system (controls). The outputs result from the system's reaction to these actions. A, B, C and D are matrices with appropriate dimensions. Knowing the transformation velocities for the creation of voxels, an attempt is generally made to reduce these two equations to one deterministic displacement like those presented in Figure 2.23. As long as the surface is free (without any connection to a support), singeing problems can be made apparent.

However, if there is a memory effect, each voxel with a behavioral function of time, for example, in the form v(t) with amortization:

$$v(t) = v(\text{infinite}) + \Delta f(t)$$

the voxels created u at time t, located at a surface position xi and yi, can interact with those that were produced before. There is therefore a dependence function of the distances defined between the voxels among themselves and the time when they were produced. In principle it is good to closely consider these aspects, which, for a complex part, can be a particularly long process (often with behavioral laws that are not totally mastered quantitatively and potentially nonlinear laws of control).

It is clear that an automation (a digital control), in order to function correctly, must be able to react to the external eventualities and overcome the shortcomings of the base model. Ensuring this ability to adapt is precisely the role of the control loops whose primary function consists of:

– observing the result of the control on the output with the support of sensors (that measure what?);

– comparing this result to the desired value for this output;

– acting robustly to counter the discrepancy established.

It would, in principle, be possible to envision output feedback (if an abnormal deformation is observed, for example), but for the reasons cited, the rejection of disturbances in this way is tricky, as their effect cannot be measured before they irreparably affect the output. If the rest times associated with v(t) are short relative to the displacement velocity, it is difficult to project a feedback that will allow the disturbance effect to be corrected (this, in fact, depends on the origin of the effect); if they are long, corrections can, in principle, be foreseen as long as an online measurement of the surface's status takes place, imposing different displacements than those presented in Figure 2.23. It is therefore difficult to foresee robust uncoupling taking into consideration these temporal rest effects affecting the voxels.

Independent of these constraints, it can be envisaged to use a monovariable adaptative control on the basis of local knowledge of disturbances (e.g. insufficient laser flow in stereolithography or powder fusion). This possible control, however, is only significant if the "delay" effects brought about by the disturbance are known.

Thus, if the production of high-quality objects can be foreseen, it is because the materials are better and better adapted to what the process demands on the one hand (example of resins with very little shrinking) or particularly to the expertise of the designers/users who, with complex and varied systems (in time due to the aging of the materials used, for example), perform technological feats that compete with the success of additive manufacturing processes.

2.4.2.3.3. Limitations for the private user

According to André-Chaignau [AND 16], 80% of innovation sales (with every possible origin) are made by 1.2% of "homes" in France. This figure is astounding because of its great deficiency and must push commercial brands to identify and concentrate their actions on the consumers likely to be interested by their innovations. It is possible that this is the same for additive manufacturing, which would then find one of its limitations in this lack of initial attractiveness. This author recalls a certain number of criteria to be followed in order to help an innovation penetrate the market:

1) Addressing the right consumer: Knowing that one in two households is not tempted by innovations, the launch of a new product must primarily target "inno-addicts" and "brand addicts". According to the product category, targeting could become hyper-targeting (young audiences, mothers, men, etc.). This hyper-targeting demands a precise knowledge of the consumer sample groups;

2) Remaining proactive: The nationwide spread of a simple innovation is often a long process (generally around 6 months, often less than the time required to possess robust training on the use of 3D machines), and when it has reached its end, it is often too late. The idea is to create much more reactive tools, even in real time to adjust the promotion campaigns in stores starting the first day (return on the new product's sales performance);

3) Developing a follow-up before and after launch: The innovation process is long. From the idea to the repeated purchase by initial testers, the strategy must be reactive to better refine the targeting (optimization of campaign performances) and react in case of results that do not match the forecasts;

4) Involving the client (e.g. Fab-Labs): To motivate the client to take the first step, it may be useful to involve him/her by offering him, for example, the possibility to test the innovation at home (or nearby, like in supermarkets). This is part of the framework of a premium offer for brand addicts. The sample can also be distributed/created in stores to recruit new clients who do not buy the brand, serving to promote the vendor;

5) Multiplying the sources of exposure: Displays, communication and the company's website and partner sites; the more the innovation multiplies its exposure, the more it will have a chance of catching the attention of potential clients at all stages of the purchasing process (from writing the grocery list to the store; moreover, some propose creating objects with demonstration machines).

In agreement with Ghilassène [GHI 16], manufacturing a 3D-printed object that requires an ill-adapted workplace (unlike classic personal printers), with hygiene and safety aspects to consider (which are too often forgotten), representing a significant cost (machine and particularly consumables) if a quality part is desired, often makes it more profitable to buy these objects from a specialist than to manufacture them [MAK 15]. It is thus necessary to foresee minimal production to make personal additive manufacturing economically attractive (or to accept limitations concerning the fun aspect of the process).

As has been shown, using a 3D printer requires mastery of computer-aided design (CAD) software insofar as all the characteristics of the object to be manufactured must be present in the digital model, thus during the design phase. "Furthermore, consumer 3D scanners today do not allow a physical object to be transformed into a digital model in a single operation. Roughly one hundred digitization acts may be necessary before reconstructing the digital version of the object" [GHI 16].

In this landscape of flexible industrial automation, there is potentially a possibility to concentrate the transition from the signifier to the signified on a single person, as he/she is capable of designing the object and creating it without third-party intervention, with the risks of reinforced individualism and the mastery of traditional language (disintermediation of society or dematerialization of intermediation). According to Giorgini [GIO 14], the exchange of perspectives via different media with different speakers cannot take place; it risks the exhaustion of knowledge due to a lack of exchange. Maybe a reduced relational path, due to its lack of physical link, will develop via the Internet to compensate for this possible effect taking shape on the medium-term horizon. Moreover, in our hierarchized society, how could an obligated reconciliation be arranged between white- and blue-collar workers, by becoming "light blue"?

The consumer can play a role as the designer here, with his/her personal computer through the design of the digital file, a real computer game with various free programs. The printing operation can then be realized in a Fab-Lab or with the help of an additive manufacturing expert. To test 3D printing, it is easy to go through a specialized 3D printing service capable of offering access to a large selection of techniques and materials without investing in the purchase of the machine that risks not being used enough or whose technology will quickly become outdated. "3D printing will seduce individuals with the possibilities that it offers them in terms of personalization or prolonging the lifetime of goods. It is, moreover, around these services that new economic models are implemented" [GHI 16].

The question of the impact of this convergence that must profoundly modify engineering methods (and engineers' training) will be dealt with later. Will there be a transition from analytical and predictive approaches to more heuristic approaches (trial/error, co-design, design thinking, etc.)? Or, on the contrary, simply adaptation?

2.4.2.4. *A transition via "makers" and Fab-Labs*

Technoscientific development is becoming more and more autonomous than other spheres; its comprehensive mastery is becoming nearly impossible. The transition from a rural civilization to a mechanic, urban and cybernetic society has standardized gestures, procedures and attitudes, and it has attenuated the micro-cultures in globalization that makes it lose its cultural specificities. "The result of this homogenization is that certain lifestyles, certain collective identities, and their hierarchy of implied values become obsolete. We have changed from a technico-industrial society to a technico-informational society that makes and will make some of our cultural and social practices obsolete. The value scales will be tipped and disrupted again, and once again, investment choices will surely emerge" [ORO 06].

The machine finds itself located in 20th Century industry, deeply imprinted in the social construct of work (visible in the economic, political, but also leisure spheres as a recreation pretext for production forces). This "instrument" from history can constitute a marker of change in civilization: from the "industrial revolution" to the "digital revolution". That said, these ruptures are not as evident as they seem. There are certainly differences in context and marked processes; however, it must be possible to identify permanences and sensitive differences which demonstrate comparable ways to proceed, including economic and social problems. Is it possible to follow the futuristic visions of certain media that exploit a "utopic idea linked to the transition where it is the person who controls the tool, to the person who controls the computer that controls the tool, with implications in terms of jobs? The risk, as in the 'RepRap' project (RepRap is the contraction of 'Replication Rapid prototype,' which came from a project by Adrian Bowyer (University of Bath, UK) aiming to enable the creation of his own largely self-replicating and free 3D printer), is that there may no longer be a need for a human in control: a device is chosen off the shelf and adapted. Human intervention then becomes minimal. It is therefore a risk to use open-source systems…". [SOD 13]. In fact, what about that?

When it is time for us to go through a large economic and societal crisis, for companies to be summoned to produce more and "better" with less means, particularly human, in homes, where hyper-connectivity overtakes conversation and friendly exchanges, if possibly deepened, certain organizations will wonder about reinforcement, even a return of inventiveness, creativity, generosity and spontaneity in cordial relationships between humans. "The Internet reestablishes collaboration at the very moment when the ideology of all-out competition is revealed for what it has always been: a truncated and narrow-minded vision of society" [DEG 15]. It comes from this position associated with "mentality 2.0", a recent research project in which society is meant to reflect "natural" tendencies and not a set of constraints and rules imposed upon its citizens.

In this orientation, this new social life of "shared dependence" would reject productivist logic, yet without searching for a return to historic village social forms. According to Arnsperger [ARN 09], the wealthiest member of society hope to maintain the current consumerist and capitalist forms, while the demand for exclusions would only concern more elementary needs like work, lodging and food; the middle class would oscillate between these two positions in a context of ambiguity and dissatisfaction. The "maker" development, with its consequences linked to sharing, introduces modifications into this somewhat anxiety-provoking landscape. However, can we expect a technology to modify behavior itself? At best, it constitutes an opportunity, probably with an impression of misappropriation, to confront the idea of a different world, of a process of transformation of the matter designed to serve the economic goal for which the technology was developed.

Indeed, this development was authorized by "mobile", extremely powerful and easily connected computers, with a relationship of widespread amazement and addiction to digital technologies (among them, additive manufacturing), leaving users with a strong impression of freedom and independence [SAD 13]. The assimilation of digital technology's potentials could then have enough power to modify cultural forms, revisit the civilizational model and work with "the emergence of a humanity that is no longer just interconnected, hyper-mobile, accessing a capital value, but henceforth hybridized with systems that orient and decide on collective and individual behaviors, under still-discrete, but already significant, modalities, called to be spread to numerous fields of society" [SAD 13].

It is also a matter of being a member of a group that displays or can display originality, a specific image, different from that of most citizens, in a world where a diversification of lifestyles and a reinforcement of autonomy can be observed with the principle of free disposal of oneself (post-conventionalist culture, according to

Lipovetsky [LIP 03]). The polarization created by the rise in forms of individualism leads to reflection on the social link and the need to open new solidarities as new forms of association and new types of social links. This is a form of social engineering that will slowly lead to changes in the habits and behaviors of a social group that previously expressed no need for them. There is only one step from there to thinking that this dysfunctional transition is intentional in order to allow access to new forms of domination and control. On this basis, behavioral induction strategies that will bring about acceptance of a new technology among makers, but also ones to make it desirable, are still needed. However, to arouse this desire, it must participate in the formatting of engagement and specific and new organizational behavior. "The meaning of our actions, our lives, is actually situational: the individual is never the 'producer of meaning'; meaning emerges from this multiple and contradictory foundation, an open set and thus nondemonstrable, infinite" [BEN 10].

In this framework, Alain Ehrenberg writes [EHR 16]: "Management consists of making individuals cooperate who must possess intelligence of social relations which allows a line of personal conduct to be adopted. Ford's concept of the organization's qualification declines to the benefit of the concept of the flexible organization's competence, and notably that of what is called emotional, social, or personal competence. Autonomy today not only means independence, but just as much cooperation and competition. It is this change in the social meaning of autonomy – liberal – that divides society, for it is, for many French citizens, tantamount to abandoning individuals to the forces of the market and thus seems to be in contradiction with our principles of solidarity." Even if it develops, there are thus still relational difficulties for the maker development, presented below, to become significant in France.

2.4.2.4.1. Hackers/Makers

People who work with 3D printing can be broken into four principal categories according to Kietzmann, Pitt and Berthon [KIE 15]. Figure 2.24 presents these four typologies, highlighting creative actors likely to innovate; others who seek to master a tool in order to create somewhat original, but ultimately rather classic products. The class of "divergent" thinkers can be enriched with contact with groups like those present in Fab-Labs, but they can display autonomous activity, as long as they have "free" access to machines. The other populations find an advantage in making equipment available, but particularly from training and the often warm dynamism that can be found there.

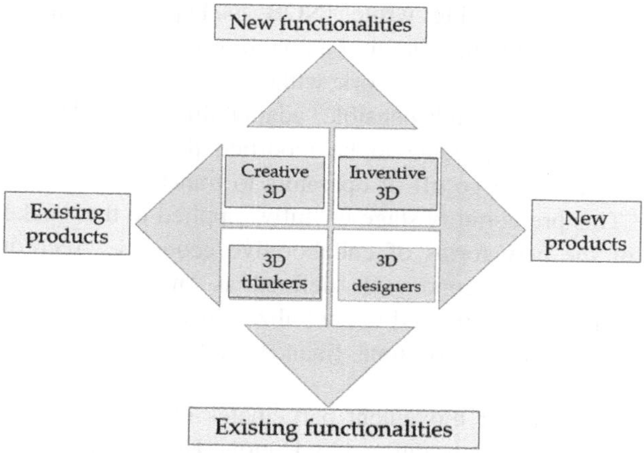

Figure 2.24. *Actors/users of 3D printing*

The "hacker" movement belongs to the development of the individual initiative, allowed through the democratization of informatics, to "play" with technologies, with forms of "technological poaching" and sometimes misappropriation from previous uses [BOS 14, AND 17]. However, it was able to be about circumventing codes, decrypting them to penetrate computer systems. These attitudes take root not only in forms of virtuosity, play or redressing wrongs in Robin Hood style, but also in the rejection of overly rigid technological (and social) norms. Beyond the mediatic aspect (see the group "Anonymous" for an example), certain hackers have invented original methodologies with an often pioneering value thanks to a creative process free of numerous constraints. The "maker" movement that grew out of this defends distributed production (it works on "making" with a generally positive vision) to personalize and lend value to our everyday objects. Michel Lallement [LAL 15] pointed out that for him, work is also always a "full social relation that engages forms of integration, shared representations, rules and identities". The makers' methods of functioning constitute original organizational forms of functioning based on cooperation values [PEC 17]. At the same time, consumers today have access to a very wide range of information (environment, resources used, social externalities, etc.) via the Internet, which, from an economic model standpoint, can be translated by the transition from the linear mode of commerce to a much more cooperative non-stationary linked mode [GIO 14, HOR 15].

In 1984/2011 David Noble wrote [NOB 84/11], "The entire difficulty of automation is rendering the machine-tool autonomous – that is, capable of following instructions specified by the framework without intervention from the workers – without compromising its indispensible adaptability [...]. Hence the role of programming, allowing a product to be modified through the use of "variable" software without having to go off of operators to transform the tool or readjust its configuration." This premonitory stage is "fully" applied to the "maker" movement, which is one of the new forms of collaborative economy. According to Rachel Botsman [BOT 14], this economy is defined as "networks of individuals and connected communities, as opposed to centralized institutions, and which transform the way in which we produce, consume, finance and learn".

In this sense, the maker movement participates in the principle of technology individuation whose dimensions defined by Hottois [HOT 94] are:

– growing integration of the technical part compared to itself;

– growing independence relative to users;

– growing integration between technology and the milieu in which it functions (and which, as a result, becomes technicized).

The promoters of the collaborative economy call upon several schools of thought [TER 16]:

– the shift from the "free" (open source) is characterized by the transition from a principle of property (property and author rights) to a logic of open, free and universal access by users to knowledge, skills and material resources via open systems; this domain is that of the makers;

– "the economy of functionality is an economy that aims to optimize the use of goods and services. The economic goal is to create the highest possible value of use for as long as possible, all while consuming the least amount of material resources and energy possible" [STA 97];

– the gift economy covers the models of exchange in which the goods and services circulating are not considered according to their value.

The collaborative economy corresponds to a wave of disintermediation through the connection of individuals and the implementation of bypasses, that is, a direct link between producers and consumers. At the same time, it is made up of both merchant and non-merchant actors. All of these actors face tension between two logics, the logic of profit and merchandising on the one hand and that motivated by social and environmental values on the other [BOR 15]. "This new conception of production introduces a totally different notion of the political economy of neoliberal globalization, that is, 'what is light, is global, but what is heavy, is local.' It is thus necessary to see the makers, even if they are still excessively in the sphere of hobbyist production, as the precursors of a new system, where design is open, global, and 'communitary' (global open design communities), but where production is done 'on demand,' on a local scale in micro-factories" [BER 15]. According to Terra-Nova, the spread of ICT reduces the concrete differences of the modes of working between the employed and the unemployed. Barthélémy and Cette [BAR 17], representatives of this organization, showed that numerous highly qualified workers benefit from autonomy that has nothing to do with that of numerous equally qualified independent workers. This use leads to largely blurring the borders between personal life and professional life. This is indeed what is observed in the makers.

Some authors, like Chen [CHE 16] or de Kerorguen and Leroy [DEK 08], saw the future embedded in a new reductive DIY (Do It Yourself) ideology. Nevertheless, even if France is not a leader in the development of open structures like Fab-Labs, in agreement with Loy [LOY 14], we can expect cultural modifications brought about by this kind of movement, which reintroduces cooperative principles into a world constrained by numerous forms of individuation.

2.4.2.4.2. Fab-Labs

Launched by MIT toward the end of the 1990s [GER 12], Fab-Labs (Fabrication Laboratories) are locations open to the public where all sorts of tools and machines, notably 3D printers, are made available for the purpose of (re)designing and creating objects. For many, this physical structure may be the origin of an inventive and singular potential in the use of these machines. Fab-Labs target entrepreneurs, designers, artists, DIYers, students or hackers who want to go from the design stage to the final object themselves [BOS 15]; for a time (interests and limitations of open systems), they bring together different populations, different age ranges and different careers. Through "construction", they are thought to be meeting places and places of personal creation in cooperation which authorize the manufacturing of unique objects/parts with the help of the more skilled and experienced.

These organizations (which are, however, not informal) seem sufficiently flexible to be capable of non-selectively aggregating varied contributions (real biodiversity), all while incarnating a collaborative model [RUM 14]. Thus, different Fab-Labs, whose number may exceed a thousand for the entire world [JAC 16], combine all five functions, in different ways, sensitively corresponding to five audiences: the discovery of the power to make, to manufacture, which targets children or DIYers; education through doing, which targets schools and universities; rapid prototyping, which targets entrepreneurs and creators; local production, which responds not only to the needs of developing countries but also to those of artists, designers or DIYers who are not seeking mass production, but innovation, the invention of objects, spaces and shapes of tomorrow [FIN 16].

In this context, it is conceivable to believe that it is possible to go from an innovation focused on industrial producers to an innovation focused on "educated" users. The educational dimension is an important element of the charter of Fab-Labs: learning, but, more importantly, doing it together. The realization of a project cannot be delegated to a more disciplined person (the principle of DIY: Do It Yourself), but must go from the acquisition of the necessary skills, facilitated by the community members, with help finding the "right" contributive people-resources of the help (and not of doing it). "These links are thought to be vectors of emancipation in which one will seek to understand and develop products by oneself instead of buying them" [EYC 12].

To earn the name Fab-Lab, a manufacturing workshop must respect the Fab-Lab charter, implemented by MIT with the need to undergo certification and validate training (see Figure 2.25, representing the Fab-Labs' missions, extracted from [GOT 16, MIT 12]), but a Fab-Lab is not necessarily affiliated with MIT [WIK 16]. According to this same resource, there were 618 Fab-Labs in the world in September 2015, 60 of which were located in France. According to FabWiki [FAB 16], these open organizations are primarily located in Western countries. Recently, a national French Fab-Lab network was created [RÉS 16].

Openness is thus a condition for success; in order to construct complex technical objects, it is necessary to have shared knowledge of all the aspects of manufacturing [BOT 12]. A form of interdisciplinary and/or collective intelligence reigns in Fab-Labs, using innovative digital technology to facilitate it. The Internet is thus central, despite gathering in "physical" locations: "It ensures permanent communication, access to useful information, even its production and spread within an evolving networking lending value to the findings of such and such group and submitting it to review by the others who can validate it, reuse it, or adapt it. From this perspective, as from the historical perspective, these collaborative locations are often connected to or stem from the world of free and open-source software of which they represent the material branch, where the theme of the common good is represented by the cultural and ecological branches" [FLO 16].

"*What is a fab lab?* Fab labs are a global network of local labs, enabling invention by providing access to tools for digital fabrication;
What's in a fab lab? Fab labs share an evolving inventory of core capabilities to make (almost) anything, allowing people and projects to be shared;
What does the fab lab network provide? Operational, educational, technical, financial, and logistical assistance beyond what's available within one lab;
Who can use a fab lab? Fab labs are available as a community resource, offering open access for individuals as well as scheduled access for programs;
What are your responsibilities? Safety: not hurting people or machines; *operations:* assisting with cleaning, maintaining, and improving the lab; *knowledge:* contributing to documentation and instruction;
Who owns fab lab inventions? Designs and processes developed in fab labs can be protected and sold however an inventor chooses, but should remain available for individuals to use and learn from
How can businesses use a fab lab? Commercial activities can be prototyped and incubated in a fab lab, but they must not conflict with other uses, they should grow beyond rather than within the lab, and they are expected to benefit the inventors, labs, and networks that contribute to their success".

Figure 2.25. *The Fab-Lab Charter (from Creative Commons [GOT 16, MIT 12])*

There is tension between local production, that responds to the needs of consumers from specific territories, and a global audience that procures the platform and which, in certain cases, can be a vector of standardization.

Organizational norms, which have cultural effects, are built in Fab-Labs that reject bureaucratic principles whose methods are considered to be de-humanizing. "They owe as much to the decision by consensus as to the designation of representatives and managers, as much to the greatest benevolence between pairs as to exchanges of vehement messages on internal networks, as much to the culture of compromise as to the authority of the most active and most experienced. The dosage and articulation of these ingredients vary from one 'hackerspace' to another, but that works and persists, even if forming a community is not exactly a walk in the park" [BER 16]. However, after having visited several of these organizations, it nevertheless seems that the presence of well-known figures (often remunerated by universities, from what the author has seen) constitutes an indispensible element of stability and legitimacy. There is thus an engagement in the profound investment of rules for successful coexistence.

For 4 years now, the Open University, an online British university, has been publishing reports on innovative pedagogies. In 2014, it emphasized massive social learning online through MOOCs, the pedagogical design founded on the basis of data analysis, the inverse class, BYOD ("Bring Your Own Data"), which invites the student to use his/her own digital work environment, the reflexive "learn to learn" movement, which allows understanding of how we learn, dynamic evaluation focusing on the students' progress and not on their level, factual learning founded on the idea that we learn better when we are engaged in a collective project, learning through narration, which, thanks to the development of a structure story:

– allows for an increase and revisiting of one's knowledge, the key concepts that determine the threshold from which a student's understanding is deeply modified;

– and, finally, allows for construction or bricolage, notably by Fab-Labs [VUE 16].

2.4.2.4.3. Industry

Global connectivity through the Internet is growing in connection to a recurrent deficiency in our exports and growth rate. According to the National Intelligence Council [NAT 17], this situation risks increasing the tensions within and among societies. According to the authors, populism today is increasing, threatening liberalism and its place in the world. Some leaders will use nationalism to reinforce control. The influence of religion will be more and more substantial and play a greater role in numerous governments. Even if the cost of manufacturing in France is 12–13% higher than that in China or the United States [BOS 17], this form of self-reflection could lead to a will to manufacture on-site, adapted to additive manufacturing each time work is done on technical products rather than those with a high added value. Moreover, in a mass-production activity, the higher the levels of performance and profitability of machines, the more work disappears and the more people see themselves thrown back into the sphere of obsolescence with a high unemployment rate (see [LA 15, LEV 04, AND 17]).

Productive development is linked to the capacity for industrial production: recent technological advances have allowed for numerous obstacles to be removed and foresee mass production using 3D printing, allowing for the release of technology that is even more associated with the rapid prototyping of parts (from 1 to 10 prototype parts) in a new economic space, simultaneously integrating the classic manufacturing system. The will of machine builders should, within the next few years, be able to compete with plastic injection or metallic foundry in terms of volume, rate and production cost. In this regard, the alliances that are created between manufacturers are a very good indicator of development and the manufacturers' desire for 3D technologies [LAM 15]. The competition has begun, but there can only be a few victors.

Independent of this aspect internal to companies, ties with Fab-Labs can constitute innovation opportunities via experimentation in these structures open to transfers to companies [AND 17].

2.4.2.5. *Industrial property and fraud*

As previously mentioned, the development of additive manufacturing processes is not without some consequences for the economy [POU 14] with, beyond the "manufacturing" aspects, the cultural transformations of consumers becoming actors in production. Thus, according to these authors, "3D printing leads to the involvement of numerous service providers in the production chain of an object, where many people are likely to be responsible: the CAD designer, the CAD modeler, the spreader of download links, download sites, the person performing the download, the provider and manufacturer of the material, printing services, and the printers themselves". This multiplicity of participants is added to the large spectrum of intellectual property rights concerned by 3D printing, such as the author's rights, patent rights, brand rights and drawing and modeling rights. As can be seen, there is a debate between partisans of industrial property and free open-source systems.

"If some makers embrace the ideals of solidary production, entrepreneurs, investors, and intellectual property lawyers throw all their weight behind the development of machines corresponding to a diametrically opposed vision. They see 'ready-to-print' products that will end up as consumer goods; the machine itself could then only manufacture objects foreseen in a catalog (the 3D printer would then function as a simple "terminal," as a photocopyer…). Again, intellectual property finds itself closely tied to the payroll stake, even if its contentious history has been suppressed, both in the maker movement and in reflections on free software" [SÖD 13].

It is rather easy to imagine that the fact of not being able to protect one's own copies, to be spied on via the Internet, pirated and so on, limits the range of globalist discourse concerning completely distributed additive manufacturing. Thus, it is extremely significant to deal with the question of intellectual property with data transmitted around the whole world via the Internet with all sorts of data-pirating possibilities for counterfeiting. 3D printing alters production because pirating allows consumers to have inexpensive copies that do not go through traditional commercial pathways. Furthermore, as Le Goff pointed out [LEG 16], "a new 'Rembrandt' painting has been created, without any man having intervened in the creation process! Thus, thanks to Big Data and Deep Learning technologies and a simple 3D printer, an AI [Artificial Intelligence] was able to produce a previously unseen painting, true to the specific practices of the painter, to the slightest details (use of light-dark, angle of the brush strokes, etc.). These examples of creation by autonomous robots are multiplying thanks to Machine Learning technologies that allow the AI to learn by itself, thanks to data analysis".

Just for computer components, the Canadian Security Intelligence Service [CAN 16] found the most underhanded form of failure to be linked to what complexity could feign, precisely because it involves the underestimation of interdependences. Thus, cascading failures take place more easily within a computer monoculture, which, with the security of personal data, constitutes the major stakes (perceived as such) linked to the development of information and communication technologies. This situation is the object of questions repeatedly brought up by the media.

Depoorter [DEP 14] found that the reduction of general product design, conception, production and distribution costs in an uncontrolled space makes additive manufacturing an "element of an industrial counter-revolution". The nature of quick and (for a lawyer) unforeseeable innovations makes the prevision of legal issues difficult. He believes that if the public perceives regulation as being excessive, this could "reinforce the weak legitimacy of the legal regime and even the consideration that it is unjust protection (particularly if the chance of being 'caught red-handed' is modest). It is enough to examine the difficulty of dealing with the questions linked to song and film pirating".

In the March 2016 report on 3D printing, the French Economic, Social, and Environmental Council (*Conseil Economique, Social et Environnemental*, CESE) displayed absolutely no concerns: "The legal rules in place are not, in their very principle, challenged by digital innovations, as it happens, by 3D printing" [ING 15]. According to Niedercorn [NIE 15], legal experts raise the question of the right to privately use copies for one's own personal needs. For the CESE, there is no urgency: "Considering the technical and cost constraints, this risk is even further from corresponding to a tangible reality".

In anticipation, certain companies sometimes hope (only) to remain the masters of the conception aspects and accept to delegate the manufacturing of objects to individuals (the current case of some toys) with, in the end, implicit delegation of responsibility if the object created does not completely conform to regulations (improper material, e.g. sharp edges). "3D printing also poses certain questions concerning the production of objects that are not compliant to security criteria. Indeed, the risks for the consumer must be anticipated, for the parts produced by a 3D printer often do not have the same physical, chemical, or mechanical properties as those of original parts. Consequently, serious risks can arise concerning shock resistance, flammability, wear and tear, or even the toxic effects connected to the object coming in contact with the skin. The manufacturing of uncertified objects for children could, moreover, lead them to run serious risks" [BER 14]. Under these open conditions, it can be recommended for one's own benefit to protect the source (digital file) or to "personalize" the object with an impossibility of reproducing it (however, makers/hackers have until now shown their ability to evade a certain number of digital protections; great care will have to be taken in order for this to be successful).

Presently, it is difficult to know what the impacts of additive manufacturing will be for companies. Pouzot, Bourguet and Jolly [POU 14] foresaw different fields of interest for them (with associated legal aspects):

– facilitating prototyping;

– evolution of the creation;

– small-scale production;

– management of replacement parts;

– product personalization;

– reorganization of the production line and logistics;

– integration of clients into the product creation and development process;

– digital marking of products to identify them beyond the traditional brand;

– preservation and immaterial development of the company.

Table 2.4, inspired in part by Blondiau [BLO 16] (see also [BRE 15, DEM 16, LE 15, LE 13, RID 12]), highlights some elements where the question of industrial property can be posed.

	Royalty	Patent	Sketch and model	Brand
CAD file	Yes	No	No	No
Upload to the Internet	Yes	Yes	No	No
Hosting	No	No	No	No
Downloading	Yes	No	No	No
Provision of 3D machines	No	No	No	No
Creation of the part	Yes	Yes	Yes	Yes
Sales	Yes	Yes	Yes	Yes

Table 2.4. *Domains in which industrial property intervenes (according to [BLO 16])*

It will be interesting in the future, with a certain stabilization of the methods of production and the relationships between manufacturers, makers and users, to examine how the law will be positioned in relation to the users of 3D printers.

2.4.2.6. *What is known about other risks?*

2.4.2.6.1. Misappropriation

In Chapter 1, several images of misappropriation were shown: weapons and (possibly) ammunitions manufacturing; in addition, copies of artworks were mentioned in the previous section. The manufacturing of sex toys is not reprehensible in and of itself. Additive manufacturing, like all technology, cannot escape these types of idiosyncrasies. Nevertheless, the possibility of not being "detected" by being capable of creating complex objects at home, sometimes with "contentious" ends in mind, may be an element to consider (see [LAV 16a, FAR 13, REG 12, DES 13]).

According to Christian Lavigne [LAV 16a], the phenomenon of misappropriation could be significant for studies from a social psychology perspective when it serves different interests, like those aiming to manufacture arms or more prosaically, objects "for pleasure", which obviously vary from one country to another, with risks of various magnitudes. In this latter case, like applications of the "Polaroid", the "made at home" objects would allow "happy" owners of said device not to be detected by the police or not to have to carry their "risqué" photos to be developed when they already existed at the neighborhood professional photographer's place. In the absence of social control authorized by "personal" 3D technologies, different types of social deviance may emerge with possible marginalizations, without this necessarily being perceptible. The use of 3D processes, which are only means of doing, thus allows sometimes limiting conventions to be done away with, its singularity to be confirmed, by developing personal cultural "tastes" that are invisible to others. As such, there is no information on the magnitude of the phenomenon, which is likely marginally relative to other forms of perversion, whether they are mild or not [AND 17].

The appearance of the Internet has led to the generalization of minor crimes, such as breaking copyright, plagiarism [CHE 16] or the illegal download of multimedia content. These crimes do not bring about any direct issues, but can lead to others. For example, the multiplication of these problems leads some organizations to implement Internet supervision policies, some of which lead to generalized censuring of online content; problems widely reported by the media include pirating (see [INT 16]) or cyber-risks for manufacturers [DEL 16].

2.4.2.6.2. Hygiene, Safety and Environment (HSE)

On the INRS Website (http://en.inrs.fr/), readers interested in additive manufacturing can find the fundamentals concerning the general risks that, in the case of 3D printing, can concern different categories: electrical, chemical (resins, powders, solvents), optics (lasers), biological (bio-printing), mechanical, screen work and so on. In particular, once particles are involved in the choice of materials and 3D processes, the maximum separation of the operators of particle emission zones must be the object of particular care (especially if the hazards of the substances in question are not known).

It is important to recall the regulation for all responsible parties that imposes taking the necessary measures to ensure the security and protect the physical and mental health of the workers at their establishment, including temporary workers (Art. L 4121-1 of the French Labor Code). The evaluation of risks for the health and security of employees has become an applicable obligation since 12/31/1992 with decree no. 2001-1016 from 11/5/2001. For this, the employer must write a "unique document" inventorying the risks identified in each workplace at the establishment and then periodically update it. This document must be filled out by the employer and can be consulted by the representative instances of the employees, the occupational physician, the labor inspector and the occupational health insurance (CARSAT in France), as well as by the employees since the modification of article R4121-4 of the labor code from the December 17, 2008 decree. This decree also foresees a warning indicating the ways that workers can access the document being displayed at a convenient and easily accessible location in the workplace, the same as for the internal regulations, if such a place exists [AST 14].

Figure 2.26 represents the different flows of matter and energy implied in additive manufacturing (according to [HUA 13]) with different effects on the environment and on humans. This figure invalidates the so-called ecological nature of 3D printing (which, we must remember, is associated with computer science, which already has its own issues).

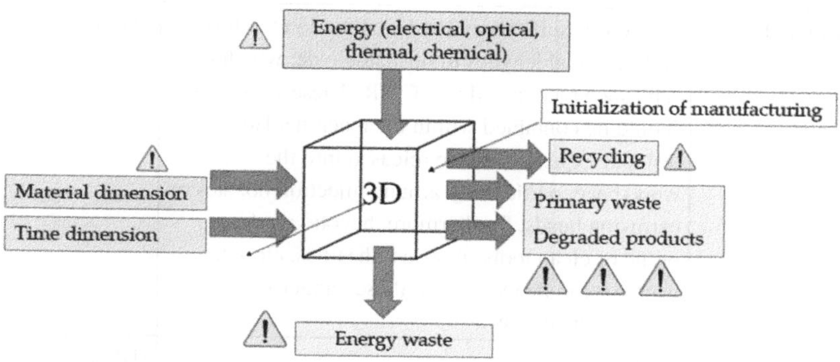

Figure 2.26. *Flow of energy and matter in additive manufacturing (with the risk components represented by triangles)*

Beyond these general considerations, some specific studies on the HSE (hygiene, safety and environment) aspects have been published (see Table 2.5).

Subject	Comments	References
Dust; particles	Metal powders with prevention propositions. There could be dust/particles other than the powders themselves.	[MEL 16]
Nanoparticles	Emission of particles from processes like in the "binder jetting printer". The essential part of the emissions in the number of particles is found in $PM_{2.5}$ with uncontrolled risks such as adequate ventilation not being implemented (~10 times more than in a "normal" atmosphere, but much less than with a fusion of polymer filaments).	[AFS 15]
Chemical risk	Wire-shaped polymer fusion processes likely to expose users to toxic gases and certain nanoparticles, especially if they are heated to a temperature above 240°C	[MER 15, STE 13, MÉL 17a]
Stereolithography	Guimon recommends using SDS (safety data sheets), which are necessary elements for good knowledge of the hazards operators may be exposed to.	[GUI 13]
Non-compliant materials	Serious risks can arise concerning shock resistance, flammability, wear and tear or even skin contact. Manufacturing uncertified objects for children could, moreover, lead to them running serious risks.	[BER 14]
Clean-up risk	Dangerous substances can be emitted in the form of gas, like aldehydes in formaldehyde, which is a substance recognized as a CMR. These substances should be contained within the machine. However, residual exposure can be released into the workspace, particularly around injection nozzles or extrusion heads. Furthermore, beyond work to purge or clean tools (nozzles, sheathes, etc.), the operator can be exposed to these dangerous gaseous substances.	[TRA 16]
Cyber-risk		[DEL 16]
Ethics	Additive manufacturing and society	[NEE 16]

Table 2.5. *Specific elements linked to hygiene and safety aspects*

"Orwellian" social control?

Because they allow information to be entered, stored, processed and spread, the computers associated with means of telecommunication can be perceived as bringing a thread in terms of "Orwellian" control over society. ICT can serve to centralize and concentrate power, or even to make all sorts of severe rules apply, no matter how strict these norms may be [ADE 85, SYS 16]. "You know, of course: Facebook does statistics on your uses of computer technology, on your surfing. They figure up the probabilities based on the assumption that two people get together based on the messages they exchange, etc. Google reads your emails. There is extensive monitoring and it is not necessarily like in '1984,' in something that will make you disappear and you will be shot in the head in a dark corner, but in a system where our lives are entirely under control, monitored, analyzed, and subsequently exploited" [BON 14].

2.4.2.6.3. Positioning relative to the lifecycle analysis and sustainable development

From an energy standpoint, according to Thomas *et al.* [THO 16], the cost of energy for a part created through 3D printing could be above 2% of the total cost of a part manufactured using traditional processes. If the material not used during the fabrication process is reusable, additive manufacturing responds rather well to the criteria for economizing on matter and energy (as long as the manufacturing of these starting materials does not cost too much on energy relative to traditional manufacturing processes and the manufacturer is not forced to repeat the part manufacturing several times).

However, it does not seem easy to find credible information concerning the potential advantages of additive manufacturing relative to traditional processes (let us not forget that, according to Flipo, Dobré and Michot [FLI 13], a simple Google search produces 7 g of CO_2, 40 million tons of electronic waste is produced per year, that reserves are running dry [BIH 14]. It would be interesting to examine the cost of transferring a large file measuring a few dozen or hundred mega-octets by electronic means, of creating an object, of possibly revisiting it to fix its shape relative to instructions, its lifetime to prove the well-founded idea of environmental comments on the revolutionary aspect of additive manufacturing processes, etc.). Certain economists, on the contrary, foresee gains linked to environmental protection with regard to additive manufacturing [TOV 16].

Le Bourhis *et al.* [LE 12, LE 13] proposed implementing a methodology that allows not only the energy consumption of the 3D machine but also the consumption of fluids and matter to be taken into consideration. This method, according to Le Bourhis *et al.* [LE 12, LE 2013] and Kerbrat *et al.* [KER 13], combines the necessary comprehensive approach in a sustainable development process and the precise evaluation of input and output flows in additive manufacturing machines. There is software (e.g. EcoInvent and Sima Pro) to allow a comprehensive process analysis as long as the following elements are available:

– the design of the part using computer methods and its digital preparation (layers, supports, etc.);

– matter and materials (including shaping them) that are involved in manufacturing an object;

– the production of the inert gas (argon) necessary in certain processes, ventilation and other security elements;

– elements for cleaning a part and their elimination;

– the energy necessary to create the 3D machine;

– recycling unused matter (lost powder, resins, etc.);

– recycling parts;

– recycling the machine at the end of its lifecycle;

– and so on.

Only some approaches are engaged in this emerging domain that seems to have been attempted [KOH 15, LE 14, MOR 07], but with conclusions that do not yet allow definitive conclusions concerning the environmental interests of 3D technologies relative to traditional processing methods. Independent of this framework, several specific studies have been published:

– Kohtala [KOH 15]: Significant synthetic study of the activities in the domain of distributed distribution, illustrating the need for further research;

– Kreiger *et al.* [KRE 14]: Recycling plastic materials used in 3D processes using filaments (this is not improving in terms of centralization);

– Wittbrodt *et al.* [WIT 13]: Technico-economic approach of "open-source" additive manufacturing systems like "RepRap";

– Huang *et al.* [HUA 13]: Analysis of the societal impact of 3D processes with a favorable orientation toward the aspects of health, durability and reducing the manufacturing stages;

– Gebler *et al.* [GEB 14]: In connection with the work of Reeves [REE 08a, REE 08b, REE 12, REE 13], these authors show the potential interest of 3D technologies in terms of reducing CO_2 production;

– Tovar [TOV 16]: This study shows that in the framework of a well-managed circular economy, there could be the possibility to save approximately 5 billion €/year.

2.4.2.7. *Liberalization of industrial property*

The report of the UK Intellectual Property Office Patent Informatics Team [UK 13] provides very well-documented information on additive manufacturing patents. Hornick and Roland [HOR 13] presented the list of patents that were filed or that are being filed in the public domain. Looking at the image with the results concerning the evolution of the number of scientific publications over time, it will be noted in Figure 2.27 that there was a quick start followed by a reduced acceleration, which could be considered a stabilization concerning the patents filed.

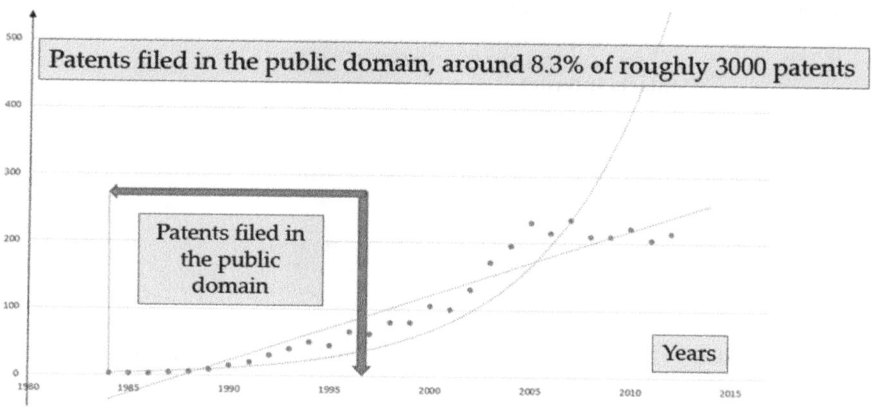

Figure 2.27. *Evolution of the number of additive manufacturing patents with two possible options: linear and exponential trends [UK 13]*

In this same report, France fails to shine because it only achieves ninth place with a "patent" coverage rate ranking at 2% relative to other developed countries (the United States makes up a considerable 45% for a population that is about five times more developed than France). What this report also shows is that the patents concern, in order of importance:

– 3D printing processes (all technologies);

– the medical, dental and pharmaceutical domain;

– material formulation (from resins to metals);

– software;

– high added value applications like micro-fluidics.

According to Google Trends[1], the number of Internet hits via "Google" undergoes the same kind of evolution with a certain decrease over time (see Figure 2.28), but strongly displaced, by approximately 7–8 years, in comparison to the results presented in Figure 2.27; in both cases, the temporal evolutions are neither linear nor exponential over time. This shift can be explained by the difference of the populations concerned: those waiting for patents; public "followers" interested in technology for "Google" hits and so on.

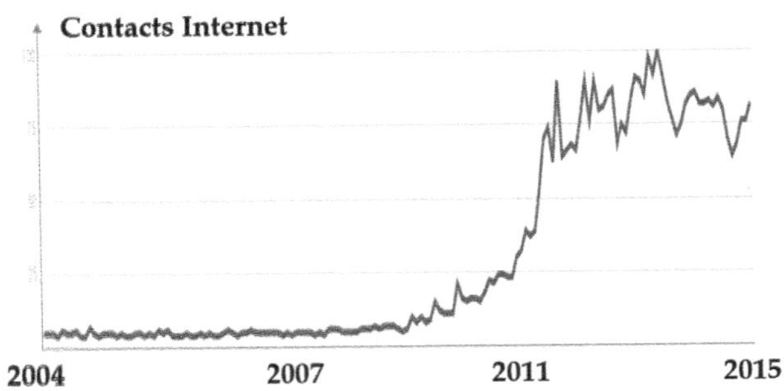

Figure 2.28. *Relative evolution of the number of Google hits on the term 3D Printing*

The period 1984–1996 is represented in Figure 2.27, strongly corresponding to the abandonment of a certain number of patents founding additive manufacturing technologies. These are the following basic patents (see Table 2.6), representing significant economic activities (but authorizing the spread of industrial activities stemming from new, very competitive companies).

1 https://www.google.co.uk/trends/explore#q=3D%20printing&cmpt=q

Data of the first submission	Concerned domain
1984	Stereolithography (André, Le Méhauté, and de Witte in France, Hull in the United States (founder of 3D Systems))
1987	Selective laser system (SLS) by DTM Corp.
1988	Submission of the melted wire or FMD (fused deposition modeling) by Stratadys
1995	Metal 3D printing or DMLS (direct metal laser sintering)
2003	3DPP (3D paper printing)
2005	First color printer by ZCorporation
2006	"Open-source" 3D printer opening the way to future home printers (see RepRap project)

Table 2.6. *Basic patents filed in the public domain (highlighted in beige)*

According to Quoniam and Reymond [QUO 14], Fabulous [FAB 15] and a bibliographical analysis concerning patents, new industrial properties are emerging, in terms of applications such as nutrition (candy, food, chocolate) and construction. However, the rupture in terms of processes comes in 2015 from the company Carbon3D with its CLIP process of space-resolved polymerization, positively exploiting a defect in resins whose kinetics is reduced by the presence of oxygen to transform it into a quality allowing the manufacturing time to be reduced by a factor of possibly seven relative to other stereolithography processes. Less innovative, but very credible in terms of application, in this same year, Hewlett Packard (HP) took up a position with a patented technology called Multi Jet Fusion. These different evolutions will be described in Chapter 3.

With this new freedom to produce machines, allowed by the loss of the property of patents blocking it, according to what numerous media outlets express, we should see an exponential explosion of the additive manufacturing market. In fact, the producers of 3D printers are the root of the effective promotion of flexible fabrication of parts in house. Thus, most of the industrial groups already have 3D machines for different ends. Nevertheless, as pointed out by Parisot [PAR 15, PAR 16], additive manufacturing has one particularity: "the manufacturers who throw themselves into mass-production first are from the most constrained sectors, like space, aeronautics, automobiles, and medical. But being a subcontractor of 3D printing in these domains is costly". Yet these sectors require a respect for quality norms and being qualified for everything concerning the process, machine, matter and even the operator himself/herself, which can lead to the availability of machines for R&D separate from the production machines.

Let us recall that the qualification process in a domain like space can last several years, and the volumes of productions on the qualified machine are sometimes limited, maybe only a few dozen parts per year with returns on investments that are difficult to ensure. "In the end, after having invested in R&D capacities, factories, and machine parks, additive manufacturing subcontractors find themselves with machines blocked for years, on which the production volumes are generally low (in aeronautics and space, at least). Certainly there is a demand on the market, for manufacturers are still searching for advanced 3D printing skills and still need to test different technologies to choose the best one for their need" [PAR 15]. There are therefore several kinds of expectations limiting development, the economy of manufacturing processes and the qualification of the aforementioned processes.

Nevertheless, as Taylor [TAY 14] reminded us in general, technological developments are displaced over time, which reinforces the idea of important industrial and applicative activity with an exponential development for this type of rather simple technology in terms of use, a "3D" photocopier. This reinforcement must be associated with the fact that a new technology, whatever it may be, does not completely destroy the preceding system of production. It must integrate itself and complete the more classic methods of producing objects and goods, even services. In particular, it allows numerous strongly personalized (at least in appearance) individual solutions to be proposed, in principle, rather than standardized solutions (at least in appearance). It is thus expected that new processes will be able to add to the advantages of existing processes whose robustness has been proven for years [GUÉ 03, CAM 11].

Yet, according to Opinion-Way presented by i-Connect [I-CO 15]: "Digital innovation is linked, above all else, to positive representations. For 47% of those interviewed in France, it is translated by 'better performing technological tools'. 44% associates this term with the idea of 'new technology' and 35% sees an R&D component or new uses. The French primarily associate digital innovation with 'better performing technological tools'. This performance assumes several shapes in their minds. More than half of all respondents (56%) feel that a product stemming from digital innovation must, as a priority, respect the environment. It is also important for them to correspond to the French way of life and the people using them daily. For example, 48% of those surveyed hope that this product is adapted to their behavior and 38% would like to control it remotely". It is indeed this tendency that is seen in the "maker" movement presented above.

The spread of an innovation, according to Gaglio [GAG 11], thus results from a "snowball effect" and is represented thanks to an S curve (slow start, exponential acceleration, then notable slowing down with normalization at an elevated level of owners; let us not forget that "norm", for Gori [GOR 13b], means "that which has a great frequency, is conform to the average, has a quantitative value deduced from a scientific description of reality, locating the functional regularities given by correlations of phenomena"). This is indeed what is observed in the two previous figures, but still with a modest global economic market relative to the only mechanical manufacturing. Everyone implied in a diffusion process would borrow the following steps:

– making knowledge available to individuals about the existence of a novelty (phase of informative knowledge). This information generally goes through the media. This phase goes beyond simple knowledge of the novelty: it also concerns the modalities of its use, "how-to-knowledge" and the fundamental principles of its internal functioning, that is, "principles knowledge";

– the persuasion phase, tributary of the relational environment. It generally can be broken into three sub-stages: stimulation from surroundings, the introduction of the novelty into the social system and its legitimization;

– the decision to adopt the novelty individually, all while being part of a collective movement (form of insurance).

The S curve allows for the spread of a novelty; this loses its status in becoming mundane. Throughout this process, five kinds of actors follow one another. They can be characterized in a reductive way (with a quantification introduced by Rogers, quoted by Gaglio in 2011) as follows:

– pioneers (2.5%) are rebellious and take risks. They go against majority views and are not always listened to or taken seriously;

– innovators (13.5 %) adopt a new behavior but have greater social prestige than pioneers. They are thus generally imitated;

– the hasty majority (34%) covers individuals described as pragmatic and moderately open to progress. They do not want to assume responsibility for the risk associated with the pioneering adoption, which is why they follow a wait-and-see policy;

– the late majority (34%) is rather conservative, skeptical and uneasy when faced with new ideas;

– latecomers (16%) only adopt the novelty at the end of the lifecycle. They are drawn to traditional values.

These elements also contribute to forms of slowing in the development of 3D technologies. However, we must remember that Alex [ALE 16a], revisiting a report by Wohler, concerning the increase in the number of "home" machines, shows continual growth of the additive manufacturing market (sales of equipment and consumer goods, printing services). This could thus have increased by 26% in 2015 to reach a value of US $5.165 billion (a particularly precise value that must be confirmed, as was shown before, with other factual data, different but also precise).

2.4.3. *Possible positioning of the industry*

2.4.3.1. *Propagandist, environmental and other overselling*

"The ultra-connected world, which is designed and reinforced daily, henceforth belongs to the engineers, a bit to the scientists, and above all else to the entrepreneurs who will be able to exploit this considerable base in order to invent new applications and promote them, new 'business models' from an almost infinite range of connected objects, a range in which additive manufacturing technologies are found" [FER 15]. However, within this world, innovations cannot be perceived in just any way, as communication creates specific stimuli filled with symbols and new values [VIN 00, JUL 96], such that the effects come about naturally, through a creation of shortage, promotion of desirabilities, intercultural exchanges, co-construction with an overabundance of opportunities. Fressoz [FRE 12] noted that new forms of direct communication participate in forms of disinhibition, the absence of interpersonal relationships, and subsequently allow forms of manipulation and formatting of social perception [AND 17].

The fact that has been reminded of throughout this chapter is that 3D printing is becoming a tool for everything, for everyone, a means of returning to national, even local, activities, all while being among the top 10 emerging technologies of the 21st Century. Presently, it has been shown that it is necessary to go beyond black-and-white systems and that the situation is not so clear, without expressing an end for this project, but it is likely necessary to reflect on the "appropriate" growth adapted to incremental innovations opened by its success.

However, let us recall that from an environmental standpoint, ICT contribute to global warming through the consumption of energy, consumption of precious (rare) materials and minerals with related exhaustion, pollution associated with processing, difficulties in recycling, with direct effects on the health of the workers in Southern Countries (to whom these dangerous products (DEEE) are "entrusted" at the end of their lives) and so on (see e.g. [BRE 08]). According to the report from the study and environmental consulting agency Bio-Intelligence Services [LAB 08], in 2005, ICT represented 2% of greenhouse gas emissions in Europe, a number that could double by 2020. It will be important to know where we are and particularly where additive manufacturing is leading us in the future by taking into consideration all the environmental elements contributing to this important issue.

2.4.3.2. Stock market activity

Upon reading the previous chapters, media pressure concerning the impact of additive manufacturing is translated by ambiguous, potentially harmful effects connected to forms of paradoxical injunction for the development of the companies that feed the market with 3D printers and adapted materials.

3D Systems, founded by Hull in 1986, was originally an SE (small enterprise), which, at that time, developed slowly. The American company now has more than 1,000 employees and has registered its revenue in 2013 as being more than US $500 million, a 45% increase from that of the preceding year. The company has developed internally and externally, for example, buying out 16 competitors or companies with specific know-how related to additive manufacturing in 2011. This acquisition process is ongoing. Similarly to its compatriot and competitor Stratasys, the company tries to dominate the additive manufacturing market (let us cite the buyout of ZCorporation – manufacturer of the ZPrinter polychrome 3D printers, of 3D Phenix Systems mastering "metal laser sintering") [NOT 14, DES 13].

Besides this aggregation of companies illustrating a rather classical process of market structuring, according to the French newspaper "Les Echos" [LES 15], this company, listed on the stock exchange, has undergone significant variations in the value of its shares, because the promised future has not been achieved. According to this source, the price of the American company Stratasys has gone down by 26%, and that of 3D Systems Corporation has gone down by 48%. It is worth mentioning that the prices of these companies were literally soaring in the months leading up to that point. "The first stock capitalization, that of the American company 3D Systems, practically multiplied by a factor of 10 from 2011 to 2013. It ended in 2013 at its highest point historically, just like that of its large rival Stratasys, to the same degree". Even if these large losses affect the image of dynamism in the domain, these companies have capitalizations "worthy of the Internet bubble, more than 60 times greater than their profits". Figures 2.29 and 2.30, inspired by Zone Bourse [ZON 16a, ZON 16b], illustrate the evolutions of the shares of 3D Systems and Stratasys.

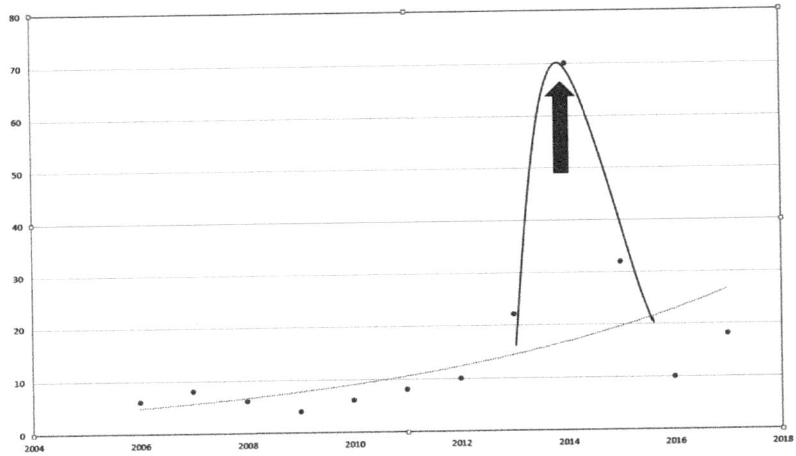

Figure 2.29. *Evolution of the price of 3D Systems' shares (exponential curve with an "Internet bubble" peak in 2014)*

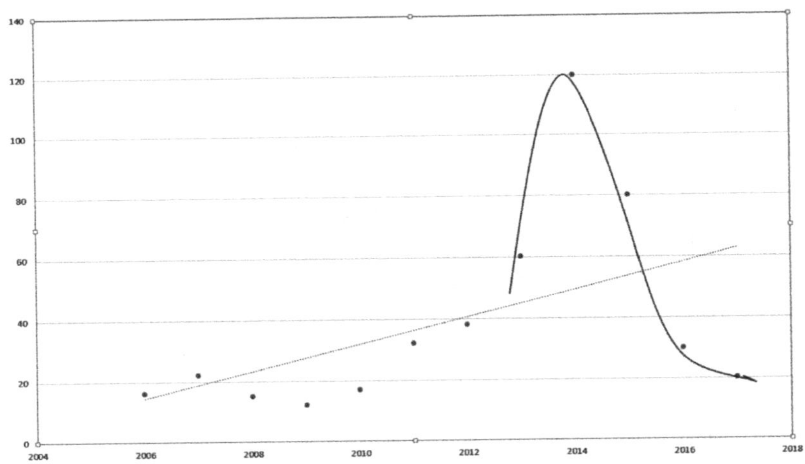

Figure 2.30. *Evolution of the price of Stratasys' shares (linear curve with an "Internet bubble" peak in 2014, followed by a large decrease)*

These disenchantments were only temporary; however, if we believe Gartner (quoted by Notteau [NOT 14]), which promises an annual growth of 20% for the next 5 years (see Table 2.1). The research firm Canalys [CAN 16] counts on a 20 billion dollar market by 2020 (with greater growth, 43% in 2013). "The story of

3D is still to follow, as with all rather high-tech industrial adventures" [JAC 14]. Nevertheless, it is necessary to remember that efficient 3D printers remain relatively expensive, rather slow in production, with certain materials costing 50–100 times more than those used for casting with plastic injections, even if it seems possible to recycle the polymer of water or milk bottles, for example [NEW 12]. Additive manufacturing still has paths of progress to explore (and this is normal).

2.4.3.3. *Governance of additive manufacturing*

3D printing, although having multiple (real or presumed) advantages, notably in industry, presents a major defect: it requires an experienced and highly qualified workforce to ensure its proper functioning. Additive manufacturing is nevertheless about to exceed the cap for producing parts in a reduced number or specific to manufacturing a prosthesis, a tool, part of a machine, an airplane, a rocket, clothing, shoes, a vehicle and so on.

For the Observatoire de la Métallurgie [OBS 17], the transfer from subtractive manufacturing processes to those of addition should be translated by evolutions in terms of operator specializations:

– design; R; R&D: researchers, metallurgists, designers, modelers;

– organization, preparation: QSE specialists, methods, lean management;

– production: machine conductors, tool makers, surface treatment specialists, quality control specialists;

– maintenance: automatician, engineers, metrologists, maintenance specialists, electromechanics and so on.

Part of these specialties already exists, at least partially, but evolutions must take place to cover the whole domain, which implies sufficient production to use all the specialized actors full time.

With a futuristic vision, additive manufacturing technology could, for some, become the cornerstone of all manufacturing processes. Considering confidentiality constraints, activity for large companies risks being internally produced according to traditional organizational methods. "Currently, we are in an interval that does not allow 1 million parts to be produced with true economic rationality, but several thousand are already projected without too many risks. When 3D printing manages to link the 'speed, volume, price' trifecta, a whole part of industry will find itself deeply upset. This horizon is probably not so far off, and as with all technological changes, it requires awareness and anticipation in terms of financial, human, and training investments, even cultural modifications" [ROC 06]. In any case, competitiveness goes through mastery of a set of factors ranging from the economy to the protection of know-how [PET 13].

3D printing can challenge the concept of the logistic chain and its consequences could be numerous: lean manufacturing, reduction of storage (except matter and materials used for 3D printing), reduction of transport and energy costs (still to be demonstrated, insofar as functional objects can be manufactured with the appropriate material(s)). According to Branche [BRA 15], "all of this chain could even become obsolete. The owners of 3D printers need only model their piece or recover a 3D file to manufacture their own object, thereby bypassing all the usual intermediaries. It is therefore in this way that the philosophy based on 'Do It Yourself', open-source, i.e. the sharing of information, together with the generalization of the digital technologies to which additive manufacturing belongs, gave birth to the maker movement and Fab-Labs or 3D manufacturing workshops" ([AND 17] and 3.2).

However, if little credit is given to the employment forecasts made by different sources [WEN 15, FRE 13, BRU 14, FOR 15, BRY 15, BOS 15], it is estimated that nearly 50% of unqualified jobs will disappear to robots in the coming years, to the degree that it is even projected today to have robots attached to 3D printers: an American start-up has developed an ultra-sophisticated robot capable of managing and controlling in-line a dozen 3D printers. This robot is in the form of a mechanical arm capable of not only starting a printer, setting the parameters, but also removing the object when it is finished. The most remarkable aspect is that this robot is compatible with all kinds of 3D printers and does not require great modifications to adapt [MOL 16]. At the same time, according to Wanted Analytics [WAN 14], the number of permanent job offers is only growing, as shown in Figure 2.31. In his survey about the industry, Platt [PLA 15] also pointed out that 35% of manufacturing activities will be impacted by additive manufacturing. As the technology is emerging, it is indeed a new job market that is opening up. "The consequences that it will have on the job market must be understood in the long term. 3D printing has created jobs and will continue to do so, but for the moment, it has no aim to replace traditional manufacturing processes" [MÉL 17].

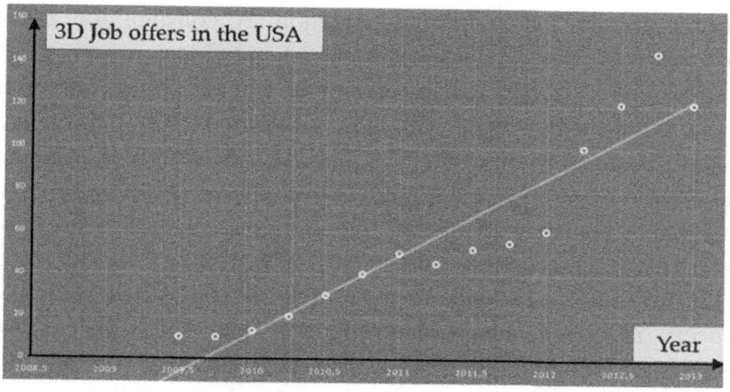

Figure 2.31. *Additive manufacturing job offers in the United States, according to Wanted Analytics [WAN 14]*

Currently as seen in Figure 2.32, this is just a drop in the ocean with regard to the considerable loss of jobs in the manufacturing sector in the United States [ITR 17]. The creation of jobs brought about by the emergence of additive manufacturing is far from compensating for the losses (currently a factor between 10^{-4} and 10^{-3}). Discussions about the re-integration of production on national soil, no matter how interesting they may be, must thus be put into perspective, despite a comparable development in terms of jobs as those offered by the classical economic sector, easier to be recognized.

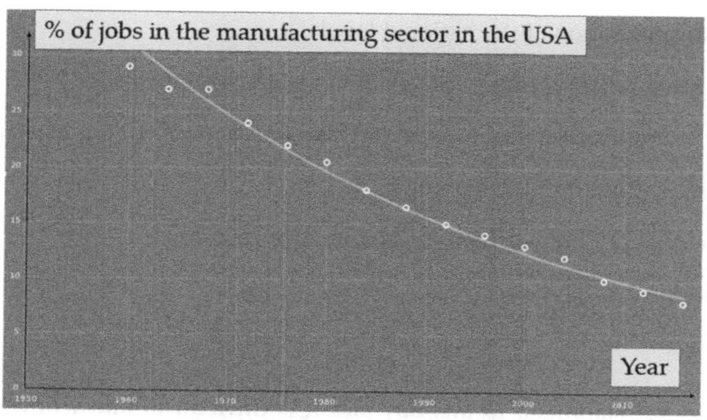

Figure 2.32. *Jobs in the manufacturing sector in the United States as a function of time*

In accordance with the maker phenomenon, if consumption turns to "self-production", this may incite a certain number of individuals to reduce the level of revenue sought after, *a fortiori* in an economic context where obtaining a stable, profitable job is becoming more and more difficult [AND 17]. In recent years, companies have become aware of the role played by creativity in re-industrialization and the elaboration of new business models. Entrepreneurship and innovations like additive manufacturing are part of the potential advantages against crises and the impetus to re-launch the economy. In fact, with 3D printing, workforce needs would no longer be the same, especially in manufacturing industries, which would lose their prospects in some areas of production and thus their *raison d'être*. Furthermore, on an individual level, if everyone can produce own objects, even at a supermarket [BLA 13], this may reduce the need to have a full-time job. We could thus experience a reduced intensity in market activities and a re-distribution of monetary flows, even a drying up of some of these. This process must thus be followed.

There is thus maintenance of Promethean optimism (according to [KUN 05]) foreseeing, thanks to additive manufacturing technologies, the pursuit of die-hard technical progress and possibly disturbances in the forms of work associated with this technology, which is not yet mature. Presently, we are only at its early stages, with a world market worth approximately 10 billion €/year, much lower than the percentage of the individual budget of most large Western countries. However, with different types of effects, it can be advantageous to follow/anticipate the evolution of the 3D printing market from the following standpoints:

– return of production at the national level;

– implications of 3D printing for sustainable development [PRI 17];

– additive manufacturing training (via Fab-Labs, schools of engineering);

– democratization of the technology;

– modifications of the work-social governance relations;

– innovations;

– and so on.

Nothing is yet stabilized to this day and it is difficult to know how 3D printing will find its place in current manufacturing processes. One particular point must be remembered, particularly stemming from the objective of this book: 3D technology is presented with a vision of technological autonomy. Yet for those who have done a little mechanics or woodworking, it is rare for only one tool to be used. Maybe the integration and association of the 3D technology integrated into every industrial production device could be accelerated with greater humility on the part of its supporters, but with a small loss of substitutional solution as a "nice" remedy in manufacturing, in the collective imaginary, and of a certain number of futurists, even of journalists sensitized by the search for the sensational.

2.4.3.3.1. Last-minute remark

The new president of the United States has planned a decree that would forbid "the commercialization of equipment, applications, and tools having a connection to additive manufacturing, also called 3D printing". This measure, according to Mélanie [MÉL 17b], should take effect in fall 2017. "To justify this new decree, President Trump explains that it 'is necessary for Americans who have lost their job due to additive manufacturing technologies to be able to find them again'. According to the president, the United States will no longer contribute to the progress of this 'false technology' that 'distances people from the American Dream". President Donald Trump has also made reference to the link between 3D printing and laziness by maintaining that "this technology is only a game that erases every notion of work that exists in our country". The White House backed Trump's standpoint by adding that "these new technologies are only beneficial if they contribute to the development of society and the economy" [MÉL 17b]. We are far from the 1933 Chicago World's Fair about the "Century of Progress", with its program: "Science finds, industry applies, Man adapts". The phenomenon is rather remarkable given that it is coming "from above" (while in France, several centuries ago, the "canut" silkweavers in France evolted against the machinism applied to weaving). Innovation indeed provokes a process of creative destruction in Schumpeter's sense, insofar as it disqualifies the production methods and products previously developed by making them obsolete and less efficient. To be continued.

2.4.3.4. *Evolutions of material choice*

In the report published by PIPAME in 2017, according to a report from IDTech, the evolutions in the choice of materials highlights the strong tendency to create metal parts for direct use. This trend was sensitive in the development of machines and confirms this progression associated with a certain use of 3D technologies by consumers or manufacturers of machine-tooling. Figure 2.33 presents these different evolutions.

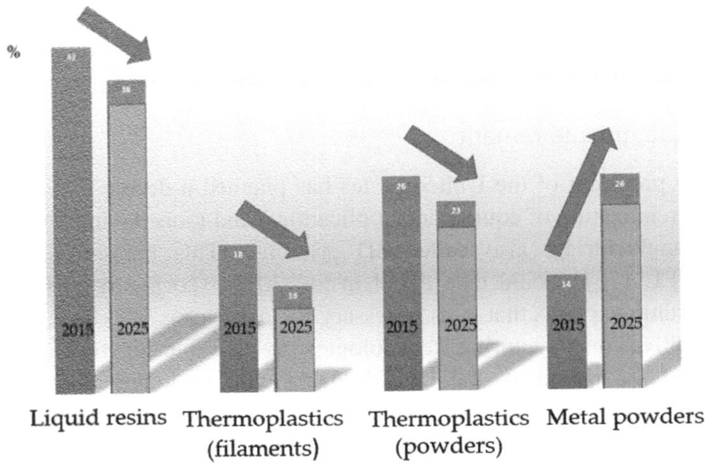

Figure 2.33. *Estimation of the parts of the market in percentage of the primary materials used in additive manufacturing*

2.5. Toward a certain stabilization: The dynamics of innovation

Schumpeter [SCH 99] wrote, "the capitalist economy is not and couldn't be stationary". Some may think and say that the principal cause of a loss of companies' competitiveness is due to a lack of appreciation for the evolution of technology and a false perception of the ephemeral character that characterizes "non-stationary" technological advances. The company questions itself and renews itself through innovation [FLI 03]. Yet, innovation is a particularly complex, badly understood process, a synonym for risk and uncertainty, actually a generator of not only fear (for a whole set of partners) but also different kinds of advantages for society (that also should not be forgotten). In this new décor, work tends to become less and less concrete, automation makes divided work more and more anonymous, monitoring tasks increase the workload without the worker at his/her company or the researcher (particularly he/she who invests in the action's science) in his/her laboratory satisfactorily mastering the complexity of the system in which they intervene and serve. The nature of the environmental, industrial and financial risks evolves; the radical (and generally incorrect) use of the precautionary principle is at our door. The company changes not only technically, but also in its organization; the number of products created evolves rapidly; mass effects change (energy, transport, information, etc.). The simultaneity of these transformations leads to fears associated with a widespread and long-lasting risk, likely to touch vast geographic areas (nuclear power, chronic pollution, massive Internet blockage, global warming, etc.).

At the start of the 20th Century, innovation was organized around the "linear" model represented in Figure 2.34 [ECR 05]. According to this classic model, the innovation process starts with fundamental research, followed by applied research, development, production and, when possible, commercialization. In this framework, "discovery" is considered rather exogenous to the innovation process. It is more or less according to this schema that 3D printing, for its stereolithography component, started its development in 1984.

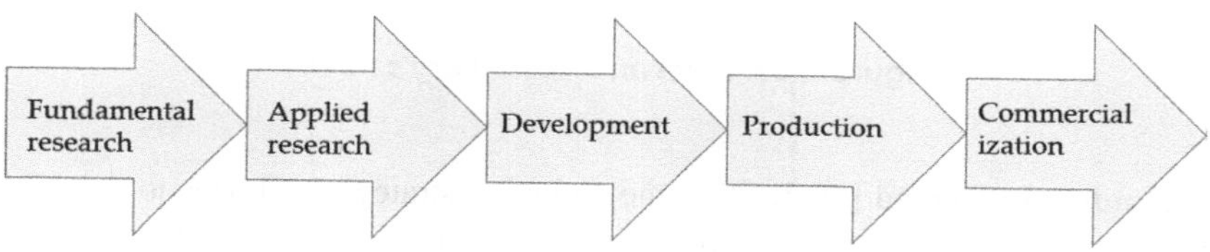

Figure 2.34. *Traditional "linear" innovation*

For economic environments, innovation finally results from the adoption of something that could have been developed in disjointed organizations. This linear model, called "science or technology push", was marginalized in favor of new slogans [MCG 10, GUC 14]. According to ECRIN [ECR 05], during the "Trente Glorieuses", the authorities understood that the situation of innovation was slightly more complex and that it was important to introduce interaction and retroaction loops. Kline and Rosenberg's model [KLI 92], which represents these interactions between societies' needs on the one hand and between economic and academic worlds on the other, is represented in Figure 2.35. This simplified model thinks of innovation in its precocious stage as a user of existing knowledge, with, when necessary, a return to basic research. Nevertheless, what this schema does not express is the origin of the idea.

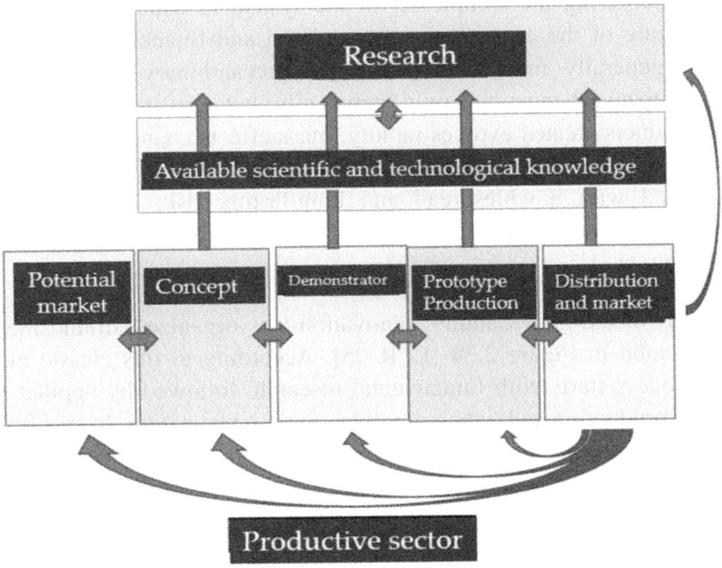

Figure 2.35. *Kline and Rosenberg's model*

According to Durand [DUR 08], "the company strategy [...] tries to resolve the paradox poxed by the relations between the freedom of creation and initiative and influence of the economic selection. It thus consists in displacing, through particular choices, the selective constraint on competitors thanks to a control of trump cards, resources, and, strategic aptitudes. For this reason, companies not only innovate to satisfy their client, but to put pressure on their rivals and push them to fail or to give up". This situation, in which the dynamic image of the company is important, allows, in principle, the links between production activity and technoscientific research (as long as it was correctly anticipated) to be reinforced. Indeed, "every organization is an immune system whose goal, as with all organisms, is to preserve a certain *status quo*. However, for the decision-makers, the adaptive value of change has become superior to that of stability" [BLA 06]. Being innovative is having clients. Innovation is a synonym for not only novelty but also progress, and it has three pillars: it is an internal dynamic of the enterprise, a common will to explore and exchange; it is a proposition that brings an obvious advantage to the consumer and finally, it requires involvement in a consumer or user culture and thus being emphasized through a communication device [RIC 15].

In their domain, the "makers" participate positively with the media in the diffusion of 3D technology with significant advantages for the user; additive manufacturing thus follows these criteria. At the same time, the company becomes more international: exchanges, personnel, legislation, normalization and ISO [ANS 17], and largely modifies its social organization: subcontracting, specialization, performance individualization, aging and so on in a rather dull context of accessing paid work. However, several phenomena (at least) are at work:

– the gigantism of industrial research centers and research centers leading, as with academic research, to a subdivision of activities. Gaudin [GAU 78] then writes that, "the object is divided into elements, and this sharing, in and of itself, guides production towards reproduction", in fact, through "fearful conformatism" [GAU 78]. We all search where things have already been found;

– creative research is only sometimes at the origin of innovation; it is the "fact of heterogeneous and improbable initiatives" [GAU 78];

– "most scientists display an incredible viscosity: the most common attitude is not openness to other explanatory schemata, but in most cases, ignorance and/or refusal to accept their existence" [MOR 12];

– innovation depends on specific events, results from history, leading to an absence of precise determinism and pre-visibility in the long term (stereotype of the butterfly wing);

– the technological systems that we use are not isolated from one another; they bounce off one another [PUE 08];

– the development of technological artifacts is regulated by their uses (appropriation); Puech [PUE 08] described adaptation of the technique to the use and adaptation of citizens to the technique (idea of co-evolution);

– the high cost of investment for new installations is likely to be an obstacle to innovation; "capitalism itself organizes its regulation of technology by choosing what must be developed and favoring profitable investments. Capitalism moderates the instatiability of technology through its own [financial] insatiability" [CHI 07];

– to understand the processes that lead to innovation, it is necessary to resituate the interdependent decisions in the context of the time, the associated irreversibilities, even the fears of undertaking [KID 82]. The conservative tendencies at every level tend toward mimicry and the *status quo* [DUR 12];

– those responsible for production do not request research whose effects are projected in a medium-term future, while they are asked for financial results and productivity in the very short term. This phenomenon is amplified by the short presence (a few years) of a number of these responsible parties at the same time. The concern with becoming engaged in a risky universe is slightly left to the successor ("Hot potato" paradigm).

However, despite these reservations, the technological movement, the means for an attractive financial result, must be the creator of new riches accessible by the citizen. At the same time, the citizen demands more assurance and regulation of the technical change through regulation, legislation aiming to protect the consumer comprehensively. Furthermore, Kaufmann [KAU 08] reminded us that "the less norms are obligatory, the more people's primary activity is producing new ones, leading to an obsession with normality" However, as Landes expressed [LAN 75], "the industrialization of the world is pursued for better and worse, and if there are people in advanced countries who retch at this cult of material realizations, it is because they have the means to indulge in a critical attitude, while the large majority of the planet's inhabitants dream of accessing this material simplicity".

According to CEFIC [CEF 14], in the area of matter and energy transformation, the evolutions of the market laws (quick financial turnaround) were felt in European R&D activity in a modest way (~2%) but with a clear trend toward decrease (~2% per year as well). Indeed, in the best of cases, R&D only produces effects after 2–3 years, if not 7–10 years. This situation, on the part of the concerned manufacturers, leads to a certain futurism in a morose and uncertain context and thus to a stabilized current vision in terms of innovations. One of the arguments, according to Grabowski [GRA 11], is that R&D expenses in the field of medical drugs are developing in a sensitively exponential way while the number of market releases is increasing little. This situation thus does not provide considerable motivation for highly risky R&D actions (see Figure 2.36).

Creativity is recognized for having an influence on the capacity to identify opportunities [LON 84, HIL 99, DET 04, CHA 14]. In this framework of active research action, individuals identify opportunities through motivated, conscious and deliberate research. Here, the opportunity comes only from the mind of the person, who, except for his/her imagination, can base this less and less on nothing or nearly nothing and transform his/her opportunity into action (necessary knowledge of the real and the possible).

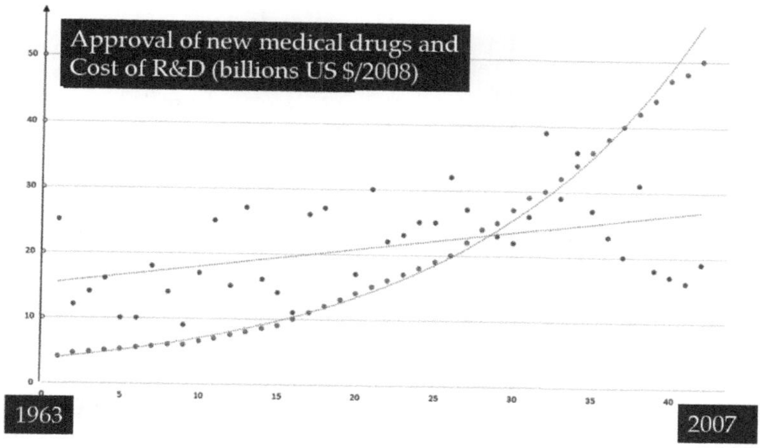

Figure 2.36. *Problem of the increased price of R&D in the particular field of pharmacy (the number of marketing authorizations would increase slightly (blue dots) while the cost of R&D increases sensitively in an exponential way (red dots). For a color version of the figure, see www.iste.co.uk/andre/printing1.zip*

Thus, once a technology is assimilated (like additive manufacturing), it can continue to develop through the knowledge acquired and as a function of the specificity of the domains in which it collects its knowledge to develop. Then, the associated specialization implies choosing people whose level of training must be more and more acute. Yet, as Lubart shows [LUB 05], creativity would follow a less favorable allure in terms of performance, according to the training time, to become more "mediocre" for very specialized trainings like doctorates, in which in-depth scientific knowledge are strongly supported (see Figure 2.37). The "dogmatization" deepening and need expanded knowledge go together, do not we risk seeing innovation lose creativity in a socially accepted domain, but still evolving?

"We cannot understand why we no longer understand. All these problems only have one cause, but one that goes unperceived: we have lived for at least three hundred years from fruits that just begged to be picked [...]. Yet, for the past forty years, these fruits have started to become rare, and we acted as if they were still there. We refused to recognize that we had reached a technological plateau and that the tree was much more stripped than we wanted to admit" [COW 11]. Could he be right? What must we "think" of this?

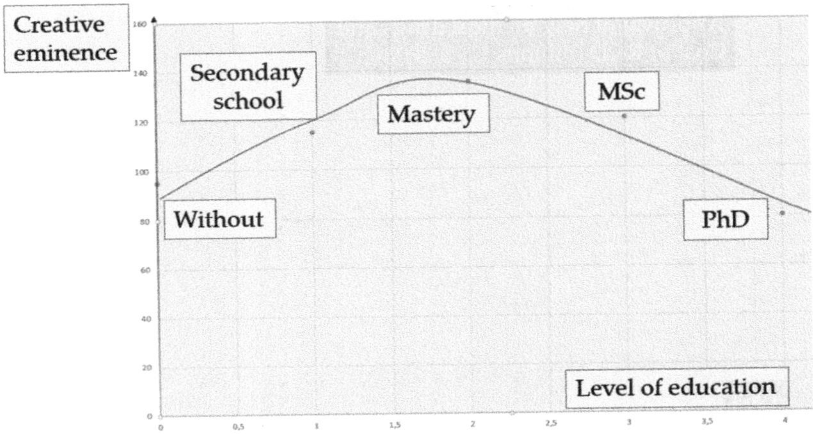

Figure 2.37. *Evolution of creativity with scholastic education*

These elements, starkly outlined, fight for a reduction in innovation in the domains that managed to lose their footing in the collective unconscious, with sufficient success to bring the economic world into a state of development, but with technological and financial limitations that slow the dynamic. These elements will be the object of the following chapters.

2.6. Conclusion

Everyone is talking about additive manufacturing; the newspapers report about new achievements in 3D technology today with potential advances (yet to be validated) in the creation of printed biological tissues, leaving the impression that everything is possible with short time gaps (within a few years, or even less). It is true that 3D technologies authorize the manufacturing of objects that were impossible to create a few decades ago and that the observation (when possible) of this manufacturing has, even for the author still, something magical about it.

If every technology finds its conditions of possibility and efficiency (which will be presented in the following chapter), it needs to find the places where it can excel. However, it must be understood that the results presented in Figure 2.5 illustrating a reduced acceleration of the development process for this domain have slightly shattered this image. Numerous little problems, presented in this chapter, slightly limit the enthusiasm for a particularly promising technology. It will have to overcome them if it wants to take the full place that it can/must expect. It is interesting to look into the mental invariants, those cognitive biases, the media's propagandist elements (even those of scientists themselves) that stop us from forming a proper idea of the pertinence of a production system.

It is particularly helpful to consider the fact that the concretization of new propositions is only effective when the researcher and/or engineer expects positive consequences from their work and when they know they can count on a network to help them to make their ideas a reality. According to Chanut-Guieu and Guieu [CHA 14], "conversely, individuals who are just as creative but without a network or motivation make little progress. These results greatly reduce the interest that should be given to the only individual creativity". The structure of the environment, which remains modest in France, can thus appear to be a less-than-favorable factor for the technological development of additive manufacturing. Let us nevertheless note the existence of recent operations carried out in industrial environments and supported by the United States to place France among the nations who matter in the area of 3D printing. At the same time, as shown by Lesage and Hurteau [LES 16], in the situation of austerity that reaches academic milieus, in their study on the effects of cutting financing for research, a higher priority is given by subsidizing organizations to support work with a high potential for concrete results, to the detriment of fundamental research. It is then possible, for a while, for the short term to be favored by reinforcing public–private partnerships. However, in this décor, where will creativity be situated with its risk-taking?

To achieve this objective in a more relaxed way in the short term, it seems useful, even necessary, to invest in the following goals:

– increasing private R&D capacities;

– increasing the potentials of economic results of public research;

– developing cooperative projects between actors;

– promoting innovative entrepreneurship;

– supporting the development of innovative enterprises.

Nevertheless, as a number of studies emphasize [DÉL 16, DEL 14, WEL 15, CUR 15, BOI 15], France has an improvable situation in terms of 3D technology development. Table 2.7, inspired by DGE's report in a SWOT analysis, gathers the strengths, weaknesses, opportunities and threats corresponding to the rise of additive manufacturing in this country.

Field	Comments
Strengths	Flexibility of 3D technologies High-level academic and R&D activity Manufacturer's strong "high-tech" demand: aeronautics, automobile, arms, etc. Sub-contracting activity Dynamic manufacturers present in niches (ceramics, metal, 3D machines, etc.) New forms of freedom in production Possibility to create complex, unique parts Vast application field: art, mechanics, medicine, research, etc.
Weaknesses	Unstructured university activity; modest legibility Difficulty financing high-risk creative research Difficulty of interdisciplinary operations (e.g. process–material pairing) Current difficulty creating multi-material parts Long manufacturing time Difficulty realizing collective manufacturing (electronics, transducers, sensors, etc.) Few materials available responding to the criteria for "good manufacturing" Cost of raw materials Few manufacturers in France Few producers of materials Insufficient human resources and skills Profitability of investments Hygiene and safety, especially for home applications
Opportunities	Additive manufacturing has covered neither all its application niches or all sizes for parts nor all materials (see Volume 2) Relocalized production or production near the buyer Mass production Process–material pairings; process–multi-materials (see Volume 2) Effective demonstration of "sustainable" aspects of this technology; material recycling Mastery of the quality of the materials used and after additive manufacturing; modeling their components Creation of consortia Bringing together the scientific community for better legibility and improved performance Scientific and technological breakthroughs to be supported (see Volume 3) Development of Fab-Labs; tendency toward the "democratization" of 3D technologies Training specialists, which creates jobs

Threats	Buyout of "young sprouts" (when they exist nationally), by foreign businesses Difficulty keeping promises in terms of the quality of parts (surface condition, mechanical strength, creation time of an object, etc.) Weakness of intellectual property Unethical applications of additive manufacturing Cyber-risk Counterfeits

Table 2.7. *Synthetic SWOT analysis applied to additive manufacturing in France*

France's neighbor Germany, on the contrary, has published a strategy aiming at strong positions adapted to its industry, as indicated in Figure 2.38 [KOH 16]. Below are presented its 14 items:

1) productivity of machines;

2) availability of materials;

3) machine design and production;

4) industrial machines;

5) normalization and standardization [ANS 17];

6) process automation;

7) digital data security;

8) quality management methods;

9) trained personnel;

10) existence of a research and an R&D strategy;

11) national position in R and R&D;

12 supports for changes in additive manufacturing (R and R&D);

13) security of productions;

14) process of integration into production devices.

The situation in France, as recently published by DGE [PIP 17], is much more difficult to pinpoint.

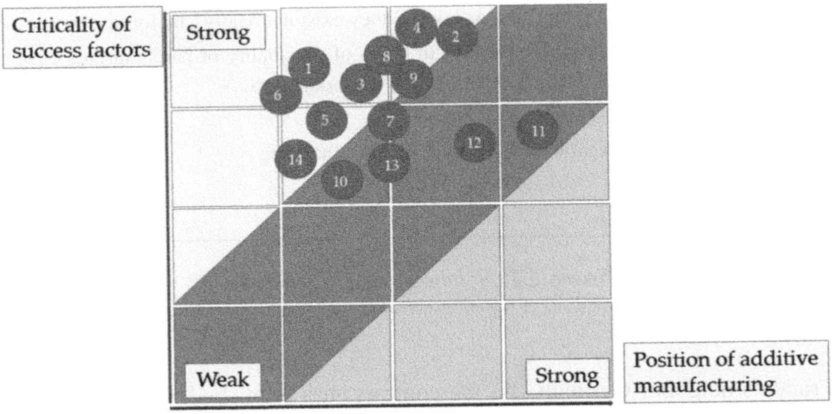

Figure 2.38. *Germany's position in additive manufacturing*

"Each of these objectives corresponds to the general ends of innovation policies and aims to compensate for a handicap, to improve incentives, to correct a fault in coordination, or to make up for the insufficient density of interactions between actors and innovation" [FRA 16]... But what can be done?

The development of 3D printing seems, beyond technique, to refer to acts of invention, object creation and original kinds of uses. These acts refer to psychological, social, rather positive factors exterior to the technology itself. In its proper way of existing, additive manufacturing must be replaced in the game of its interactions and its solidarity with what surrounds and supports it. It would thus not do for 3D processes' image of success to be marred by abusive propaganda that could eventually slow the process of the domain's inclusion among industrial devices (see e.g. [BOU 16, VIC 16a]). Possibly, counter-productively, this is a matter of new research programs in France aiming to develop an entire production chain: from the elaboration of materials to the equipment and the printing processes.

It is generally accepted that scientific research produces predictive scientific theories and theories of applications. These theories must help define strategies for intervening in reality. They can also create objects that are never seen before and new representations, ideas and theoretical tools that can go beyond their original domain (e.g. optical application to stereolithography in 1984). As long as they have been produced, 3D objects have become objects of knowledge and analysis of possible openings that could be associated with them. However, between reality and promises, the representations, fantasies, constructs or images created in the absence of knowledge escape the scientific domain. They prematurely lead to the manufacturing of standpoints capable of providing (certainly questionable) points of reference to give a view on a complex domain that is yet to be constructed for various reasons explained in this chapter.

To conclude this chapter, let us remember this text from Tinland [TIN 94]: "We must situate our desires and our practices, our representations and our decisions in the framework of this technological ensemble, made not only of objects and *modi operandi*, but also of offered or suggested possibilities, constraints with which make concessions, of choices, of acceptances, of refusals, and also of dreams [...]. Men are no longer faced with this ensemble that they could have as a function of representations and desires from elsewhere".

> *"But what changes in Man is not his/her biophysical nature; it is his/her symbolic being and what this articulates (subjectivity, inter-subjectivity, affectivity, the collective...). Under the effect of the physical milieu that it technically transforms, humanity is led to evolve symbolically by producing a new culture, a new society that will allow an appropriate articulation of itself [...] with the evolutional dynamic of technology [...]. Schematically, there is Man who technically modifies the natural milieu and who is modified by reflection via the symbolic-cultural creation" [HOT 94].*

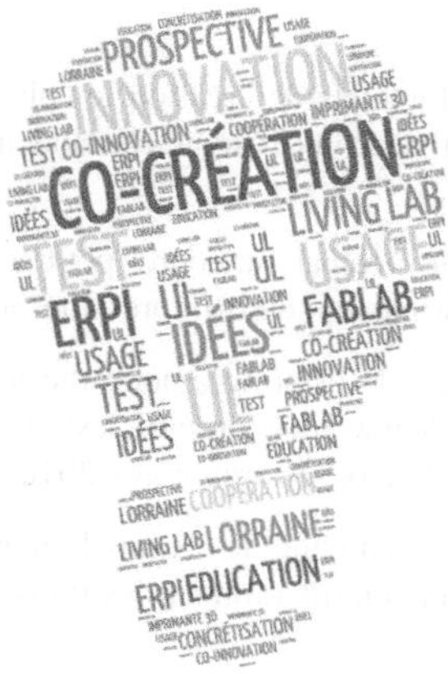

Figure 2.39. *"Word cloud" from the Lorraine Fab Living Lab® in Nancy (Photo credit: Laboratoire ERPI / ENSGSI, University of Lorraine, 2014)*

2.7. Bibliography

[3D 16a] 3D NATIVES, "Scanners 3D", available at: http://www.3dnatives.com/category/scanners-3d/, 2016.

[3D 16b] 3D NATIVES, "Des aides fiscales pour les PME intéressées par l'impression 3D", available at: http://www.3dnatives.com/aides-fiscales-pme-impression-3d-20012016/, 2016.

[ACA 15] ACADÉMIE DES TECHNOLOGIES, "Les enjeux stratégiques de la fabrication additive: positionnement de l'Académie des Technologies", available at: http://academie-technologies-prod.s3.amazonaws.com/2015/06/18/08/31/49/22/Avis_de_l_AT_sur_Fabrication_additive_juin_2015_1_.pdf, 2015.

[ADE 85] ADER M., *Le choc informatique*, Folio-Actuel, Paris, 1985.

[ADD 16] ADDIVERSE, "Impression 3D: 7 prédictions pour 2015 par additiverse.com", available at: http://www.additiverse.com/impression-3d-7-predictions-en-2015-par-additiverse-com/, 2016.

[AFS 15] AFSHAR-MOHAJER N., WU C.Y., LADUN T. *et al.*, "Characterization of particulate matters and total VOC emissions from a binder jetting 3D printer", *Building and Environment*, vol. 93, pp. 293–302, 2015.

[AFT 06] AFTU (Association Pour Une Fondation Travail-Université), "Technologies convergentes, évaluations divergentes", Education Permanente, "De nouveaux enjeux pour la politique scientifique", available at: http://www.ftu.be/documents/ep/EP-24.pdf, 2006.

[AGA 12] AGARWALA M. (ed.), "3D Printing and Additive Manufacturing State of the Industry", Annual Worldwide Progress Report – Wohler Associates, Fort Collins, USA, 2012.

[ALE 15a] ALEX, "TOP 25 des logiciels de création les plus utilisés pour l'impression 3D", available at: http://www.3dnatives.com/25-logiciels-modelisation-impression-3d-07082015/, 2015.

[ALE 15b] ALEX, "Un nouveau format standard pour l'impression 3D?", available at: http://www.3dnatives.com/format-standard-impression-3d/, 2015.

[ALE 15c] ALEX, "Gartner places 3D printing among the top 10 strategic technology trends for 2016", available at: http://www.3ders.org/articles/20151224-gartner-places-3d-printing-among-the-top-ten-strategic-technology-trends-for-2016.html, 2015.

[ALE 15d] ALEXANDRE L., "Comment les trans-humanistes ont gagné nos cœurs", available at: http://www.wedemain.fr/Comment-les-transhumanistes-ont-gagne-nos-coeurs_a903.html, 2015.

[ALE 16a] ALEX, "Le marché de l'impression 3D estimé à 26,7 milliards de dollars en 2019", available at: http://www.3dnatives.com/marche-impression-3d-2019-idc-25012016/, 2016.

[ALE 16b] ALEX, "Le marché de l'impression 3D a augmenté de 25.9% en 2015", available at: http://www.3dnatives.com/marche-de-limpression-3d-2015-06042016/, 2016.

[AND 94] ANDRÉ J.C., CORBEL S., *Stéréo-photo-lithographie laser*, Polytechnica, Paris, 1994.

[AND 12a] ANDERSON C., *Makers: La nouvelle révolution industrielle*, Pearson-France, Paris, 2012.

[AND 12b] ANDERSON C., "L'imprimante 3D aura plus d'impact que le web", available at: http://rue89.nouvelobs.com/2012/12/25/chris-anderson-limprimante-3d-aura-plus-dimpact-que-le-web-238098, 2012.

[AND 16a] ANDRÉ J.C., "Bio-printing and NBIC Convergences: Promises, ethics, epistemological approaches", *Nanoscale 3D Printing Workshop, Conference*, Singapore, 1–2 December 2016.

[AND 16b] ANDRÉ J.C., GALLAIS L., AMRA C., "Procédé pour la réalisation d'un objet tridimensionnel par un processus de photo-polymérisation multi-photonique et dispositif associé", French Patent no. 16-59211, 28 September 2016.

[AND 16c] ANDRÉ-CHAIGNEAU D., "Innovation produit: Qui sont les acheteurs?", available at: http://www.toute-la-franchise.com/vie-de-la-franchise-A24581-quels-acheteurs-pour-les-innovations-produits.html, 2016.

[AND 17] ANDRÉ J.C., "Fabrication additive: Révolution ou simple évolution sociétale?", *Techniques de l'Ingénieur*, ref. AG115, V1, p. 21, Paris, 2017.

[ANI 17] ANIWAA, "Les technologies de scan 3D et le processus de scan 3D – Qu'est ce que le scan 3D? Comment fonctionne le scan 3D? Quelles sont les technologies de scan 3D disponibles?", available at: http://www.aniwaa.fr/les-technologies-de-scan-3d-et-le-processus-de-scan-3d/, 2017.

[ANS 17] ANSI, "Standardization Roadmap for Additive Manufacturing", ANSI, New York, available at: https://www.ansi.org/news_publications/news_story?menuid=7&articleid=6a8f99db-84d9-40f8-b70b-78e692f77361, 2017.

[ARC 16] ARCAND P., "Le chantier sur l'acceptabilité sociale: une démarche innovante tournée vers l'avenir", *Éthique publique*, vol. 18, 2016, available at: http://ethiquepublique.revues.org/2428; doi: 10.4000/ethiquepublique.2428 available at: http://ethiquepublique.revues.org/2428.

[ARN 09] ARNSPERGER C., *Ethique de l'existence post-capitaliste; pour un militantisme existentiel*, Cerf, Paris, 2009.

[AST 14] AST74, "Evaluation des risques professionnels", available at: http://www.ast74.fr/fr/reglementation/id-4-evaluation-des-risques-professionnels, 2014.

[BAR 15] BARLIER C., BERNARD A., *Fabrication additive: du prototypage rapide à l'impression 3D*, Dunod, Paris, 2015.

[BAR 17] BARTHÉLÉMY J., CETTE G., "Travailler au XXIème Siècle; L'uberisation de l'économie?", available at: http://tnova.fr/system/contents/files/000/001/287/original/04012017_-_Travailler_au_XXI_me_Si_cle.pdf?1483524639, 2017.

[BAU 72] BAUDRILLARD J., *Pour une critique de l'économie politique du signe*, Gallimard, Paris, 1972.

[BAU 12] BAUMERS M., Economic Aspects of Additive Manufacturing: Benefits, Costs and Energy Consumption, Doctoral Thesis, Loughborough University, Leicestershire, 2012.

[BAU 16] BAUMERS M., DICKENS P., TUCK C. *et al.*, "The cost of additive manufacturing: machine productivity, economies of scale and technology-push", *Technological Forecasting and Social Change*, vol. 102, pp. 193–201, 2016.

[BEC 07] BECKERT B., BLÜMEL C., FRIEDEWALD M., "Visions and realities in converging technologies; Innovation", *The European Journal of Social Sciences*, vol. 20, pp. 375–394, 2007.

[BEN 10] BENASAYAG M., *Organismes et artéfacts*, La Découverte, Paris, 2010.

[BER 14a] BERCHON M., LUYT B., *L'impression 3D*, Eyrolles, Paris, 2014.

[BER 14b] BERDAH P., "L'impact de l'imprimante 3D sur la propriété intellectuelle", available at: http://glorieusefrance.fr/wp-content/uploads/2014/03/Imprimante-3D.pdf, 2014.

[BER 14c] BERNARD A., MANIL P., "IRFU-CEA – Fabrication additive", Paris, available at: http://www.google.fr/url?sa=t&rct=j&q=&esrc=s&source=web&cd=1&ved=0ahUKEwjfpZCt8rzQAhUHtxoKHXseDwEQFggiMAA&url=http%3A%2F%2Firfu.cea.fr%2FPhocea%2Ffile.php%3Ffile%3DSeminaires%2F3387%2FBERNARD_Journee_CEA-IRFU_01122014_V2.pdf&usg=AFQjCNEHpWBS_avTzdgc5T2iYDfa7KH5Mw&bvm=bv.139250283,d.d2s, 2014.

[BER 15a] BERREBI-HOFFMANN I., BUREAU M.C., LALLEMENT M., "Entretien avec Michel Bauwens", *Recherches sociologiques et anthropologiques*, vol. 46, available at: http://rsa.revues.org/1546, 2015.

[BER 15b] BERRY G., "L'informatique modifie-t-elle notre manière de penser?", available at: http://www.lefigaro.fr/sciences/2012/04/20/01008-20120420ARTFIG00572-l-informatique-modifie-t-elle-notre-maniere-de-penser.php, 2015.

[BER 16] BERGÈRE J.M., "Quand les hackers réinventent le travail", available at: http://www.metiseurope.eu/quand-les-hackers-reinventent-le-travail_fr_70_art_30085.html, 2016.

[BIH 14] BIHOUIX P., *L'âge des low-tech; vers une civilisation techniquement soutenable*, Anthropocène/Le Seuil, Paris, 2014.

[BIO 16a] BIOTECH-FINANCES, "Global Market for Bio-printing Market Should Reach $1.8 Billion by 2021 – Key Vendors are Autodesk, Celenys & Digilab Inc. – Research and Markets", available at: http://www.eei-biotechfinances.com/global-market-for-bio-printing-market-should-reach-1-8-billion-by-2021-key-vendors-are-autodesk-celenys-digilab-inc-research-and-markets/, 2016.

[BIO 16b] BIOTECH-FINANCES, "Global 3D Bio-printing Market Research and Analysis 2015-2022 – Research and Markets", available at: http://www.eei-biotechfinances.com/global-3d-bio-printing-market-research-and-analysis-2015-2022-research-and-markets/, 2016.

[BLA 13] BLANC S., "Chéri, achète de la lessive et imprime un porte-clé", Slate, 30/10/2013. available at: http://www.slate.fr/story/79484/imprimante-3D-supermarche-auchan, 2013.

[BLA 14] BLACKMAGIC3D, "The 8 best 3D printing stories of 2014", available at: http://blog.blackmagic3d.com/8-best-3d-printing-stories-2014/, 2014.

[BLA 06] BLANQUART O., MALMAISON P., SOULIER E., *Le storytelling: concepts, outils et applications*, Hermès-Lavoisier, Paris, 2006.

[BLO 16] BLONDIAU G., La propriété intellectuelle face à la démocratisation de l'impression 3D, le titulaire de droits face à l'utilisateur; Un équilibre instable?, Masters Thesis, Université catholique de Louvain, 2016.

[BOE 09] BOENISCH G., "Jean-Michel Cornu, Nouvelles technologies, nouvelles pensées? La convergence des NBIC", *Questions de communication*, vol. 15, pp. 490–491, 2009.

[BOI 04] BOIFFIN J., "Territoire: agronomie, géographie, écologie, où en est-on? Le point de vue d'un chercheur agronome", *Natures Sciences Sociétés*, vol. 12, pp. 307–309, 2004.

[BOI 15] BOILLAT E., "Techniques de fabrication additive", available at: http://www.swissmem.ch/fr/presentations-fabrication-additive.html, 2015.

[BON 14] BONNET C., "CRéacteur; Résilience par l'action créative", available at: https://head.hesge.ch/communicationvisuelle/wp-content/uploads/2015/03/BONNET_Claire_BachelorTh.pdf, 2014.

[BOR 15] BOREL S., DEMAILLY D., MASSÉ D., "L'économie collaborative, entre utopie et big business", Esprit, available at: http://www.eurozine.com/pdf/2015-07-28-borel-fr.pdf, 2015.

[BOS 14] BOSQUÉ C., NOOR O., RICARD L., *Fablabs, etc.; les nouveaux lieux de fabrication numérique*, Eyrolles, Paris, 2014.

[BOS 15a] BOSQUÉ C., "Enquête au cœur des FabLabs, hackerspaces, makerspaces; Le dessin comme outil d'observation", *Techniques & Culture*, vol. 64, pp. 168–185, "Essais de bricologie. Ethnologie de l'art et du design contemporains", available at: http://media.virbcdn.com/files/39/49248a9dffa2fb29-Ledessincommeoutildobservation.pdf, 2015.

[BOS 15b] BOSTON CONSULTING GROUP, "Man and machines in industry 4.0 – How will technology transform the industrial workforce through 2025?", available at: https://www.bcgperspectives.com/Images/BCG_Man_and_Machine_in_Industry_4_0_Sep_2015_tcm80-197250.pdf, 2015.

[BOS 15c] BOSTON CONSULTING GROUP, "Job seekers trends 2015; channels, search and income change", available at: https://www.bcgperspectives.com/Images/BCG-Job-Seeker-Trends-December-2015_tcm80-202726.pdf, 2015.

[BOS 17] BOSTON CONSULTING GROUP, "Honing us manufacturing's competitive edge", available at: https://www.bcgperspectives.com/Images/BCG-Honing-US-Manufacturings-Competitive-Edge-Jan-2017_tcm80-218567.pdf, 2017.

[BOT 12] BOTTOLIER-DEPOIS F., *FabLabs, Hakerspaces: nouvelles formes d'innovation et militantisme libertaire*, Observatoire du Management Alternatif, HEC, Paris, 2012.

[BOT 14] BOTSMAN R., "The sharing economy lacks a shared definition" quoted by DEMAILLY D., NOVEL N.S. in "Economie du partage: enjeux et opportunités pour la transition écologique", Institut du développement durable et des relations internationals, Etude "Nouvelle prospérité" no. 03/14, July 2014.

[BOU 14] BOUFFARON P., "Impression 3D: Les prémisses d'une nouvelle (r)évolution industrielle", French Embassy in the US, available at: http://sf.france-science.org/wp-content/uploads/2014/11/SMM14_025.pdf, 2014.

[BOU 16] BOURSORAMA, "CERINNOV: lancement d'un projet R&D collaboratif dans l'impression 3D", available at: http://www.boursorama.com/actualites/cerinnov-lancement-d-un-projet-r-d-collaboratif-dans-l-impression-3d-9228ea7d6935751a43d084c34bf5ff00, 2016.

[BRA 15] BRANCHE P.M., "Impression 3D; la fin annoncée des transports", available at: http://www.voxlog.fr/dossier/13_2/logistique-et-transport-limpression-3d-remet-en-cause-la-supply-chain, 2015.

[BRE 08] BREUIL H., BURETTE D., FLURY-HERARD B., "TIC et développement durable", available at: http://www.ladocumentationfrancaise.fr/var/storage/rapports-publics/0940 00118.pdf, 2008.

[BRE 15] BRESSÉ P., "Impression 3D de pièces détachées et contrefaçon: enjeux et perspectives", *Propriétés intellectuelles*, vol. 57, 366 pp., 2015.

[BRU 14] BRUEGEL, "Chart of the week: 54% of EU jobs at risk of computerisation", available at: http://bruegel.org/2014/07/chart-of-the-week-54-of-eu-jobs-at-risk-of-computerisation, 2014.

[BRY 14] BRYNJOFSSON E., MCAFEE A., *Le deuxième âge de la machine; travail et prospérité à l'heure de la révolution technologique*, Odile Jacob, Paris, 2014.

[BRY 15] BRYNJOFSSON E., MCAFEE A., "Will humans go the way of horses? Labor in the second machine age", *Foreign Affairs*, vol. 94, pp. 8–14, 2015.

[BÜH 06] BÜHLERA E.A., CAVAILLÉ F., GAMBINO M., "Le jeune chercheur et l'interdisciplinarité en sciences sociales – Des pratiques remises en question", *Natures Sciences Sociétés*, vol. 14, pp. 392–398, 2006.

[BUT 16] BUTTIN N., SAFFRÉ B., *Activer l'économie circulaire*, Eyrolles, Paris, 2016.

[CAH 14] CAHN J.G., "Impression 3D: Perspectives industrielles et/ou utopie sociétale", *Lettre de prospective (CCI Ile-de-France)*, vol. 43, pp. 1–5, available at: http://www.cci-paris-idf.fr/sites/default/files/etudes/pdf/documents/friedlandpapers-201402-43.pdf, 2014.

[CAM 11] CAMPBELL T., WILLIAM C., IVANOVA O. *et al.*, "Could 3D printing change the world? Technologies, potential and implications of additive manufacturing" available at: https://info.aiaa.org/SC/ETC/MS%20SubCommittee/Alice%20Chow_3D%20Printing%20Change%20the%20World_April%202012.pdf, 2011.

[CAN 16a] CANALYS, "3D printing market to be worth US$22.4 billion in 2020", available at: https://www.canalys.com/static/press_release/2016/media-alert-17052016-3d-printing-market-be-worth-us224-billion-2020.pdf, 2016.

[CAN 16b] CANTENEUR P., "Impression 3D, simple gadget ou révolution de l'industrie?", available at: http://www.atelier.net/trends/articles/impression-3d-simple-gadget-revolution-de-industrie_442862, 2016.

[CAR 61] CARNAP R., *Le mythe moderne du progrès; décortiqué et démonté par le philosophe*, Agone, Marseille, 1961.

[CCI 15] CCI PARIS-IDF/CONSEIL GÉNÉRAL DE L'ARMEMENT, "L'impression 3D: porte d'entrée dans l'industrie du 21ème siècle", available at: http://www.adverbe.com/wp-content/uploads/2015/09/rapport-impression3d.pdf, 2015.

[CEF 14] CEFIC, "Evolution of competitiveness in the European chemical industry: historical trends and future prospects", available at: http://www.cefic.org/Documents/RESOURCES/Reports-and-Brochure/Oxford-Study-2014.pdf, 2014.

[CHA 14] CHANUT-GUIEU C., GUIEU G., "Créativité, PME et entrepreneuriat: des zones d'ombre et de lumière", *Revue internationale P.M.E.: économie et gestion de la petite et moyenne entreprise*, vol. 27, pp. 35–63, 2014.

[CHE 12] CHENAVAZ R., "Dynamic pricing, product and process innovation", *European Journal of Operational Research*, vol. 222, pp. 553–557, 2012.

[CHE 16] CHEN Z., "Research on the Impact of 3D Printing on the International Supply Chain", *Advances in Materials Science and Engineering*, Article ID 4173873, available at: http://dx.doi.org/10.1155/2016/4173873, 2016.

[CHE 07] CHEVASSUS-AU-LOUIS B., *L'analyse des risques: l'expert, le décideur et le citoyen*, Quae, Paris, 2007.

[CHE 16] CHEVASSUS-AU-LOUIS N., *Malscience; de la fraude dans les labos*, Le Seuil, Paris, 2016.

[CHI 07] CHIAPELLO E., "Dynamique du capitalisme et régulation de la technique", in FELTZ B., GOUJON P., LAVELLE S. *et al.* (eds), *Ethique, technique et démocratie*, Brulant-Academia, Louvain-la-Neuve, 2007.

[CHI 09] CHIFFLET S., "L'imaginaire technoscientifique: cas de la convergence NBIC", available at: http://www-llb.cea.fr/Phocea/Vie_des_labos/Seminaires/index.php?id_type=19&type=19&y=2009&id=1663 and "Le récit de la convergence NBIC: vers une nouvelle cosmogonie?", Thesis, Université Stendhal, Grenoble, 2009.

[CLA 15] CLAPAUD A., "2015: Le marché de l'impression 3D dans la tourmente", available at: http://www.4erevolution.com/2015-impression-3d/, 2015.

[COL 15] COLIN N., VERDIER H., *L'âge de la multitude – entreprendre et gouverner après la révolution numérique*, Armand-Colin, Paris, 2015.

[CON 15] CONTEXT, "L'impression 3D personnelle s'invite dans les foyers", available at: http://www.3dnatives.com/ventes-imprimantes-3d-04082015/, 2015.

[COT 14] COTTELEER M., HOLDOWSKY J., MAHTO M., *The 3D opportunity primer: the basics of additive manufacturing*, Deloitte University, Westlake, 2014 (quoted by Verhulst J. (2015)).

[COW 11] COWEN T., *The Great Stagnation: How AMERICAN are All the Low-hanging Fruit of Modern History, Got Sick and Will (eventually) Feel Better*, Dutton, New York, 2011.

[CRO 12] CROCKER G., *A managerial philosophy of technology*, Palgrave – McMillan, London, 2012.

[CUR 15] CURTET M., MORESCHI M., "Impression 3D par dépôt de matière fondue", available at: http://cerig.pagora.grenoble-inp.fr/memoire/2015/impression-3d.htm, 2015.

[DAS 16] DASTI, "An OECD horizon scan of megatrends and technology trends in the context of future research policy", available at: http://ufm.dk/en/publications/2016/files/an-oecd-horizon-scan-of-megatrends-and-technology-trends-in-the-context-of-future-research-policy.pdf, 2016.

[DEF 04] DEFFONTAINES J.-P., HUBERT B., "Un regard sur l'interdisciplinarité à l'Inra – Point de vue de deux chercheurs du département Sciences pour l'action et le développement (SAD)", *Natures Sciences Sociétés*, vol. 12, pp. 186–190, 2004.

[DEG 15] DE GRAVE A., "La fin des Léviathans", in FILLIPOVA D. (ed.), *Société collaboratrice; la fin des hiérarchies*, Rue de l'échiquirer, Paris, 2015.

[DEH 16] DEHU E., "L'impression 3D révolutionne déjà l'industrie et l'artisanat: la preuve en images", available at: http://www.wedemain.fr/L-impression-3D-revolutionne-deja-l-industrie-et-l-artisanat-la-preuve-en-images_a1756.html, 2016.

[DEK 08] DE KERORGUEN Y., LEROY E., *Vivre en 2028; notre futur en 50 mots clés*, Lignes de repères, Paris, 2008.

[DEL 14] DELOITTE, "Du nouveau dans le secteur manufacturier: les répercussions de l'impression tridimensionnelle", available at: http://www2.deloitte.com/content/dam/Deloitte/ca/Documents/insights-and-issues/ca-fr-perspectives-et-enjeux-daffaires-du-nouveau-dans-le-secteur-manufacturier.pdf, 2014.

[DEL 16] DELOITTE, "Cyber Risk in Advanced Manufacturing", available at: https://www.mapi.net/system/files/attachments/files/CyberRisk_Manu_FullStudy_Landscape_Brochure_LPXCIC07_06_17_7101_FINALfor.pdf, 2016.

[DEM 16] DE MICHAUX B. (ed.), *L'impression 3D, défis et opportunités pour la propriété intellectuelle*, Larcier, Brussels, 2016.

[DEP 14] DEPOORTER B., "Intellectual Property Infringements & 3D Printing Decentralized Piracy", *Hastings Law Journal*, vol. 1483, available at: http://repository.uchastings.edu/faculty_scholarship/1011, 2014.

[DES 13a] DESAI D.R., MAGLIOCCA G.N., "Patents, Meet Napster: 3d Printing And The Digitization Of Things", in Indiana University Robert H. McKinney School of Law Research Paper No. 2013-37; Thomas Jefferson School of Law Research Paper No. 2338067, available at: https://ssrn.com/abstract=2338067, pp 7–55, 2013.

[DES 13b] DESMEDT P., "L'américain 3D Systems s'empare du numéro un français de l'impression 3D", available at: http://www.usinenouvelle.com/article/l-americain-3d-systems-s-empare-du-numero-un-francais-de-l-impression-3d.N218519, 2013.

[DET 04] DE TIENNE D.R., CHANDLER G., "Opportunity identification and its role in the entrepreneurial classroom: a pedagogical approach and empirical test", *Academy of Management Learning and Education*, vol. 3, pp. 242–257, 2004.

[DGE 16] DGE, "Technologies clés – Préparer l'industrie du futur 2020", available at: http://www.entreprises.gouv.fr/files/files/directions_services/politique-et-enjeux/innovation/technologies-cles-2020/technologies-cles-2020.pdf, 2016.

[DIR 14] DIRECCTE, "L'impression 3D: Etat des lieux et perspectives", available at: http://www.centre.direccte.gouv.fr/IMG/pdf/L_impression_3D_etat_des_lieux_et_perspectives_-_DIRECCTE_Centre_en_partenariat_avec_CCI_Centre_-_10_decembre_2014-2.pdf, 2014.

[DUP 17] DUPONT L., "Lorraine Fab Living Lab", available at: https://laboratoire-erpi.wikispaces.com/Lorraine+Fab+Living+Lab?responseToken=021644860a63f191bab4570bf70bdd6e9, 2017.

[DUR 12] DURAND D., *Des outils, des machines et des Hommes*, Presses Universitaires de Provence, Aix-en-Provence, France, 2012.

[DUR 08] DURAND-RICHARD M.J. (ed.), *L'analogie dans la démarche scientifique – perspective historique*, L'Harmattan, Paris, 2008.

[ECR 05] ECRIN, *Technologies du futur; enjeux de société*, OMNI Science, Paris, 2005.

[EHR 16] EHRENBERG A., "Le drame français des grands souvenirs: un culte réactionnaire du passé?", available at: http://www.metiseurope.eu/le-drame-francais-des-grands-souvenirs-un-culte-reactionnaire-du-passe_fr_70_art_30392.html, 2016.

[ELL 88] ELLUL J., *Le bluff technologique*, Hachette, Paris, 1988.

[ETT 92] ETTIGHOFFER D., *L'entreprise virtuelle ou les nouveaux modes de travail*, Odile Jacob, Paris, 1992.

[EYC 12] EYCHENNE F., *Fab-Lab; l'avant-garde de la nouvelle revolution industrielle*, FYP, Paris, 2012.

[FAB 15] FABULOUS, "Le guide de l'impression 3D", available at: http://www.fabulous.com.co/guide-impression-3d/en-bref/histoire/, 2015.

[FAB 16] FabWiki, "Portal: Labs", available at: http://wiki.fablab.is/wiki/Portal:Labs, 2016.

[FAR 13] FARIVAR C., "'Download This Gun': 3D-printed Semi-automatic Fires over 600 Rounds" Ars Technica, available at: http://arstechnica.com/techpolicy/2013/03/download-this-gun-3d-printed-semi-automatic-fires-over-600-rounds/, 2013.

[FER 15] FERRÉOL V., BRETONES L., "Bienvenue dans l'ère de l'imprévu", available at: http://www.lesechos.fr/idees-debats/editos-analyses/021287519685-bienvenue-dans-lere-de-limprevu-1149082.php, 2015.

[FLI 03] FLICHY P., *L'innovation technique*, La Découverte, Paris, 2003.

[FIN 16] FING, "Faire émerger et connecter des FabLabs en France", available at: http://fing.org/?Le-Fab-Lab-lieu-d-artisanat, 2016.

[FLI 13] FLIPO F., DOBRÉ M., MICHOT M., *La face cachée du numérique; l'impact environnemental des nouvelles technologies*, Ed. de l'Echappée, Paris, 2013.

[FLO 16] FLOSSMANUAL, "Fablab, Hackerspace, les lieux de fabrication numérique collaboratif", available at: http://fr.flossmanuals.net/fablab-hackerspace-les-lieux-de-fabrication-numerique-collaboratif/introduction/, 2016.

[FOR 15] FORD M., *The Rise of the Robots: Technology and the Threat of a Jobless Future*, Basic Books, New York, 2015.

[FRA 16] FRANCE STRATÉGIE, "Quinze ans de politiques d'innovation en France", available at: http://www.strategie.gouv.fr/publications/quinze-ans-de-politiques-dinnovation-france, 2016.

[FRE 13] FREEDONIA, *World 3D Printing: Additive Manufacturing*, Cleveland, 2013, (quoted by Verhulst (2015)).

[FRE 12] FRESSOZ J.B., *L'apocalypse joyeuse: une histoire du risque technologique*, Le Seuil, Paris, 2012.

[FRE 13] FREY C., OSBORNE M., *The Future of Employment: How Susceptible are Jobs to Computerisation?*, Oxford Martin School, Oxford, 2013.

[GAG 11] GAGLIO G., *Sociologie de l'innovation*, PUF, Paris, 2011.

[GAL 68] GALBRAITH J.K., *La paix indésirable*, Calmann-Lévy, Paris, 1968.

[GAO 15] GAO, "3D Printing: opportunities, challenges and policy; implications of additive manufacturing", available at: http://www.gao.gov/assets/680/670960.pdf, 2015.

[GAU 78] GAUDIN T., *L'écoute des silences*", Collection 10/18, Paris, 1978.

[GEB 14] GEBLER M., SCHOOT-UITERKAMP A.J.M., VISSER C., "A global sustainability perspective on 3D printing technologies", *Energy Policy*, vol. 74, pp. 158–167, 2014.

[GER 12] GERSHENFELD N., "How to make almost anything; the digital fabrication revolution", *Foreign Affairs*, vol. 91, pp. 43–57, 2012.

[GÉN 16] GÉNEAU D., CAULIER S., "Le 3D imprime sa marque sur l'industrie; Construire localement, à moindre coût, avec un service après-vente amélioré", available at: http://www.ledevoir.com/economie/actualites-economiques/465475/le-3d-imprime-sa-marque-sur-l-industrie-construire-localement-a-moindre-cout-avec-un-service-apres-vente-ameliore, 2016.

[GIB 15] GIBSON I., ROSEN D., STUCKER B., *Additive manufacturing technologies: 3D printing, rapid prototyping and direct digital manufacturing*, Springer, New York, 2015.

[GIL 78] GILLE B., *Histoire des techniques*, La Pléiade, Paris, 1978.

[GHI 16] GHILASSENE F., "L'impression 3D et la question de la redevance pour copie privée", INPI available at: https://www.inpi.fr/sites/default/files/impression_3d_redevance_pour_copie_privee.pdf, 2016.

[GIO 14] GIORGINI P., *La transition fulgurante; vers un bouleversement systémique du monde*, Bayard, Paris, 2014.

[GOM 16] GOMEZ P.Y., *Intelligence du travail*, Desclée de Brouwer, Paris, 2016.

[GOR 13a] GORE A., *The Future: Six Drivers of Global Change*, First Editions, New York, 2013.

[GOR 13b] GORI R., *La fabrique des imposteurs*, Babel, Paris, 2013.

[GOR 16] GORDON R., HARROP J., "3D Printing of Metals 2015–2025; Pricing, properties and projections for 3D printing equipment, materials and applications", available at: http://www.idtechex.com/research/reports/3d-printing-of-metals-2015-2025-000441.asp, 2016.

[GOT 16] GOTRONIC, "Carte des Fab Labs", available at: http://www.gotronic.fr/ins-carte-des-fablabs-50.htm, 2016.

[GRA 01] GRANGIER G.G., *Science et réalité*, Odile Jacob, Paris, 2001.

[GRA 11] GRABOWSKI H., "The evolution of the pharmaceutical industry over the past 50 years: a personal reflection", *Journal of the Economics Business*, vol. 18, pp. 161–176, 2011.

[GRU 14] GRUNEWALD S., "Is 3D printing priced out of reach of 60% of business who want it?" available at: http://3dprintingindustry.com/2014/12/12/3d-printing-priced-60-percent-businesses/, 2014.

[GUC 14] GUCHET X., *Philosophie des nanotechnologies*, Herrmann, Paris, 2014.

[GUÉ 03] GUÉRIN-TALPIN G., *Communication de crise*, Préventique, Bordeaux, 2003.

[GUI 11] GUILLAUD H., "La programmation du monde: où nous mène la convergence technologique?", available at: http://www.Internetactu.net/2011/07/19/la-programmation-du-monde-ou-nous-mene-la-convergence-technologique/, 2011.

[GUI 13] GUIMON M., "La fabrication additive, un empilement de risques?", *Hygiène et sécurité du travail*, vol. 233, pp. 88–92, 2013.

[HAU 14] HAUSMAN K.K., HORNE R., *L'impression 3D pour les nuls*, Frist Edition, Paris, 2014.

[HEI 13] HEINRICH P., "L'impression 3D – ou fabrication additive", available at: http://www.heinrich-consultant.fr/wp-content/uploads/2013/02/Impression-3D.pdf, 2013.

[HIL 99] HILLS G.E., SHRADER R.C., LUMPKIN G.T., "Opportunity recognition as a creative process" in REYNOLDS P.D., BYGRAVE W.D., MANIGART S. *et al.* (eds), *Frontiers of Entrepreneurship Research*, Wellesley, Babson College, 1999.

[HLA 14] HLAVIN M., "3D printing: the next industrial revolution", *Appliance Design*, vol. 62, pp. 22–23, 2014.

[HOR 13] HORNICK J., ROLAND D., "Many 3D Printing Patents Are Expiring Soon: Here's A Round Up & Overview of Them", available at: http://3dprintingindustry.com/news/many-3d-printing-patents-expiring-soon-heres-round-overview-21708/, 2013.

[HOR 15] HORNICK J., ROSARIO C., "Why the Memjet v. HP lawsuit will be a footnote in the 3D printing patent war", available at: http://3dprintingindustry.com/2015/08/25/why-the-memjet-v-hp-lawsuit-will-be-a-footnote-in-the-3d-printing-patent-wars/?utm_source=3D+Printing+Industry+Update&utm_medium=email&utm_campaign=409c8ce214-RSS_EMAIL_CAMPAIGN&utm_term=0_695d5c73dc-409c8ce214-64557777, 2015.

[HOS 14] HOSKINS S., *Le guide de l'impression 3D pour artisans et designers*, Marabout, Paris, 2014.

[I-CO 15] I-CONNECT, "La perception de l'innovation numérique en France", available at: http://www.filrfid.org/2015/09/la-perception-de-l-innovation-numerique-et-technologique-en-france.html, 2015.

[HOT 94] HOTTOIS G., "Gilbert Simondon; entre les interfaces techniques et symboliques", in TINLAND F. (ed.), *Ordre biologique; ordre technologique*, Champ Vallon, Paris, 1994.

[HUA 13] HUANG S.H., LIU P., MOKASDAR A. *et al.*, "Additive manufacturing and its societal impact: a literature review", *International Journal of Advanced Manufacturing Technology*, vol. 67, pp. 1191–2103, 2013.

[IND 15] INDUSTRIE ET TECHNOLOGIE, "Médicaments imprimés, famille robot, Google Science Fair: les innovations qui (re)donnent le sourire", available at: http://www.industrie-techno.com/medicaments-imprimes-famille-robot-google-science-fair-les-innovations-qui-re-donnent-le-sourire.39303, 2015.

[ING 15] INGELAERE R., "Innovation technologique et performance industrielle: l'exemple de l'impression 3D", available at: http://www.lecese.fr/sites/default/files/pdf/Avis/2015/2015_07_innovations_techno_impression_3d_.pdf, 2015.

[INT 16] INTELLECTUAL PROPERTY OFFICE, "Annual IP crime report: 2015 to 2016", available at: https://www.gov.uk/government/uploads/system/uploads/attachment_data/file/557539/ip-crime-report-2015-16.pdf, 2016.

[ISO 16] ISO, "ISO/ASTM 52915:2016(fr)", available at: https://www.iso.org/obp/ui/fr/#iso:std:iso-astm:52915:ed-2:v1:fr, 2016.

[ITR 17] ITRI (Information Technology& Innovation Foundation), "Trade vs. Productivity: What Caused U.S. Manufacturing's Decline and How to Revive It", available at: http://www2.itif.org/2017-trade-vs-productivity.pdf?_ga=1.86832733.491750456.1487121244, 2017.

[JAC 16] JACKSON B., "Fab Labs take over the world – who will be 1000th?" available at: https://3dprintingindustry.com/news/fab-labs-take-world-will-1000th-97568/, 2016.

[JAC 14] JACQUOT G., "En bourse, l'impression 3D fait un peu moins fantasmer", available at: http://www.lemonde.fr/technologies/article/2014/04/21/en-bourse-l-impression-3d-fait-un-peu-moins-fantasmer_4404706_651865.html, 2014.

[JEO 15] JEONG S., KIM J.C., CHOI J.Y., "Technology convergence: What developmental stage are we in?" *Scientometrics*, vol. 104, pp. 841–871, 2015.

[JEO 15] JEONGA S., LEE S., "What drives technology convergence? Exploring the influence of technological and resource allocation contexts", *Journal of Engineering and Technology Management*, vol. 36, pp. 78–96, 2015.

[JIN 17] JIN Y., HE Y., FU G. *et al.*, "A non-retraction path planning approach for extrusion-based additive manufacturing", *Robotics and Computer-Integrated Manufacturing*, vol. 48, pp. 132–144, 2017.

[JUL 96] JULIEN F., *Traité de l'efficacité*, Grasset, Paris, 1996.

[JUT 14] JUTHIER J.E., "Le marché de l'impression 3D va croître de 20% par an", available at: http://www.journaldunet.com/economie/industrie/marche-de-l-impression-3d-selon-xerfi.shtml, 2014.

[KAZ 14] KAZIUMAS A., *Pratique de l'impression 3D*, available at: https://books.google.fr/books?isbn=2212270631, 2014.

[KAU 08] KAUFMANN J.C., "Quand Je est un autre. Pourquoi et comment ça change en nous", Armand Colin, Paris, 2008.

[KER 13] KERBRAT O., LE BOURHIS F., PONCHE R. *et al.*, "Evaluation de la performance environnementale des procédés de fabrication", *21st French Mechanics Congress*, 26–30 August, Bordeaux, available at: https://hal.archives-ouvertes.fr/hal-00914003/document, 2013.

[KID 82] KIDDER J.T., *Project Eagle*, Flammarion, Paris, 1982.

[KIE 15] KIETZMANN J., PITT L., BERTHON P., "Disruptions, decisions, and destinations: Enter the age of 3-D printing and additive manufacturing", *Business Horizons*, vol. 58, pp. 209–215, 2015.

[KIM 13] KIM K.H., MOON Y.H., "A new measurement to understand convergence phenomenon", *Asian Journal of Innovation and Policy*, vol. 2, no. 1, pp. 37–62, 2013. doi: http//dx.doi.org/10.7545/ajip.2013.2.1.037.

[KIM 15] KIM H.C., "Acceptability engineering: the study of user acceptance of innovative technologies", *Journal of Applied Research and Technology*, vol. 13, pp. 230–237, 2015.

[KLE 04] KLEIN E., *Les Tactiques de Chronos*, Champs Flammarion, Paris, 2004.

[KLI 92] KLINE S.J., ROSENBERG N., *La technologie et l'économie: les relations déterminantes*, OECD, Paris, 1992.

[KOH 15] KOHTALA C., "Addressing sustainability in research on distributed production: an integrated literature review", *Journal of Cleaner Production*, vol. 106, pp. 654–668, 2015.

[KOH 16] KOHLHUBER M., KAGE M., KARG M., "Additive Fertigung", available at: http://www.acatech.de/fileadmin/user_upload/Baumstruktur_nach_Website/Acatech/root/de/Publikationen/Kooperationspublikationen/3Akad_Stellungnahme_AdditiveFertigung_2016_web.pdf, 2016.

[KRE 14] KREIGER M.A., MULDER M.L., GLOVER A.G. *et al.*, "Life cycle analysis of distributed recycling of post-consumer high density polyethylene for 3-D printing filament", *Journal of Cleaner Production*, vol. 70, pp. 90–96, 2014.

[KUN 05] KUNDERA M., *Le Rideau*, Gallimard, Paris, 2005.

[LA 15] LA FABRIQUE DE L'INDUSTRIE, "Automatisation, emploi et travail", available at: http://www.la-fabrique.fr/uploads/telechargement/Le_robot_tue_t_il_l_emploi.pdf, 2015.

[LAB 08a] LABOUZE E., DES ABBAYES C., IYAMA S. *et al.*, "Impacts of information and communication technologies on energy efficiency", EU-DG INFSO, 2008.

[LAB 08b] LABOUZE E., DES ABBAYES C., IYAMA S. *et al.*, "Bio intelligence SERVICE, Impacts of Information and Communication Technologies on Energy Efficiency, Final Report", available at: http://www.ifap.ru/library/book373.pdf or the preceeding http://, 2008.

[LAL 14] LALY J.M., "Premiers pas dans l'impression 3D: Cas pratique n°1", available at: http://www.3dnatives.com/impression-3d-cas-pratique-1/, 2014.

[LAL 15] LALLEMENT M., "L'âge du faire. Hacking, travail, anarchie", in *La couleur des idées*, Le Seuil, Paris, 2015.

[LAM 15] LAMIGEON V., "L'impression 3D, nouvelle révolution industrielle", available at: http://www.challenges.fr/challenges-soir/20150908.CHA9140/l-impression-3d-nouvelle-revolution-industrielle.html, 2015.

[LAN 75] LANDES D.S., *L'Europe technicienne : Révolution technique et libre essor industriel en Europe occidentale de 1750 à nos jours*, Gallimard, Paris, 1975.

[LAU 05] LAURENT L., "Nanotechnologies et technologies convergentes: quel avenir à vingt ans?", Le Banquet, available at: http://www.revue-lebanquet.com/nanotechnologies-et-technologies-convergentes-quel-avenir-a-vingt-ans/, 2005.

[LAV 16a] LAVIGNE C., "Modèles numériques dans l'art et la Nature", available at: http://www.arsmathematica.org/IS2016-PDV/index.html, 2016.

[LAV 16b] LAVIGNE C., Personal communication, 2016.

[LE 12a] LE BOURHIS F., KERBRAT O., HASCOËT J.Y. *et al.*, "Évaluation et modélisation des impacts environnementaux en fabrication directe, application à la projection de poudre", *Proceedings of AEPR'12, 17th European Forum on Rapid Prototyping and Manufacturing*, 12–14 June, Paris, available at: https://hal.archives-ouvertes.fr/hal-00844885/document, 2012.

[LE 12b] LE MAREC J., FAURIE M., "Communication et réflexivité dans l'enquête par des chercheurs sur des chercheurs", *Symposium Proceedings, "Réflexivité en contexte de diversité: un carrefour des sciences humaines"*, Limoges (01/02-12-2010), available at: https://halshs.archives-ouvertes.fr/hal-00671219/document, 2012.

[LE 13a] LE BOURHIS F., KERBRAT O., HASCOËT J.Y. *et al.*, "Sustainable manufacturing: evaluation and modeling of environmental impacts in additive manufacturing", *International Journal of Advanced Manufacturing Technology*, vol. 69, pp. 1927–1939, 2013.

[LE 13b] LE STANC C., "Impression en 3D et Propriété Intellectuelle", *Propriété Industrielle*, pp. 1466–1467, 2013.

[LE 14] LE BOURHIS F., Modélisation de la performance environnementale des procédés de fabrication par ajout de matière; application à la projection de poudres métalliques, Thesis, Ecole Centrale de Nantes, Nantes, France, 2014.

[LE 15] LE GOFFIC C., "Contrefaçon dans le cadre de l'impression 3D, responsabilités et remèdes", *La propriété intellectuelle*, pp. 227–239, 2015.

[LE 16] LE GOFF T., "IA et propriété intellectuelle, un enjeu clef du 21ème siècle", available at: http://www.journaldunet.com/economie/expert/65903/ia-et-propriete-intellectuelle--un-enjeux-clef-du-21eme-siecle.shtml, 2016.

[LEC 17] LECOMPTE-BOINET G., "L'aéronautique parée pour basculer dans la fabrication additive", available at: http://www.journal-aviation.com/actualites/36093-l-aeronautique-paree-pour-basculer-dans-la-fabrication-additive, 2017.

[LES 15] LES ECHOS (BOURSE), "Graphique historique", available at: http://bourse.lesechos.fr/bourse/details/graphique_histo.jsp?code=US88554D2053&place=XNYS&codif=ISIN, 2015.

[LES 16] LESAGE S.E., HURTEAU P., "L'innovation en recherche universitaire: Austérité et arrimage aux intérêts du secteur privé", available at: http://iris-recherche.s3.amazonaws.com/uploads/publication/file/web_NouveauTexteEtude_RechercheUniversitaire_vli_02.pdf, 2016.

[LEV 04] LEVY F., MUMANE R.J., *The New Division of Labor: How Computers are Creating the Next Job Market*, Princeton University Press, 2004.

[LIN 12] LINDEMANN C., JAHNKE U., MOI M. *et al.*, "Analyzing product lifecycle costs for a better understanding of cost drivers in additive manufacturing", *Conference Paper: Solid Freeform Fabrication Symposium 2012*, Austin, Texas, 2012.

[LIP 03] LIPOVETSKY G., "Luxe éternel, luxe émotionnel", in LIPOVETSKY G., ROUX E. (eds), *Le luxe éternel; de l'âge du sacré au temps des marques*, Gallimard, Paris, 2003.

[LIP 06] LIPOVETSKY G., *Le bonheur paradoxal; essai sur la société d'hyper-consommation*, Gallimard, Paris, 2006.

[LON 84] LONG W., MCMULLAN W.E., "Mapping the new venture opportunity identification process", in HORNADAY D.J., TARPLEY F., TIMMONS J. *et al.* (eds), *Frontiers of Entrepreneurship Research*, Wellesley, 1984.

[LOY 14] LOY J., "eLearning and eMaking: 3D Printing Blurring the Digital and the Physical", *Education Sciences*, vol. 4, pp. 108–121, 2014. doi: 10.3390/educsci4010108.

[LUB 05] LUBART T., *Psychologie de la créativité*, Armand Colin, Paris, 2005.

[LUB 16] LUBINEAU P., "La Fabrication Additive en France: cartographie des acteurs", available at: http://aif.diwi.org/wp-content/uploads/2016/07/20160719-Etude-Carnot-FIMP-Fab-Add-Cartographie-des-compétences-Diff-AIF.pdf, 2016.

[MAD 15] MADDYNESS, "Prospective: Impression 3D; état des lieux, chiffres clés et perspectives de la fabrication additive", available at: http://www.maddyness.com/prospective/2015/05/26/impression-3d-additive/, 2015.

[MAD 17] MADEELI, "Fabrication Additive : Enjeux et Impacts en Occitanie Pyrénées-Méditerranée", available at: http://www.madeeli.fr/wp-content/uploads/2017/01/Fabrication-Additive-en-Occitanie-Etude-2016.pdf, 2017.

[MAI 16] MAILLARD D., "Le consommateur, voilà l'ennemi!", available at: http://www.metiseurope.eu/le-consommateur-voil-l-ennemi-thinsp_fr_70_art_30468.html, 2016.

[MAK 15] MAKERSHOP, "L'achat d'une imprimante 3D est-il rentable face aux services d'impression?", available at: http://www.makershop.fr/blog/rentabilite-imprimante-3d-rentable-service-impression/, 2015.

[MAN 13] MANYLKA J., CHUI M., BUGHIN J. *et al.*, "Disruptive technologies: advances that will transform life, business, and the global economy", vol. 180, Mac Kinsey Global Institute, CAL, USA, 2013.

[MAR 15] MARKETS AND MARKETS, "4D Printing Market worth $537.8 Million by 2025", available at: http://www.marketsandmarkets.com/PressReleases/4d-printing.asp, 2015.

[MAR 17] MARTI J.F., "Design Thinking: un outil qui favorise l'innovation et remet en cause les certitudes dans l'entreprise", available at: http://www.hbrfrance.fr/chroniques-experts/2017/01/13811-design-thinking-un-outil-qui-favorise-linnovation-et-remet-en-cause-les-certitudes-dans-lentreprise/, 2017.

[MCG 10] McGRAIL S., "Nanodreams and nightmares: emerging technoscience and the (re)interpreting the future, present and past", *Journal of Future Studies*, vol. 14, pp. 23–48, 2010.

[MCK 13] McKINSEY GLOBAL INSTITUTE, *Disruptive Technologies: Advances That Will Transform Life, Business and the Global Economy*, McKinsey Global Institute & Company, Seoul, 2013.

[MÉA 09] MÉADEL C., "La réception", *Revue Hermès – CNRS*, Paris, 2009.

[MEL 16] MELLIN P., JÖNSSON C., ÅKERMOA M. *et al.*, "Nano-sized by-products from metal 3D printing, composite manufacturing and fabric production", *Journal of Cleaner Production*, vol. 139, pp. 1224–1233, 2016.

[MÉL 17a] MELANIE, "L'impression 3D, futur eldorado du marché du travail?", available at: http://www.3dnatives.com/metiers-impression-3d-25012017/, 2017.

[MÉL 17b] MELANIE, "Zimple, l'impression 3D sans émissions toxiques", available at: http://www.3dnatives.com/zimple-interview-140320173/, 2017.

[MÉL 17c] MELANIE, "Donald Trump veut interdire l'impression 3D aux Etats-Unis!", available at: http://www.3dnatives.com/trump-impression-3d-01042017/, 2017.

[MER 15] MERLOT F., MAZZONI S., "Gas evolution during FDM 3D printing and health impact", available at: http://www.3dsafety.org/3dsafety/download/mf2015_eng.pdf, 2015.

[MIC 09] MICHEAU Y., *Qu'est-ce que le mérite?*, Gallimard, Paris, 2009.

[MIR 16] MIRZENDEHDEL A.M., SUREH K., "Support structure constrained topology optimization for additive manufacturing", *Computer-Aided Design*, vol. 81, pp. 1–13, 2016.

[MIT 12] MIT, "The Fab Charter", available at: http://fab.cba.mit.edu/about/charter/, 2012.

[MOL 16] MOLITCH-HOU M., "Tend.ai proposes cloud robotics for 3D print farm automation", available at: http://www.engineering.com/3DPrinting/3DPrintingArticles/ArticleID/12500/Tendai-Proposes-Cloud-Robotics-for-3D-Print-Farm-Automation.aspx, 2016.

[MOR 07] MORROW W.R., QI H., KIM I. *et al.*, "Environmental aspects of laser-based and conventional tool and die manufacturing", *Journal of Cleaner Production*, vol. 15, p. 932e943, 2007.

[MOR 12] MORANGE M., *Les secrets du vivant; contre la pensée unique en biologie*, La Découverte, Paris, 2012.

[MOR 14] MORA A., "L'impression 3D: L'industrie du futur est en marche", *Management*, vol. 217, pp. 64–71, March 2014.

[MOU 15] MOUSSET L., "L'impression 3D entre fiction et réalité", available at: http://information.tv5monde.com/info/l-impression-3d-entre-fiction-et-realite-48944, 2015.

[NAT 17] NATIONAL INTELLIGENCE COUNCIL, "Global Trends Report – The Paradox of Progress", available at: https://www.dni.gov/files/images/globalTrends/documents/GT-Full-Report.pdf, 2017.

[NEE 16] NEELY E.L., "The risks of revolution: ethical dilemmas in 3D printing", *Science and Engineering Ethics*, vol. 22, pp. 1285–1297, 2016.

[NEU 86] NEUMAN W.R., "Paradox of mass politics – knowledge and opinion in the American electorate", Harvard University Press, Cambridge, 1986.

[NEW 12] NEW ECONOMIST, "A third-world dimension", available at: http://www.economist.com/news/science-and-technology/21565577-new-manufacturing-technique-could-help-poor-countries-well-rich-ones, 2012.

[NFI 16] NFI, "Nouvelle France Industrielle; Construire l'industrie française du futur", available at: http://proxy-pubminefi.diffusion.finances.gouv.fr/pub/document/18/20960.pdf, 2016.

[NIE 15] NIEDERCORN F., "La déferlante de l'impression 3D questionne la propriété intellectuelle", available at: http://www.lesechos.fr/thema/brevet-marque/021116764893-limpression-3d-cauchemar-de-la-propriete-intellectuelle-1127666.php, 2015.

[NOB 84/11] NOBLE D.F., *Forces of Production: a Social History of Industrial Automation*, Transaction Publishers, Piscataway, 1984/2011.

[NOR 04] NORDMANN A., *Converging Technologies: Shaping the Future of European Societies*, European Commission, Brussels, 2004.

[NOT 14] NOTTEAU Y., "Histoire de l'impression 3D", available at: http://www.additiverse.com/actualites/2014-09-29/histoire-de-l-impression-3d, 2014.

[OBA 16] OBATON A.F., "La maîtrise du processus de fabrication additive est un véritable enjeu pour les industriels", available at: http://www.usinenouvelle.com/article/tribune-la-maitrise-du-processus-de-fabrication-additive-est-un-veritable-enjeu-pour-les-industriels.N385082, 2016.

[OBS 17] OBSERVATOIRE DE LA METALLURGIE, "L'élaboration et la transformation des métaux par forge, fonderie et fabrication additive métallique – Rapport complet, 2017", available at: http://www.observatoire-metallurgie.fr/sites/default/files/uimm_documents/etude_metallurgie_2017.pdf, 2017.

[OEC 92] OECD, *La technique de l'économie*, OECD, Paris, 1992.

[OEC 16] OECD, "Measuring GDP in a digitalised economy", available at: http://www.oecd-ilibrary.org/economics/measuring-gdp-in-a-digitalised-economy_5jlwqd81d09r-en, 2016.

[OEC 17] OECD, *3D Printing – What the Next Production Revolution Means for the Environment*, LIVEstream, Paris, France, 27 February 2017.

[OLL 15] OLLA P., "Opening Pandora's 3D printed box", *IEEE Technology and Society Magazine*, vol. 34, pp. 74–80, 2015.

[ORO 06] OROZCO E.L., Le processus de transmission des valeurs chez les jeunes: etude comparative de trois configurations colombiennes, Thesis, University of Rennes, France, available at: https://halshs.archives-ouvertes.fr/tel-00199402/document, 2006.

[PAN 13] PANETTA K., "Need something? Print-it; 3D printing in engineering", *Electronic Component News*, vol. 57, pp. 16–19, 2013.

[PAR 99] PARTOUNE C., "L'approche systémique", available at: http://www.lmg.ulg.ac.be/competences/chantier/contenus/cont_syst.html, 1999.

[PAR 15] PARISOT F., "Sous-traitant en impression 3D, une position délicate", available at: http://www.usinenouvelle.com/article/sous-traitant-en-impression-3d-une-position-delicate.N350845, 2015.

[PAR 16a] PARISOT F., "Impression 3D: baisse d'intérêt du grand public", available at: http://www.usine-digitale.fr/article/impression-3d-baisse-d-interet-du-grand-public.N375797, 2016.

[PAR 16b] PARISOT F., DENIEL P., "Impression 3D, la discrète révolution", available at: http://www.usinenouvelle.com/article/impression-3d-la-discrete-revolution.N375791, 2016.

[PEC 17] PECKHAM T.K., BAKER M.G., CAMP J.E. *et al.*, "Creating a Future for Occupational Health", *Annals of Work Exposures and Health*, vol. 61, pp. 3–15, available at: https://academic.oup.com/annweh/article/61/1/3/2762736/Creating-a-Future-for-Occupational-Health, 2017.

[PET 13] PETRICK I.J., SIMPSON T.W., "3D printing disrupts manufacturing: how economies of one create new rules of competition", *Research-Technology Management*, vol. 56, pp. 12–16, 2013.

[PIP 17] PIPAME, *Prospective – Futur de la fabrication additive*, DGE, Paris, available at: http://www.entreprises.gouv.fr/etudes-et-statistiques/futur-fabrication-additive-pipame, 2017.

[PLA 15] PLATT J.R., "Thirty-five percent of engineering jobs now require 3-D printing skills", available at: http://theinstitute.ieee.org/career-and-education/career-guidance/thirtyfive-percent-of-engineering-jobs-now-require-3d-printing-skills, 2015.

[POR 09] PORTER A.L., YOUTIE J., "How interdisciplinary is nanotechnology?", *Journal of Nanoparticle Research*, vol. 11, pp. 1023–1041, 2009.

[POT 14] POTSTADA M., ZYBURA J., "The role of context in science fiction prototyping: the digital inustrial revolution", *Technological Forecasting and Social Change*, vol. 84, pp. 101–114, 2014.

[POU 06] POUTEAU S., "Biotechnologies: la vie en morceaux et l'éthique en péril", *Journal International de Bioéthique*, vol. 17, pp. 15–27, 2006.

[POU 14] POUZOT R., BOURGUET F., JOLLY L., "La protection des impressions 3D dans les différents domaines de la propriété intellectuelle", available at: http://www.lagbd.org/index.php/La_protection_des_impressions_3D_dans_les_diff%C3%A9rents_domaines_de_la_propri%C3%A9t%C3%A9_intellectuelle_(fr), 2014.

[PRI 17] PRIARONE P.C., INGARAO G., "Towards criteria for sustainable process selection: on the modelling of pure subtractive versus additive/subtractive integrated manufacturing approaches", *Journal of Cleaner Production*, vol. 144, pp. 57–68, 2017.

[PUE 08] PUECH M., *Homo-sapiens technologicus: philosophie de la technologie contemporaine; philosophie de la sagesse contemporaine*, Le Pommier, Paris, 2008.

[PWC 16] PwC, "Tech breakthroughs megatrend – How to prepare for the technological breakthroughs megatrend, and the eight technologies to start with", available at: http://www.pwchk.com/webmedia/doc/636081726468991499_tech_megatrend_aug2016.pdf, 2016.

[QUO 14] QUONIAM L., REYMOND D., "Etat de l'art des brevets existants sur l'impression 3D", *Le Club Reverse Engineering*, Marseille, France, 2014.

[RAP 16] RAPOPORT D., "Trans-humanisme: à qui appartient notre corps?", available at: http://www.lesechos.fr/idees-debats/cercle/cercle-158086-transhumanisme-a-qui-appartient-notre-corps-2008706.php?ezOL9P149ZO6F8P2.99, 2016.

[RAT 12] RATTO M., REE R., "Materializing information: 3D printing ans social change", First Monday, 17, available at: http://firstmonday.org/ojs/index.php/fm/article/view/3968/3273, 2012.

[RAY 16] RAYNAL J., "L'impression 3D: un marché mondial à plus de 35 milliards de dollars en 2020", available at: http://www.industrie-techno.com/l-impression-3d-un-marche-mondial-a-plus-de-35-milliards-de-dollars-en-2020.45234, 2016.

[REC 16] RECROSIO E., "Know your CAD files: Nine Facts about STL", available at: https://www.sculpteo.com/blog/2016/11/16/know-your-cad-files-nine-facts-about-stl/?utm_medium=email&utm_campaign=NL%20US%2016112016&utm_content=NL%20US%2016112016+CID_51f638db7a2628bf11206172ae2d3f2a&utm_source=Campaign%20Newsletter&utm_term=Learn%20More, 2016.

[REE 08a] REEVES P., "Additive Manufacturing – A Supply Chain Wide Respons to Economic Uncertainty and Environmental Sustainability", available at: http://www.econolyst.co.uk/resources/documents/files/Paper___2008__AM_a_supply_ chain_wide_response.pdf, 2008.

[REE 08b] REEVES P., "How the Socio-economic benefits of rapid manufacturing can be used to off-setthe tchnological limitations", Additive Manufacturing Platform Europe, available at: http://www.rm-platform.com/index2.php?option=com_doc man&task=doc_view&gid=626&Itemid=5, 2008.

[REE 12a] REEVES P., "Example of Econolyst Research – Understanding the Benefits of AM on CO_2", available at: http://www.econolyst.co.uk/reso- urces/documents/files/Presentation_2012_AM_and_carbon_footprint.pdf, 2012.

[REE 12b] REEVES P., "Putting 3D printing into value stream: opportunities for new business models", available at: http://www.econolyst.co.uk/resources/documents/files/Presentation%20-%20Oct%202012%20-%203D%20Printshow%20London%20UK%20-%20putting%203D%20Printing%20into%20the%20value%20stream.pdf, 2012.

[REE 13] REEVES P., "3D Printing in Aerospace and Automotive" http://www.econolyst.co.uk/resources/documents/files/Inside%203D% 20printing%20industry%20conference%20New%20York%20-%203D%20Printing %20in%20aerospace%20&%20automotive.pdf, 2013.

[REG 12] REGALADO A., CUNNING N.M., "Plan to Prevent 3-D Printer Piracy" *MIT Technology Review*, available at: http://www.technologyreview.com/view/429566/nathan-myhrvolds-cunningplan-to-prevent-3-d-printer-piracy/, 2012.

[REN 14] RENAN E., *L'Avenir de la science – Pensées de 1848*, Flammarion, Paris, 2014.

[RES 16] RESEAU FRANÇAIS DES FAB LABS, available at: http://www.fablab.fr/, 2016.

[RIC 15] RICHARDIN A., "Innovation: 55% des français sont en attente de produits rupturistes", available at: http://www.maddyness.com/prospective/2015/09/09/innovation-mediacom/, 2015.

[RID 12] RIDEOUT B., "Printing the impossible triangle: the copyright; implications of three-dimensional printing", *Journal of Business, Entrepreneurship & the Law*, vol. 161, pp. 163–64, 2012.

[ROC 06] ROCHER S., MATHE J.C., "Contrôler la rupture technologique pour rester leader: une stratégie gagnante? Une illustration au travers du marché des consoles de jeux vidéo", *Management & Avenir*, vol. 4, pp. 9–24, 2006.

[ROD 16] RODOLPHE, "L'optimisation topologique pour l'impression 3D", available at: http://www.3dnatives.com/optimisation-topologique-3d-07032016/, 2016.

[RUA 17] RUANO-BORBALAN J.C., "Techno-sciences en société: les voies multiples de la légitimation des savoirs", *Innovations*, vol. 1/2017, pp. 5–15, 2017.

[RUM 14] RUMPALA Y., "'Fab labs', 'makerspaces': entre innovation et émancipation?", *Revue Internationale de l'Economie Sociale*, vol. 334, pp. 85–97, 2014.

[SAA 15] SAAM C., "La fabrique du futur", available at: http://www.patentattorneys.ch/2015/10/17/la-fabrique-du-futur/, 2015.

[SAD 13] SADIN E., *L'humanité augmentée: L'administration numérique du monde*, L'échappée, Paris, 2013.

[SCH 99] SCHUMPETER J.A., *Théorie de l'évolution économique*, Dalloz, Paris, 1999.

[SCU 15a] SCULPTEO, "Apprendre l'impression 3D", Elephorm, available at: www.elephorm.com, 2015.

[SCU 15b] SCULPTEO, "Your guide to the 4 areas where 3D printing is unstoppable", available at: http://www.sculpteo.com/en/get/ebook/unstoppable_3D_printing/, 2015.

[SCU 15c] SCULPTEO, "3D printing and the tiny house movement in the US", available at: http://www.sculpteo.com/blog/2015/08/11/3d-printing-and-the-tiny-house-movement-in-the-us/?utm_medium=email&utm_campaign=Newsletter+12082015+US&utm_content=Newsletter+12082015+US+CID_66d4c457b681ed42b4dd366d94b6d4ce&utm_source=Campaign%20Newsletter, 2015.

[SCU 16a] SCULPTEO, "STL (Standard Tessellation Language): Format de fichier pour l'impression 3D", available at: https://www.sculpteo.com/fr/glossaire/stl-definition-fr/, 2016.

[SCU 16b] SCULPTEO, "The state of 3D printing 2015 vs 2016", available at: http://www.sculpteo.com/blog/2016/02/17/the-state-of-3d-printing-2015-vs-2016/?utm_medium=email&utm_campaign=Newsletter%202172016%20US&utm_content=Newsletter%202172016%20US+CID_6038000b1c0963ea3c725ccb68a200a5&utm_source=Campaign%20Newsletter&utm_term=Predictions%20for%20the%20State%20of%203D%20Printing%20Report%202016, 2016.

[SCU 16c] SCULPTEO, "The state of 3D printing", available at: http://www.sculpteo.com/static/0.30.0-64/download/report/Sculpteo_State_of_3D_Printing.pdf, 2016.

[SCU 17] SCULPTEO, "3D Printing Tutorials: Prepare your model for 3D printing", available at: https://www.sculpteo.com/en/tutorial/?utm_medium=email&utm_campaign=NL%20US%2008022017&utm_content=NL%20US%2008022017+CID_5fe54d50af7de4f4aecb018ef5ec90b1&utm_source=Campaign%20Newsletter&utm_term=Learn%20More, 2017.

[SER 16a] SERRAT D., "En quoi la révolution digitale est-elle une nouvelle révolution?", available at: http://theconversation.com/en-quoi-la-revolution-digitale-est-elle-une-nouvelle-revolution-62395, 2016.

[SER 16b] SERVICE CANADIEN DU RENSEIGNEMENT DE SECURITE, "Perspectives sécuritaires 2018 – Risques et menaces éventuels", available at: https://www.csis-scrs.gc.ca/pblctns/ccsnlpprs/2016/2016-06-03/GLOBAL_SECURITY_POST-CONFERENCE_FRANCAIS.pdf, 2016.

[SÖD 13a] SÖDERBERG J., "Automating amateurs in the 3D printing community: connecting the dots between 'deskilling' and 'user-friendliness'", *Work Organisation, Labour & Globalisation*, vol. 7, pp. 124–139, 2013.

[SÖD 13b] SÖDERBERG J., "Illusoire émancipation par la technologie", *Le Monde Diplomatique*, available at: https://id-libre.org/file/presse/articles%20divers/MonDiplo_2013_01_p._3_imprimante_3D.pdf, 2013.

[STA 08] STAR S.L., GRIESEMER J.R., "Ecologie institutionnelle, traductions et objets-frontières: des amateurs et des professionnels au musée de zoologie vertébrée de Berkeley 1907–1939", in LAHIRE B., ROSENTAL C. (eds), *La cognition au prisme des sciences sociales*, Ed. des Archives Contemporaines, Paris, 2008.

[STA 97] STAHEL W.R., *The Functional Economy: Cultural and Organizational Change*, in NATIONAL ACADEMY OF ENGINEERING (ed.), *The Industrial Game: Implications for Environmental Design and Management*, The National Academies Press, 1997.

[SUC 16] SUCHAUD N., "Prospective: Quel impact l'innovation numérique peut-elle vraiment avoir sur l'économie?", available at: http://www.maddyness.com/prospective/2016/01/04/g9plus-idate-innovation-economie/, 2016.

[SYS 16] SYS C., "Scandale: Yahoo! espionne les comptes de ses utilisateurs depuis 2015", available at: http://dailygeekshow.com/yahoo-espionnage-mails/, 2016.

[TAG 01] TAGUIEFF P.A., *Resister au bougisme*, Mille et Une Nuits, Paris, 2001.

[TAN 16] TANCOGNE-DEJEAN T., SPIERINGS A.B., MOHR D., "Additively-manufactured metallic micro-lattice materials for high specific energy absorption under static and dynamic loading", *Acta Materialia*, vol. 116, pp. 14–28, 2016.

[TAY 14] TAYLOR S., "Trend evolution: 3D printing trends", available at: http://3dprintingindustry.com/2014/02/24/trend-evolution-3d-printing-trends-part-1/, 2014.

[TCH 17] TCHIEHE D.N., GAUTHIER F., "Classification of risk acceptability and risk tolerability factors in occupational health and safety", *Safety Science*, vol. 92, pp. 138–147, 2017.

[TEC 15] TECHNOPOLIS, "Regional Innovation Monitor Plus 2015 - Mapping advanced manufacturing networks and exploring new business opportunities", available at: https://ec.europa.eu/growth/tools-databases/regional-innovation-monitor/sites/default/files/report/RIM%20Plus_Mapping%20Advanced%20Manufacturing%20Networks_Thematic%20Paper.pdf, 2015.

[TER 16] TERRASSE P., BARBEZIEUX P., HERODY C., "Rapport au premier ministre sur l'économie collaborative" available at: http://www.gouvernement.fr/sites/default/files/liseuse/6421/master/projet/Rapport-de-Pascal-Terrasse-sur-le-développement-de-l-économie-collaborative.pdf, 2016.

[THO 16] THOMPSON M.K., MORONI G., VANEKER T. *et al.*, "Design for additive manufacturing: trends, opportunities, considerations, and constraints", *CIRP Annals – Manufacturing Technology*, vol. 65, pp. 737–760, 2016.

[TIN 94] TINLAND F., "Le site de la technique; éclairages théoriques et enjeux pratiques", in TINLAND F. (ed.), *Ordre biologique; ordre technologique*, Champ-Vallon, Paris, 1994.

[TOV 16] TOVAR I., "In a circular economy 3D printing can help unlock $5bn of waste", available at: https://3dprintingindustry.com/news/circular-economy-3d-printing-can-help-unlock-5bn-waste-97792/, 2016.

[TRA 16] TRAVAIL ET SECURITE, *L'industrie de la plasturgie*, INRS, Paris, available at: http://www.travail-et-securite.fr/ts/dossier.html, 2016.

[TUR 15] TURKLE S., *Seuls ensemble; de plus en plus de technologies; de moins en moins de relations humaines*, Ed. de l'Echappée, Paris, 2015.

[UK 13] UK INTELLECTUAL PROPERTY OFFICE PATENT INFORMATICS TEAM, "3D printing industry: a patent overview", available at: https://www.gov.uk/government/uploads/system/uploads/attachment_data/file/445232/3D_Printing_Report.pdf, 2013.

[VER 15] VERHULST J., Analyse de l'état et de l'avenir du marché de l'impression 3D, Masters' Thesis, Université Catholique de Louvain, Belgium available at: http://dial.uclouvain.be/memoire/ucl/en/object/thesis%3A2710/datastream/PDF_01/view, 2015.

[VER 16] VERDO Y., "Une société malade de sa peur du risque ?", available at: http://www.lesechos.fr/idees-debats/sciences-prospective/0211040768661-une-societe-malade-de-sa-peur-du-risque-2007473.php?HQPiR300MKKzqWSB.99, 2016.

[VIC 16a] VICTOR, "La bio-impression: Un marché en croissance de 36% d'ici à 2022", available at: http://www.3dnatives.com/bio-impression-marche-15112016/, 2016.

[VIC 16b] VICTOR, "SOFIA, le plan français dédié à la fabrication additive métal", available at: http://www.3dnatives.com/sofia-fabrication-additive-29112016/, 2016.

[VIN 00] VINSONNEAU G., *Culture et comportement*, Armand Colin, Paris, 2000.

[VOL 15] VOLLE M., "Du virtuel au réel et vice versa", available at: http://www.volle.com/opinion/dedale.htm, 2015.

[VUE 16] VUES D'ENSEMBLE, "Dossier prospective: penser l'avenir", available at: http://www.univ-catholille.fr/ebooks/VE-n65-oct2016/files/publication.pdf, 2016.

[WAN 14] WANTED ANALYTICS, "Demand For 3D Printing Skills Soars", available at: http://3dprintingfromscratch.com/2014/09/wanted-analytics-demand-for-3d-printing-skills-soars/, 2014.

[WEL 15] WELLER C., KLEER R., PILLER F., "Economic implications of 3D printing: market structure models in light of additive manufacturing revisited", *International Journal of Production Economics*, vol. 164, pp. 43–56, 2015.

[WEN 15] WENT R., KREMER M., KNOTTNERUS A., "Netherlands Scientific Council for Government Policy – Mastering the Robot – The Future of Work in the Second Machine Age", available at: http://test6.wrr.nl/fileadmin/en/publicaties/PDF-Verkenningen/Mastering_the_Robot_Web.pdf, 2015.

[WIK 16] WIKIPÉDIA, "Fab lab", available at: https://fr.wikipedia.org/wiki/Fab_lab, 2016.

[WIT 13] WITTBRODT B.T., GLOVER A.G., LAURETO J. *et al.*, "Life-cycle economic analysis of distributed manufacturing with open-source 3-D printers", *Mechatronics*, vol. 23, pp. 713–726, 2013.

[WOR 17] WORLD ECONOMIC FORUM, "The Global Risks Report 2017", available at: http://www3.weforum.org/docs/GRR17_Report_web.pdf, 2017.

[XER 16] XERFI, "L'impression 3D en France et dans le monde", available at: http://www.xerfi.com/presentationetude/limpression-3D-en-France-et-dans-le-monde_6EEE25, 2016.

[ZON 16a] ZONE BOURSE, "3D Systems Corporation", available at: http://www.zonebourse.com/3D-SYSTEMS-CORPORATION-8053422, 2016.

[ZON 16b] ZONE BOURSE, "Stratasys Ltd" available at: http://www.zonebourse.com/STRATASYS-LTD-12111547/, 2016.

PART 2

3D Processes

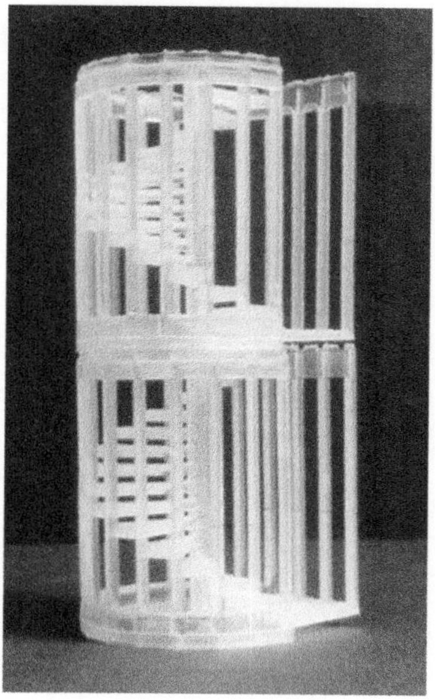

Introduction to Part 2

"The scientific process tends to assemble all the parts it has at its disposal at a given moment. As new parts appear endlessly, this leads to frequent challenges". [DEP 86]

"It is more difficult to escape the attraction of our myths than the universal attraction". [PIC 09]

"A man who tackles science with the perspective of a police inspector will spend the most of his time thwarting tricks that would have never been played on him, would follow suspects who would voluntarily respond to a direct question, and, in a general manner, would play the game of the constables and thieves, such as it is played now in the realm of official and military science. I have not the slightest doubt that the state of mind, worthy of an inspector, presented by the barons of scientific administration is one of the major reasons for the sterility of so much scientific research today". [WIE 14]

"Individuals who are just as creative but without a network or motivation make little concrete. These results greatly reduce the interest that should be given to the individual creativity. This, if it is not framed in a motivational organizational device, risks being quite unproductive in the end. Furthermore, individualist readings, considering that it is the individual who is creative, are substituted by organizational readings that are interested in the management of creativity in a collective framework". [CHA 14]

> *"For discipline to be born, the intellectual conditions are not sufficient. It is necessary, for example, for 'candidate' research that will become the center of a new discipline to be at the crossroads of various social groups, in such a way that, faced with these, they are led to assume a certain autonomy". [FOU 96]*

> *"For Kuhn, in science, change – from one paradigm to another – is a mystical conversion that is not governed by the laws of reason and cannot be, and it comes entirely from the realm of the (social) psychology of the discovery. Change in science is a sort of religious change". [LAK 94]*

> *"If the centers, the orthodox, the continuers try to monopolize positions in the university institution, the heterodox, on the other hand, believe in new trajectories beyond disciplines". [BOU 01]*

> *"The procedure ends up taking precedence over the objective, the mission is forgotten in favor of the rule. Our societies are thus being disincarnated, and the reality of social and political work, as with the mind and the values that guide it, are masked in favor of the method". [BEI 12]*

> *"Modeling is the main thing, and maybe the only tool at our disposal to study the behavior of the complex systems that we must consider". [SIM 13]*

> *"The calculations dissimulate the ignorance". [BOU 13]*

> *"Modeling and interdisciplinarity are, and particularly become, crucial subjects of scientific practice on the heuristic level as on that of its social utility". [MAT 14]*

In a random way, technological choices result from the happenstance of history, associated with creativity on the one hand, allowing the emergence of new technical systems, and the existence of a potential application market, conditioned by the cultural models and technological habits of society, on the other [BUN 83]. This creativity, for the most part, does not set itself apart in its premises of a fine distinction between ends and means. It is only when a demonstrator is realized that applicational ends and, later, certain mis-appropriations can be imagined. However, once the ends of a technology are, for a time, stabilized, the choice of means is no longer indifferent, a function of the options retained. These are not decided at the moment of emergence of the generic idea. Thus, we enter into the space of incremental, rather boxed in, innovation, without evident breakthroughs.

These different steps recorded and observed for more than 30 years (see Part 1 of this volume) highlight breaks to innovation that are not specific to additive manufacturing. According to Miraton [MIR 13], "Innovation requires [...] something unreasonable: wanting to create that which does not exist. It demands fighting against all of those convinced that we could not take the risk of changing a product that is selling well or who think that everything possible has already been imagined". However, for all that, the power of information and communication technologies is such that partner technologies, like additive manufacturing, which are likely to respond to aspirations of industrial and home trends, have finally managed to emerge. They even allow forms of recomposition of certain cognitive functions and production activities to be foreseen (see Part 1 and Giorgini, 2014).

The initial model system that consists in bringing and transforming matter locally has illustrated a conceptual foundation, very simple at the end of the day. As a function of the possible, in the current evolutions that will be described in Chapter 3, it is shown that the surplus of descriptive content for each additive manufacturing technology remains minimal in relation to that which is indispensible to advancement (maximal empirical shoring according to Nicolescu, 2002). There is a multitude of viewpoints, although with successive coherence arrangements stemming from incremental evolutions supported by industrial demand, in terms of both processes and materials. These evolutions are the objectives of Chapter 1 in Volume 2.

Tulips [GOR 17] - Copyright ©Prodways ©Hamilton de Oliveira

Bibliography

[BEI 12] BEIGBEGER C., *Puisque c'est impossible, faisons-le*, J.C. Lattès, Paris, 2012.

[BOU 13] BOULEAU N., "L'excessive mathématisation; symptomes et enjeux", in BOURG D., JOLY P.B., KAUFMAN A., *Du risque à la menace; penser la catastrophe*, PUF, Paris, 2013.

[BOU 01] BOURDIEU P., *Science de la science et réflexivité*, Raisons d'Agir, Paris, 2001.

[BUN 83] BUNGE M., *Epistémologie*, Maloine, Paris, 1983.

[CHA 14] CHANUT-GUIEU C., GUIEU G., "Créativité, PME et entrepreneuriat: des zones d'ombre et de lumière", *Revue internationale P.M.E.: économie et gestion de la petite et moyenne entreprise*, vol. 27, pp. 35–63, 2014.

[DEP 86] DE PRACONTAL M., *L'imposture scientifique en dix leçons*, La Découverte, Paris, 1986.

[FOU 96] FOUREZ G., *La construction des sciences*, De Boeck, Brussels, 1996.

[GIO 14] GIORGINI P., *La transition fulgurante; vers un bouleversement systémique du monde*, Bayard, Paris, 2014.

[LAK 94] LAKATOS I., *Histoire et méthodologie des sciences*, PUF, Paris, 1994.

[MAT 14] MATHIEU N., SCHMID A.F., "Reconsidérer le lien entre modélisation et interdisciplinarité", in MATHIEU N., SCHMID A.F. (eds), *Modélisation et interdisciplinarité; six disciplines en quête d'épistémologie*, Quae, Paris, 2014.

[MIR 13] MIRATON D., "Culture de l'innovation et éducation: retour d'expérience d'industriel", in CHEVALLIER-LE GUYADER M.-F. (ed.), *Partager la science; l'illettrisme scientifique en question*, Actes Sud, Paris, 2013.

[NIC 02] NICOLESCU B., "Niveaux de réalité", in D'ESPAGNAT B. (ed.), *Implications philosophiques de la science contemporaine, vol. 2*, PUF, Paris, 2002.

[PIC 09] PICQ P., *Le monde a-t-il été créé en sept jours?*, Champs Science, Paris, 2009.

[SIM 13] SIMON H.S., cited by LE MOIGNE J.L., "Sur l'intelligence de la complexité en entrelaçant épistémologie et pragmatisme", in LE MOIGNE J.L., MORIN E. (eds), *Intelligence de la complexité*, Hermann, Paris, pp. 11–27, 2013.

[WIE 14] WIENER N., *Cybernétique et société*, Le Seuil, Paris, 2014.

3

Processes, Machines and Materials

"The development of technical means is the result of a constantly accumulated collective experience. Each generation continues to inherit experience from all those who preceded it. In the technical field, progress is a summet". [DAU 79]

"Once formed, a perfectly organized system of opinions, closed upon itself, possessing numerous particularities and numerous relations, lasts over time, resisting everything that contradicts it". [FLE 08]

"If systematic technology is an indispensible base, the tool only exists in the gesture that makes it technically efficient". [LER 65]

"Domination of the machine presupposes a society in the final stages of growing entropy, where the probability is negligible and where the statistical differences between individuals are inexistent. We luckily have not reached such a state". [DUB 90]

"The problem that is posed in the reality experienced by some research is not sinking into the intoxication of the technical and of acquisition with endless data, without once again entering into the field of hypothesis and its verification through experimentation". [KOU 14]

"A subtle balance is to be found between creativity and routine to ensure hyper-growth. The notion of balance is often put forth to explain and justify hyper-growth. Managers are aware that they must take the time to find the proper balance to follow their dominant logic. Some managers even believe that they must constantly manage a 'dynamic imbalance'". [CHA 14]

"Chance, far from taking the form of a universal law of Nature, will instead have the allure of a contingent time command, of an arrow of time that can cease not to make sense". [MIQ 00]

"There is colonization of the future that consists in living at its expense, an imperialism of the present absorbs and feeds off of future times... which substitutes the short time with the long term, the immediate with duration". [INN 08]

"This could be called the Sirius paradox: it is futile to remain outside reality and cultivate a certain haughty view of the proper analysis. Nevertheless, by coming too close to reality, this becomes even denser and seems to saturate the horizon with micro-meanings that end up clouding our view and making any analysis impossible". [MEI 14]

"There are also artificial fogs that we take for ramparts; they constitute the principal obstacles, for they block every process of alternative imagination at the root itself". [VIV 12]

"Among all human activities, technology is the only one that never returns to its point of departure". [LER 83]

"The primary difficulty [...] consists in trying to imagine something that you have never seen, which agrees in its most minute details with what has already been seen, and which nevertheless differs from what has already been thought – this, all while formulating a proposition that is not vague, but precise". [FEY 98]

3.1. Introduction

To return to a more scientific and technological standpoint, even to exit from a slightly (too) philosophical approach, concerning additive manufacturing, the observation of nature shows that this tends to be organized into ordered structures. Then, in principle, two options may be at work: the first consisting in mastering a set of physicochemical and energetic systems so that the desired object appears without returning to classical principles of construction (see Volume 2) and the second, more reasonable one, aiming to manufacture an object through the addition of matter (which, from a manufacturing standpoint, is even slightly original, because normally, processing methods are practiced through the elimination of matter).

Although the name "fourth industrial revolution" may perhaps be exaggerated in the current trend, the idea of Industry 4.0 into which additive manufacturing is integrated goes beyond the simple development of industry, furthering automation and integration into the space of information and communication technologies. It could be possible to mention the future fate of industry in the manner of economist Jean-Louis Levet [LEV 12] by emphasizing that it does not disappear, but that it transforms and that its borders become vaster. The progress of digital technologies and their availability at reasonable costs actually have the potential to make industrial production even more flexible and more agile, to increase its productivity and to develop new offers with a high added value [KOL 15], with more supple production infrastructures, capable of reacting practically in real time to changes (demand, providers, repairs, etc.), the ability to manufacture personalized or customized products on a large scale and in the most economic way possible, the development of new business models (marketing, sales, service, exploitation, entertainment) and supported technological innovation.

Nevertheless, as pointed out by OECD [OEC 16], by implying research that is the object of Volume 2, and as was recalled in the previous chapters, additive manufacturing is part of the 10 key presently emerging technologies. Figure 3.1, from this report from the Organization, nevertheless presents additive manufacturing on an axis corresponding to advanced materials, while logic would wish for it to be placed in the applications of the digital. Furthermore, underlined in red are the application domains that go far beyond the "materials" component, where they advanced. As will be shown in Volume 3, the "biotechnologies" component emerges significantly alongside more classical applications in the domains of robotics, drones and so on.

This situation in fact places additive manufacturing as a generic technology that, in this figure, could have better been placed at its center (see orange arrow in Figure 3.1) even more so as the set of application niches that are drawn are not yet stabilized.

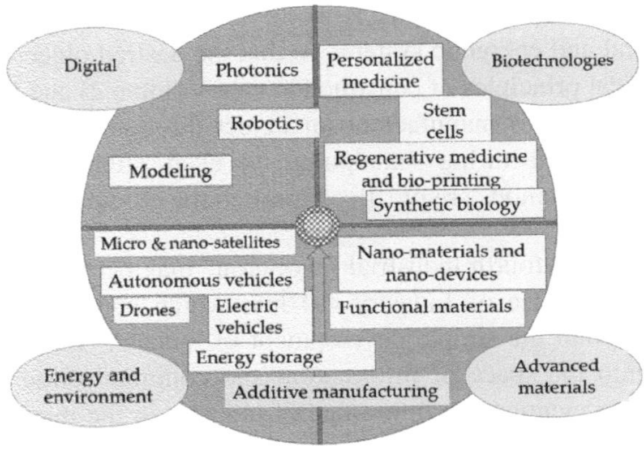

Figure 3.1. *Position of additive manufacturing in the priorities of emerging technologies (only the domains concerned with additive manufacturing are presented) (according to [OEC 16])*

However, between 20 and 30 years have been absolutely necessary to achieve this renowned position, now accepted for additive manufacturing, but it is likely temporary. It could only be a flash in the pan followed by collapse, but the sum of the economic and social advantages presented in the previous chapters has made it an integrated technology in the market and now in the collective unconscious. One of the reasons for this position is linked to the fact that we rationalize, with minimal energy, emerging technologies with conservative, outdated and other ways of thinking, which stops us foreseeing new organizations and effective supports to occupy the space accessible by the technology as fast as possible. Furthermore, Wiener wrote in 1960, "the suppositions that were axiomatic for the preceding generation cannot be transferred to new machines. I find myself faced with a public whose attitude concerning the machine was formed on the basis of an imperfect understanding of the structure and *modus operandi* of modern machines" (at that time, this was cybernetics; but this reflection can be applied to other emerging technologies, like 3D printing).

Additive manufacturing technologies are developing with the following five principal concerns present and engaged toward more prospective aspects:

– showing that additive manufacturing belongs to an industrial process allowing the "time-to-market" to be reduced;

– finding new prospects for the technology: from children's games (Mattel®) to printing organs;

– innovating to maintain the dynamic: an incremental approach toward new processes and particularly materials and multi-materials; a breakthrough approach by revisiting the concept (example of 4D printing in Volume 3);

– performing design tests and communicating; the initial demonstrators are developed in light structures with modest investment, but strong creativity;

– manufacturing finished products in the fewest stages possible.

Considering this road map, the area covered is very vast, and clearly a single technology will not be able to cover it. Thus, as presented in the introduction of this chapter, stereolithography at first served as a demonstrator and, to illustrate the interest of the concept, other technologies have emerged, each with advantages and disadvantages, not only among themselves, but also vis-à-vis other complementary or competitive technologies. At present, the technological palette is more mixed than ever with services and applications that beg to be put together in a specific way following the creativite capacity of each individual or of "simple" industrial needs.

The object of this chapter is to present these different technologies and to try to position them in the current space of object production. According to Vaezi, Seitz and Yang [VAE 13], 3D technologies (through the addition of layers, or more generally, of matter) are significant in number and can be represented by different themes as gathered in Figure 3.2 (see also [OBA 15, OBA 16, GUO 14, DEH 16, GAR 16, MAD 17, MUR 17, GOH 17, BHU 17, SOL 16, LOW 17, CHE 17b]) and in Table 3.1, interpreted from Hague, Reeves and Jones [HAG 16], Bouffaron [BOU 14] and Brethome *et al.* [BRE 16]. The principles are sensitively identical, and the processes essentially differ in the methods of material adhesion to the object being created (see e.g. [CAH 14, MIZ 15, PRÉ 15, ROB 15, GRO 14, CAS 14, VER 15]). Nevertheless, if the rule of the game is indeed to have either localized energy or a localized positioning of the material, or both, it seems necessary to engage a somewhat finer analysis of these different processes to then examine their application specificities.

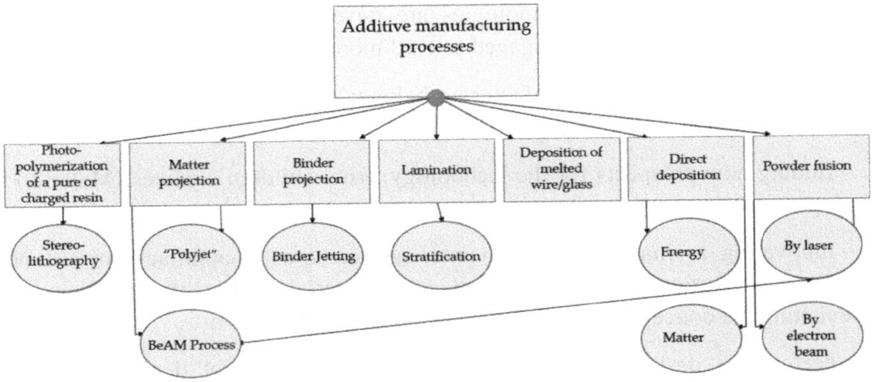

Figure 3.2. *Different additive manufacturing processes*

Process	Energy	Matter	Principle	Nature of the connections	Post-treatment	Specificities
Stereolithography	Light	Monomer	Chemical reaction	Space resolved photo-polymerization	Post-polymerization Clean-up Elimination of supports	Mono- and biphotonic absorption Supports From the µm to the m Aging and removal problems Possibility of working with charged materials
Matter jetting (1)	Light	Drop of liquid monomer; of fusible powder	Chemical reaction	Photo-polymerization Powder fusion	Deburring Post-polymerization	Broad choice of materials
Matter jetting (2)	Without	Drop of a melted material	Solidification	Collage performed on the surface	See above	See above

Binder jetting	Binding agent	Powdered material	Collage reaction	Collage performed on the surface	Elimination of excess powder	No supports Multiple colors Object fragility
Shaving	Welding bond	Sheet	Cutting performed on the surface	Collage	Separation of the sheets	Anisotropic object Inexpensive
Deposit of melted wire/glass	Without	Melted material	Solidification	Collage performed on the surface	Elimination of supports	Precision Surface quality Inexpensive
Direct deposit Powder fusion (1)	Laser	Powdered material	Solidification	Collage performed on the surface	Elimination of excess powder	Improvable surface condition Broad choice of materials
Direct deposit Powder fusion (2)	Electronic beam source	Powdered material	Solidification	Collage performed on the surface	Elimination of excess powder	See above
Direct deposit Powder fusion (3)	IR + Mask	Powdered material	Solidification	Collage performed on the surface	Elimination of excess powder	See above

Table 3.1. *Different additive manufacturing methods: some differentiation elements*

The issue is then no longer that of knowing how 3D technologies attained their legitimacy or how their emergence was "negotiated" with society, but now describing how these different technologies imposed themselves among the competition in their market space. This activity implied the exploitation of a minimum of creativity, know-how and innovation on the one hand and some certitudes and convictions translated into facts on the other hand, with uses that respond to certain questions or that allow the "secretion" of new demands.

3.2. Stereolithography

3.2.1. *History of 2D1/2 processes [AND 15, AND 17, LAV 15]*

In 1984, although very extensive progress was made in the field of computer science, the actuator and shaping materials aspects are still a little disharmonious to create personalized shapes, for biomedical applications and so on. There is thus an industrial attractiveness for parts active in a considerable global market. Other, less futuristic applications are already in place thanks to still "rustic" methods of manufacturing parts with inert matter, in very diverse domains.

These creations of objects without machine-tools (which remove matter) are developing more and more from processes that are no longer called "additive manufacturing", stemming from a process using light developed in the 1980s in Nancy, France. It depended on the knowledge of the coordinates of the object to be created, recorded in the memory of a computer monitoring galvanometric mirrors and controlling the movement of light flux to transform a liquid oligomer (or resin) into a solid through the polymerization of a layer, "voxel" by "voxel" (see Figure 3.3). The addition of a second layer, then a third and so on thus allowed the prototype part to be created, like the Neolithic potters as they created objects like containers for doves. This basis still serves as the founding principle of additive manufacturing technologies, which will be described in the following paragraphs.

In 1984, on the basis of the idea of "playing" with a one-proton process (see Figure 3.4) via amplification linked to a free-radical chain reaction, photo-polymerization brought about by a continuous UV laser, the transformation of a liquid monomer (e.g. a dimethyl acrylate) into a solid is produced where light reaches the surface, as long as the light energy absorbed exceeds a certain threshold (it is, in fact, slightly more complex, but this is not essential for the purpose of understanding). Concerning the penetration of the light, it is best to have an optical thickness (μ) linked to the concentration of a photochemical initiator slightly larger than that of the layer placed on the object being constructed and to adapt the power of the laser beam to polymerize the fluid profoundly in order to also connect the layers to one another.

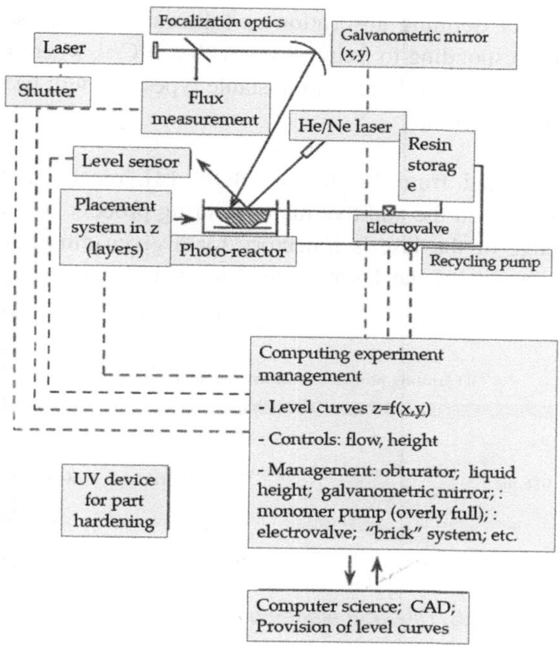

Figure 3.3. *Initial system for creating an object by laser stereolithography [AND 94]*

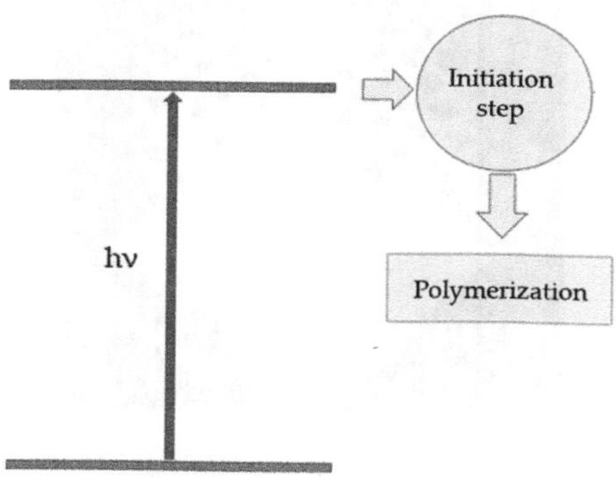

Figure 3.4. *Monophoton absorption process*

It was a matter of defining absorption as follows: a chromophore absorbs the wavelength (λ) (corresponding to frequency ν with $\lambda = C/\nu$, where C is the speed of light in vacuum), which leads directly to unstable types that will be the origin of the polymerization chain reaction.

Figure 3.5, extracted from Bouffaron [BOU 14], recalls and completes the different stages implied in the additive manufacturing process. The object is created using the knowledge of the laser's trajectory for a given dimension based on the computerized movement of the laser beam. Closer and closer, layer by layer, the object is constructed.

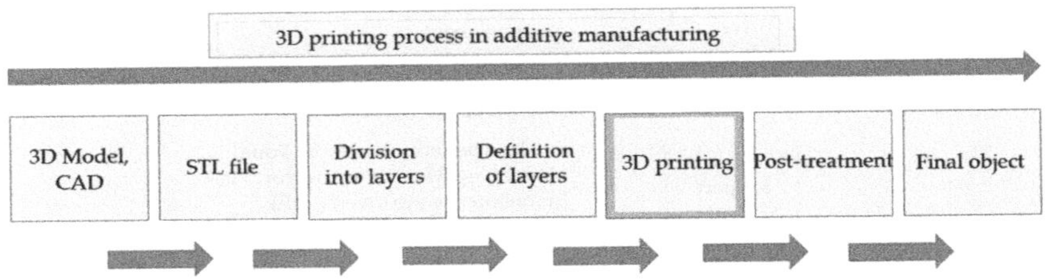

Figure 3.5. *Different stages of additive manufacturing*

A typical application example corresponding to a simple object that is difficult to create by other means is presented in Figure 3.6.

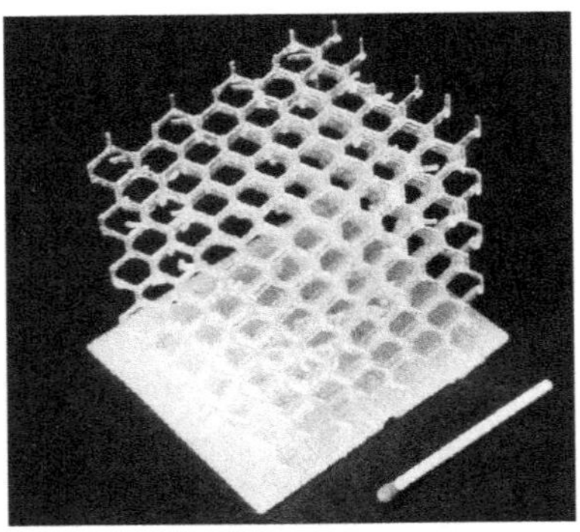

Figure 3.6. *Example of the interest of additive manufacturing in creating complex objects: 3D mesh*

But not everything is so simple in real life! Thanks to masks, initial experimentation has been attempted using linear acrylic monomers but the polymer formed was soluble in the liquid that had led to its emergence; the failure was bitter and hence led to the idea of using multi-functional monomers and thus a material that was insoluble in its monomer. The enthusiasm lasted as long as convex objects (with bases larger than the part that was under construction) were created using masking. However, the creation, with modest light flux, of a small bridge with its deck spanning liquid was translated by two piles connected by an imprecise volume: the polymer of the deck, denser than the monomer, "flowed" during transformation. Thus, two difficulties were quickly targeted, that of increasing the density of the photo-product in relation to the precursor (thus provoking removal) and that of the need to place temporary supports "at the right place" [STR 13] (even if it meant later trimming them) so that the part will be possible to create.

There was thus a need to do engineering work, a bit less rustic than this presentation might leave one to believe, by exploiting everything accessible in our "toolbox". Playing with the process; the choice of light sources and the optimized direction of the light beams, materials and their formulations; the choice of photochemical initiators and the computerized control of the galvanometric mirrors, which precisely position the light beam on a reactive surface and then pilot the creation of a part (with its supports), was an exercise of intellectual acrobatics or "learned" bricolage that went slightly beyond "classical" interdisciplinarity, because the disciplinary supports (both human and financial) were spent with moderation, but that is not all ... The activity was immediacy, urgency management and, at least at the start, very artisanal, but it was "learning" research. Thus, it was done with UV, visible or IR lasers. More and more complex parts started to result, to our great satisfaction (it was magic!), and more than 20 patents were filed.

This "3D photocopying" process, based on a quickly mediatized digital file, was immediately attractive for socioeconomic milieus wishing to create more directly costly, complex prototype parts impossible to create otherwise (see previous chapters). Currently, the annual market amounts to several billion euros. However, earlier, few heads of CILAS (a French subsidiary of Alcatel Enterprises) and CNRS had believed in it when this operation took shape, which means that the know-how that could have been acquired, at least in France, is practically no longer present on national soil. Even start-ups supported by researchers of the CNRS-Nancy team were bought by the American manufacturer who had patented the idea of the device 15 days after us, leading to a bitter little lesson for the support of creativity and for possible industrialization stemming from research. This topic will be discussed again in Volume 2!

3.2.2. *Other techniques developed since 1984*

Figure 3.7 presents a variant of the stereolithography process (SLA or more simply SL) using a dynamic computer-controlled mask to create micronic objects.

By defining the areas outside the surface to be polymerized through darkening the mask, the whole surface of interest can be transformed from a liquid to a solid at the dimension in question. Then, each layer can be created in a single operation using a laser or a more conventional light source [BER 97, BER 96, BER 97a, SUN 05]. Figure 3.7, from Wiederkehr, Goetschy and Wunschel [WIE 08], illustrates a variant that is better adapted to "home" applications. In these different cases, gains are made in the beam's movement time, as was considered in section 1.1, but compromise obliges: there is a much lower density of light flow (for the same source, the relation of the object surface to that of a voxel; the voxel being the transformed elementary volume), which may increase polymerization time.

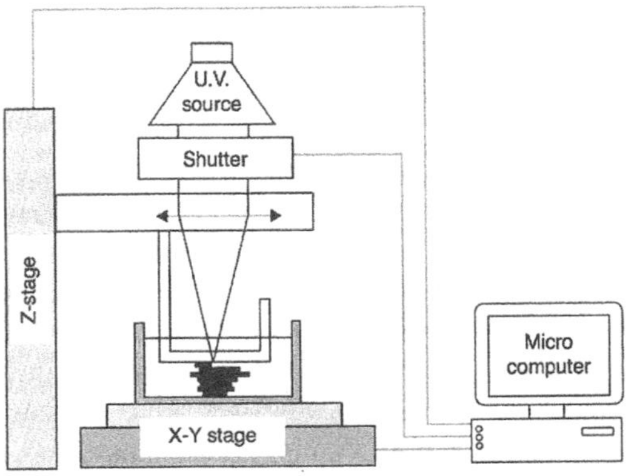

Figure 3.7. *Micro-stereolithography [BER 11]*

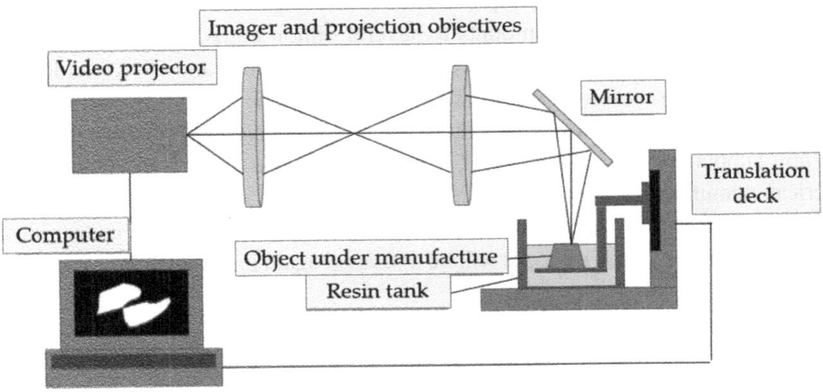

Figure 3.8. *Optical assembly using an overhead projector*

Hybrid systems have been developed [EMA 15] where dynamic masks are used for portions of the surface to have a larger number of pixels, a more significant flow density and, in principle, to make gains in resolution (briefly, as long as the flow density is constant over the entire surface accessible to polymerization). Other recent articles, including Zhou, Ye and Zhang [ZHO 15] and Kang, Park and Cho [KAN 12], revisited the use of dynamic masks. Published approximately 15 years after the first papers on the subject, these publications deal with refinements in the light–matter interactions.

To the author's knowledge, the idea of creating a large number of identical parts has not been the object of great interest in research environments to date, and yet this method of collective manufacturing that has long been exploited in electronics has shown its full interest in terms of reducing manufacturing costs. In the bibliography created, we cite an example (unrelated to stereolithography processes) concerning the near future. The vision is to explore the possibility of printing circuits with 3D motifs measuring 10 nm on silicon chips with a diameter of 200 mm and measuring 100 nm on 300 mm chips [LAN 15]. Through a pilot line, the "Inspire" project from CEA-LETI (a department of the French Atomic Energy and Alternative Energy Center) looks into the integration of lithography through nano-printing in an automated manufacturing set. It would be interesting to examine if optical technologies like stereolithography are not capable of escaping the confinement of the single part, as this was only practiced in the University of San Diego in California using radiation via micro-mirrors to create micronic objects capable, according to the authors, of exploring (in the near future) the human body [KOS 15a].

3.2.3. *Light–matter interaction and space-resolved polymerization*

In transparent environments, monophotonic absorption follows the Beer–Lambert law. Locally, for a given concentration c, the intensity of light absorbed is proportional to the product c. exp(-ε.c.x), where x is the distance from the light-input window to the vessel, c is the concentration of the absorber and ε is a coefficient known as the molecular extinction coefficient. It is expressed as:

$$Ia(x) = \varepsilon.c.F(0).\ \exp(-\varepsilon.c.x)$$

where F(0) is the density of photon flow at the considered wavelength. The optical thickness is also designed by μ with the equation:

$$\mu = 1/\varepsilon.c$$

From this basis alone, it is imaginable that if e is the thickness of a layer to be polymerized, it will be necessary to choose μ as close to e; ε depends on the absorbent substance, ranging from a few dozen units to 10^5 in the classical units of physical chemists and spectroscopists ($M^{-1}.l.cm^{-1}$). This finding illustrates that for typical thicknesses of 0.1 mm, the initiator concentrations will generally be more than 10^{-3} $M.l^{-1}$. Under these conditions, the essence of the reactive mix will be made up of a pure monomer or a mix of monomers engaged in the desired liquid–solid transformation.

NOTE.– *Charged environment*: Numerous applications imply charged environments (e.g. creation of ceramics); under these conditions, light is the object of diffusion, which leads to modifications in optical thickness (even if, at first approximation, the transmission of light energy follows a law close to Beer–Lambert's [BAL 84, BRA 85]).

This penetration thickness (Ep) is, in the absence of absorption by the resin, approximately defined by:

$$Ep = 2.\, D_{50}/(3.Q.C)$$

where

– D_{50}: the average diameter of the particles (assumed to be spherical);

– Q: light diffusion factor defined by $h.\Delta n^2/\lambda$ with:

- h: inter-particle distance;

- λ: radiation wavelength;

- Δn: the difference in refractive indices between the resin and the organic or the mineral charge (with refractive indices approximately 2).

– C: charge density.

With the initiator, if μ is the optical thickness in the absence of a charge, the thickness μ_c in the presence of solid additives is expressed by the following approximated relation:

$$1/\mu_c = 1/\mu + 1/Ep$$

Second aspect, the resins contain groupings absorbing far-off ultraviolet light, which imposes molecule choices that will initiate the polymerization reaction in a "free" spectral window. For chain reactions, generally ketones are used and electrically excited in the middle UV light domain (~350 nm), rather than more complex compounds that are absorbed in visible light, due to parasitic polymerization risks brought about by natural light (even if it was possible to create objects in visible light more than 20 years ago as demonstrated by Bertsch [BER 96]).

These electronically stimulated compounds can be fragmented into free radicals R, which will be susceptible to initiating a radical polymerization. Figure 3.9 [SUS 10, DEC 02, AND 94] schematically represents a radical polymerization chain reaction (with its mechanism) allowing, through a radical created through a photonic process, a large number of molecules (monomers and/or oligomers) to be transformed.

Initiation stage (photochemical)	A → 2 R.
Propagation (M for monomer)	R. + M → R. (k_1)
Termination (dimer)	R. + R. → R_2
Termination (entrapment)	R. → (R.) (k_2)

Figure 3.9. *Radical chain polymerization mechanism*

Considering the concentration of the monomers (a nearly pure environment of monomers upon initiation), polymerization takes place step by step at the molecular level, as indicated by Figure 3.10. If excess multi-functional monomers are available, the polymer constructed is normally insoluble in the monomer that created it (see [AND 94]).

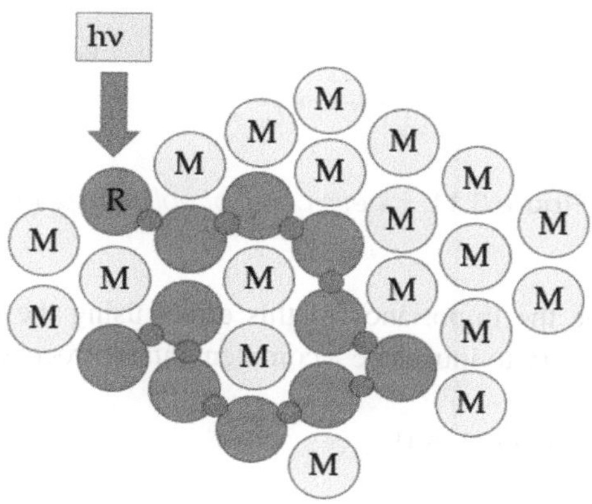

Figure 3.10. *Polymerization through spatial proximity (hν represents the photon's energy, where ν is the radiation frequency and h is the Planck constant)*

From a kinetic standpoint, the transformation follows a sigmoidal law as presented in Figure 3.11 with a slow initial kinetic linked to the presence of reaction inhibitors/quenchers (e.g. hydroquinone to allow the preservation of monomers, oxygen in solutions, etc.) that are slowly consumed [ALL 85, ALL 89, LUC 15, AND 01] in a "free" polymerization reaction, but more and more aggravated by the three-dimensional polymer that forms, limiting the space accessible to the reaction of macro-radicals with the leftover monomers and thus the reaction kinetics [IRM 09, AND 01]. It is possible to show through experimentation that free radicals are then present, but trapped in a polymeric matrix.

If the initiator takes part in the inhibitor consumption reaction, it is necessary to have a concentration of this substance greater than that of the inhibitors (if not, there could be a risk of polymerization not taking place or of a different degree of polymerization than required). In radical polymerization reactions, oxygen is often the primary inhibitor; its concentration in organic fluids is about 10^{-2} M.l^{-1}. Under these conditions, if a μ value of 0.1 mm is desired, for example, with $c > 0.01$ M.l^{-1}, the value of ε must be below 10^4 M^{-1}.l.cm^{-1}, which allows the choice of the "right" photochemical initiator relative to the desired manufacturing to be calibrated (see [MOK 15] for initiator selections; [STA 08, NEU 05, DEL 16]).

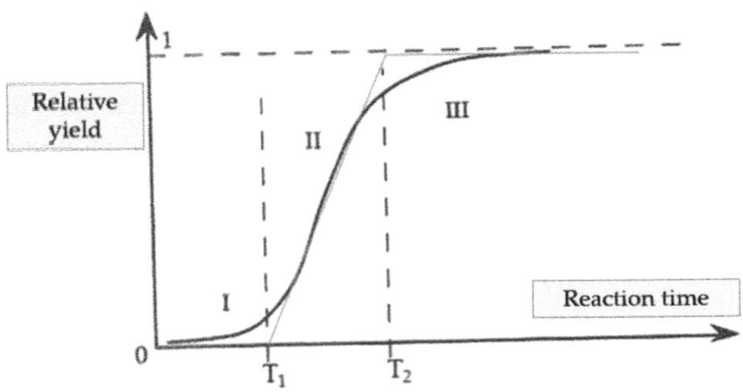

Figure 3.11. *Kinetic curve of polymerization*

In order to define the importance of this effect using the method presented in Figure 3.6, it is necessary to introduce a complementary process, namely:

$R + O_2 \rightarrow$ Termination (kq)

Thus, by hypothesizing that kq is controlled by diffusion (kq >> k_1), for a given intensity of light absorbed Ia, with this mechanism (which is reductive, but allows the problems at hand to be understood), the value of T_1, corresponding to the effective initiation of the polymerization, is equal to:

$$d(O_2)/dt = - kq(O_2).(R.) = Ia,$$

leading to $T_1 = (O_2)/Ia$ (the change in the concentration of the inhibitor has a linear effect on T_1).

For $t > T_1$ and $t < T_2$, the reaction kinetics is defined by:

$$R = k_1(M).Ia/k_2$$

corresponding to approximation by an oblique right segment presented in Figure 3.11. In this process, a significant share of the monomers is trapped in the polymeric network that is gradually created (see Figure 3.10). As soon as the gelling point is exceeded, the environment becomes practically solid with reaction yield close to 40%. Thus, in phase 3, presented in Figure 3.11, trapped free radicals and a large amount of monomers remain in the transformed material. As a function of time, of molecular agitation, the polymerization reaction can be followed with a lower kinetic rate, leading to sometimes harmful aging in the parts manufactured through stereolithography.

Thus, the time required for the transformation corresponds to an estimated yield of 40% (approximately), or, with $M = 5\ M.l^{-1}$, its approximate value is:

$$T = T_1 + 0.4.k_2/(k_1.Ia) = ((O_2) + 0.4.\ k_2/(k_1))/Ia$$

3.2.4. *Consequences*

3.2.4.1. *Degree of polymerization*

For a flow density F(0,r) of monochromatic light arriving at the resin's surface, hypothesizing a Gaussian beam (cylindrical symmetry), the intensity of light absorbed locally is defined by:

$$Ia\ (z,r) = -dF(z,r)/dz \text{ with } F(r,z) = F(0,r).\exp(-\varepsilon.c.z)$$

T_l is then defined by:

$$T_l = \exp(\varepsilon.c.z).(O_2)/(F(0,r).\varepsilon.c) \text{ with } z_l = A. \text{Ln } (T_l/T_{l,0})$$

where A is a constant and $T_{l,0}$ is the value of T_l for z = 0 and $T_l > T_{l,0}$. Under these conditions, for values above T_l, the polymerization reaction can begin.

A Monte Carlo simulation with a Gaussian laser beam reaching the resin's surface orthogonally, presented in Figure 3.12, illustrates the complex influence of exposure time to light on the shape of the voxel.

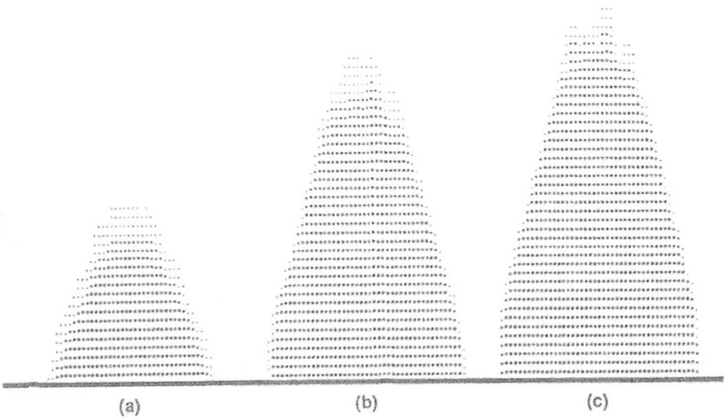

Figure 3.12. *Polymerization of a monomer by radical reaction (a: 50,000 lancers, b: 1,000,000, c: 2,750,000).*

An important effect is then brought about, which is that after T1, the evolution of the polymerization sensitively follows a logarithmic law as a function of radiation time, which leaves something "lacking" in the choice of polymerization time, as long as the inhibitors have been consumed.

3.2.4.2. *Solution*

With a pure resin (without solvents), the diffusion of oxygen can be taken into consideration for modest voxel sizes (a few micrometers). With a resin having a viscosity of 100 cP, a 1 s period for polymerization, the oxygen can spread over a distance measuring 3 μm (classical hypothesis of a transport over the distance d defined by $d = (D.t)^{1/2}$, where D is the coefficient of the molecular diffusion of oxygen). On this basis, the presence of the inhibitor can lead to a reduction of the size of the polymerized volume. To find this volume precisely, the following equation would need to be solved:

$$\partial(O_2)/\partial t = D.(\partial^2(O_2)/\partial x^2 + \partial^2(O_2)/\partial y^2 + \partial^2(O_2)/\partial z^2) - Ia$$

by hypothesizing the stability of D as a function of the advance of the polymerization reaction. However, without even seeking to perform a precise calculation, it can be shown that oxygen can spread from an air–resin interface (hence, the sticky feature of the parts created by stereolithography) and in liquid (see Figure 3.13). Then, if a second voxel is manufactured in contact with the first, the transport of oxygen is modified, which can have effects on the spatial resolution for voxel sizes of a few micrometers (no major effect in classical conditions with a layer thickness of 0.1 mm).

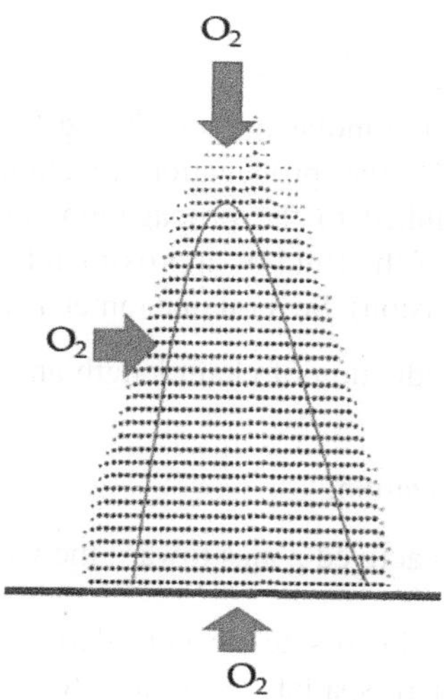

Figure 3.13. *Improvement of the spatial resolution due to the spread of oxygen*

3.2.4.3. *Strengthening*

In a chain reaction, several monomer molecules can be transformed through an absorbed photon. By hypothesizing a quantum yield of 1 for initiation, the chain length λ is expressed (for t > T1) by:

$$\lambda = k_1 . (M)/k_2$$

With acrylic components, k_1 is $k_{1,0}$.(exp(-E/RT) with (according to [PER 05, SAG 11, WAN 09, NIK 10, BEU 03, SIE 10, ROB 50, MAE 98, DOU 04, BAR 10, BAR 11, QUI 14]):

$k_{1,0} = 4.10^7$ $M.l.s^{-1}$

E = 20 $kJ.M^{-1}$ (activation energy); R: constant of perfect gases; T: temperature (°K)

(M) = 5 $M.l^{-1}$

$K_2 = 10^3\ s^{-1}$

$\lambda = \sim 120$

3.2.4.4. *Possible consequences*

For a chain length of 120, a molar mass of 200 g for each monomer unit and a heat capacity of 1.5 J/g/°C, the propagation reaction generates approximately 20 $kJ.M^{-1}$ through a step linked to the propagation process, that is, a maximum increase in the temperature of the fluid of approximately 80 °C (heat that dissipates quickly through thermal diffusion). Several phenomena can modify this situation:

– the use of oligomers (reduction of chain length and increase in the molar mass of the oligomeric units);

– the kinetics of polymerization;

– the heat transfer from reactived zones toward the whole reactor.

These elements are thus factors to be considered to master the process and, positively, for the associated spatial resolution (obviously, differences can be expected between the polymerization brought about by a sphere, a line or a surface, as well as through a presence near volume elements in the shape of different writings on the influence of matter or heat transfer processes). This result shows that the method of tracing the surface to be polymerized can have effects because unstabilized (unrelaxed) thermal transfers can play a role on the kinetics of polymerization (voxel size, degree of polymerization, etc.).

3.2.5. *Families of materials used*

Figures 3.14 and 3.15 and Table 3.2, taken from André and Corbel [AND 94], Davis and Mitchell [DAV 11] and Nguyen-Tri [NGU 04], give some examples of the monomers used to create parts through photo-polymerization (see also [STA 05, COR 03, PRO 02, SAL 05, DUF 05, KEL 93, ALL 89, VAR 01, BAR 11]).

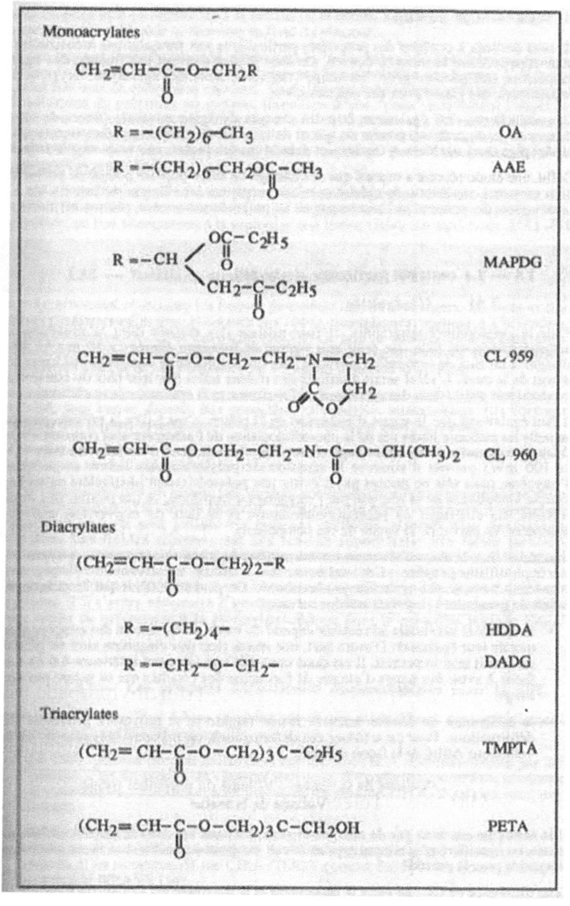

Figure 3.14. *Examples of acrylic monomers that can be used to create parts through photo-polymerization*

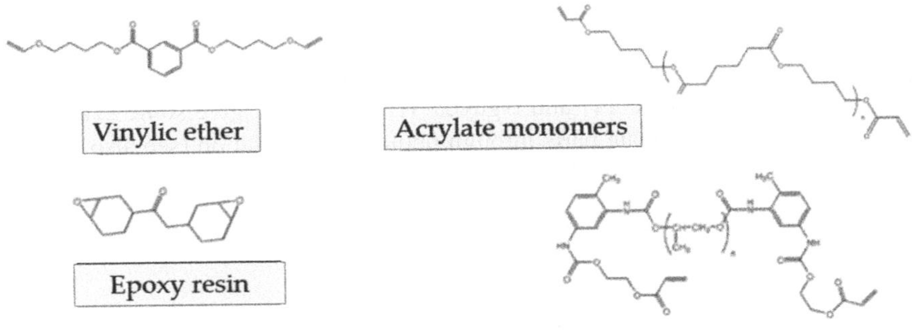

Figure 3.15. *Other examples of monomer formulas that can be used to create parts through photo-polymerization*

Radical monomer		
Formulation	Functionality	Common name
HEA	1	Hydroxyethylacrylate
HPA	1	Propanol monoacrylate
HDDA	2	Hexanediolacrylate
TPGDA	2	Tripropyleneglycoldiacrylate
TEGDE	2	Triethyleneglycoldiacrylate
PTIEA	3	Pentaerythritoacrylate
TMPTA	3	Trimethylpropane triacrylate
PETA	4	Pentaerythritotetraacrylate

Telechelic oligomer

$$CH_2{=}CH{-}\underset{\underset{O}{\|}}{C}{-}R{-}\underset{\underset{O}{\|}}{C}{-}CH{=}CH_2$$

With R: polyester, polyether, polysiloxane, polyurethane

Table 3.2. *Examples of monomers that can be used to create parts through photo-polymerization*

Because it is of the utmost importance to create a material that is strictly insoluble in the monomer that created it, it is necessary to use, at least in part, polyfunctional monomers to allow reticulation. Furthermore, it is also possible to use pre-polymers (oligomers) that lead to less significant volumetric decreases than that with their monomers (but with a higher viscosity, which, in the process proposed, can constitute a constraining piece of data due to the time required to place the layers).

Concerning the photochemical initiators, the substances classically used for radical polymerizations are ketone components (like benzophenone, whose decomposition into free radicals after the absorption of an appropriate photonic energy is represented in Figure 3.16). A rather recent publication by Yagci, Jockusch and Turro [YAG 10] presents an exhaustive list of the photo-initiators used in radical and ionic photo-polymerization (with 293 references). Interested readers can benefit from this source (which is still current; see also [DEL 16]).

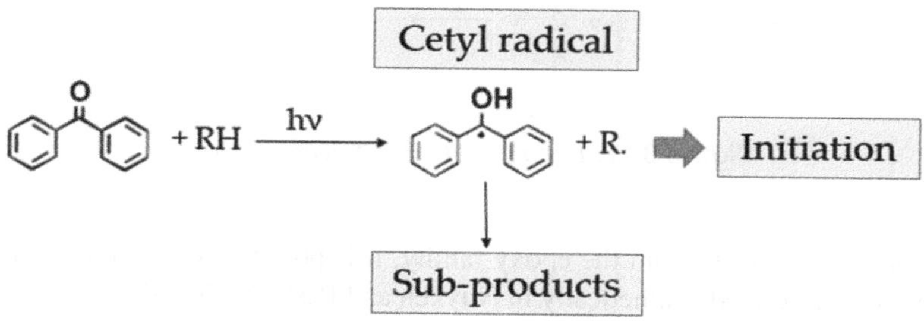

Figure 3.16. *Illustration of a photochemical initiation process in the case of benzophenone*

NOTE 1.– Other polymerizations:

Using an epoxy (cationic monomer) and a photo-initiator composition, polymerization under the effect of light radiation leads to the desired liquid–solid transformation. Certain components adapted to stereolithography applications of a photon are available on the market. These compositions include monomers, typically epoxies and the initiator. Figure 3.17 illustrates a classical ionic polymerization mechanism with the following stages: initiation, propagations (1) and (2) and transfer.

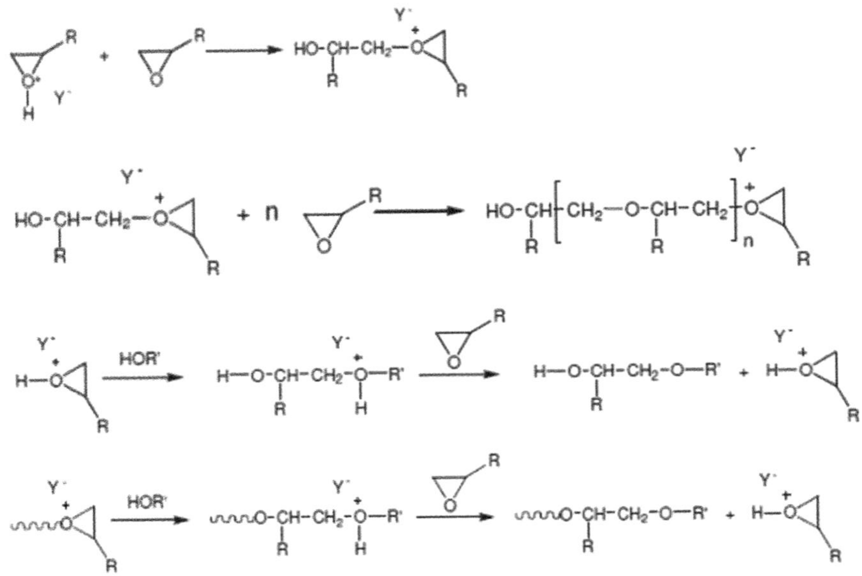

Figure 3.17. *Example of ionic polymerization*

Beyond components from the epoxy family, it is possible to use a large number of monomers described synthetically in Nuyken and Pask [NUY 13].

NOTE 2.– Shrinkage linked to polymerization:

The reduction of the polymerization of resins composed of an acrylic matrix base is inherent to the polymerization reaction itself and depends on their chemical composition, the volumetric fraction of the charges and the degree of conversion (measurement of the degree of polymerization) during the polymerization, which is never total and uniform. Indeed, the conversion of monomers into a network of polymers leads to a replacement of the Van der Waals bonds with shorter covalent bonds between the molecules [WAT 91, YAP 00]. Before polymerization, the electrostatic bonds between the monomers measure 0.3–0.4 nm. Once the polymerization has been effected, the monomers are connected by covalent bonds measuring 0.15 nm [PEU 97]. With monomers displaying a lower molecular mass, the volumetric reduction can reach more than 10%. With oligomers, this reduction can be less than 5%; this is thus a piece of data that must be taken into consideration because it has an influence on the quality of the parts manufactured using stereolithography.

Karrer *et al.* [KAR 92], by mastering polymerization reactions and cycle opening in the monomer, were able to find practically no reductions, but produced a defect linked to the very high viscosity of the resin; the addition of certain polymeric charges could lead to the same result, but with the same defect. The exploitation of the process patented by André, Gallais and Amra [AND 16], which avoids transition through the placement of layers, could once again lend interest to this kind of work.

NOTE 3.– Charged resins:

Most manufacturers provide "home-made" resins, resins that have photochemical and rheological characteristics, as well as specific characteristics of deformation (shrinkage) during photo-transformation under the indispensible condition of optical and/or mechanical properties adapted to their applications. The resins used are generally mixes of mono- and multi-functional acrylate monomers (to ensure insolubility of the polymer in the monomer that created it) or epoxy monomers (cationic polymerization). However, the most interesting applications in terms of the industrial market of functional parts concern the creation of parts with charged resins using diverse powdered materials (alumina, zircon, hydroxyapatite, etc.). These resins, as a function of their charge, display a lower volumetric shrinkage than for the resin by itself (function of the percentage of added charge). After stereolithographic manufacturing, these objects will be transformed by heating, de-binding, sintering and other processes to obtain objects exempt from resins (ceramics in particular), as illustrated in Figure 3.17 from France 3 (French TV station) about the creations of 3D Ceram [3DC 16] (a French company involved in 3D ceramics machining). These types of processes are now industrialized (see e.g. [3DC 16, CHA 02, CHE 15]). The significant problem of the volumetric reduction linked to the resin is essentially transferred to the cohesive assembly of powders, which, after sintering, can undergo volumetric reductions measuring 40–50%. The expectation of the reduction, which has two origins, is complex and may often require a prior "draft" during the manufacturing process.

In these operations, the light penetration depth can be affected by the presence of charge, which may require modification during the process of concentrating initiators dissolved in the resin and/or of the incident light flux and/or the length of voxel manufacturing.

Concerning such matters, the possibilities are numerous. As examples, we could cite:

– epoxy resins charged with multi-layer carbon nanotubes, allowing the mechanical properties of the objects manufactured to be reinforced [SAN 07];

– acrylic resins containing nanoparticles of magnetite (Fe_3O_4) (diameter ~50 nm) to manufacture sensors [LEI 11];

– hybrid resins containing acrylic and epoxy or vinyl ether components, used by Bartolo amd Gaspar [BAR 08], resins in which metallic particles (cobalt and tungsten carbide) are introduced; the objects are de-bound and sintered in a later phase;

– acrylic resins containing different types of charge (alumina, metals, alloys, etc.) as proof of concept [CHE 93, GRI 96, CHA 03, DE 16], today with high-tech industrial applications by 3DCeram with ceramics, as shown in Figure 3.18;

– epoxy resins associated with strongly charged suspensions of hydroxyapatites for bio-printing operations [SCA 14];

– phenolic-epoxy-acrylate resins, phenolic, triethylene glycol to shape pores and benzoin dimethyl ether as a photo-initiator, used by Tian *et al.* [TIA 12]. The object created underwent pyrolysis and was then treated with silicon to shape parts in silicon carbide, a particularly hard material; similar methods with SiH_4 leading to silicon carbide objects have been exploited by Eckel *et al.* [ECK 16]. However, it is not yet possible to know if the quality of the compound meets industrial demands.

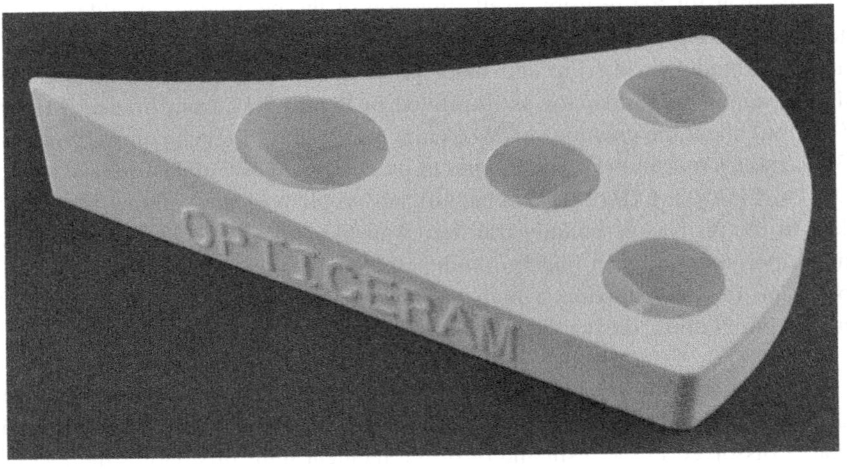

Figure 3.18. *Ceramic part after sintering (photograph generously provided by 3DCeram)*

3.2.6. *Layer implementation*

A partially created object is immersed into the vessel containing the reactive fluid and taken out so that the fluid exceeds the object in construction by Δz in the most horizontal way possible. If the reactive fluid is slightly viscous, balancing for a quick ascent of the object under construction will also be quick; for a more viscous monomer, however, this balancing time is much longer, sometimes outright unacceptable. Under conditions of preliminary flow, with no external artifice, this height decreases exponentially with relaxation times that may exceed 1 min, which leads to the highlighting of the problem concerning the implementation of layers, a significant factor to be considered in this stereolithography process. In the presence of total anchorage, it may be useful to take into account Tanner's law. This law states that in a free relaxation time, a relaxation is proportional to the time at the power -3/10 [DE 05]. With very viscous environments, this relaxation may be unacceptable in an industrial process in which financial investment, applied to an object's manufacturing time, can become a production cost criterion.

According to André and Corbel [AND 94], a first path had been found to accelerate this temporal relaxation by bringing the object up beyond the final expected position. Currently, other techniques using mechanical scrapers are used successfully; however, for layers with modest thicknesses (measuring a few dozen micrometers), it is best not to destroy the object under construction with the scraper (avoiding critical shear). Thus, any process reducing or eliminating the implementation of layers will be accepted, for this would allow more relaxed work with oligomers, which are possibly charged and highly viscous.

3.2.7. *Coupling of polymerized surface generation and volumetric reduction*

This aspect was already described in Chapter 2. In addition to the methods described concerning the avoidance of reduction problems, we must remember that with this technique, there is anisotropy of thickness polymerization, strongly independent of the method of irradiation. This may, however, be slightly affected by the thermal effects that were mentioned, which can lead to working with voxels with more complex shapes than expected (see [CHE 16b]).

3.3. Process of wire fusion

Figure 3.19, from the Lorraine Fab-Lab Lorraine [DUP 17] shows how, in principle, a melted polymer can be used to solidify an object. The accomplishment of this principle with computerized movements corresponds to this henceforth well-established process.

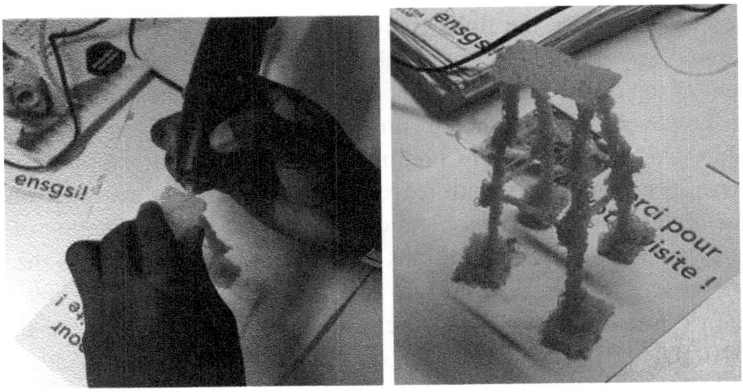

Figure 3.19. *Nearly the first steps: "Manual" additive manufacturing with melted polymers (Photo credit: Laurent Dupont, ERPI Laboratory, ENSGSI, Lorraine University, France, 2015 – www.lf2l.fr)*

In 1988, Scott and Lisa Crump founded the "Stratasys" company for the industrial development of the FDM (fused deposition modeling) process, which would be patented the following year. The company released its first printers using the "layer-by-layer" deposition of all types of melted material, thanks to a moving extrusion head. The FDM process would give rise to personal printers, used by a number of "makers" and Fab-Labs (see Part 1). These different FDM sub-technologies can be used to create prototype parts and parts for shaping lost wax (see e.g. [VOX 15]).

The object being constructed is placed on a fixed support; the extruder is connected to a computer-controlled system that can be moved according to the three spatial axes and/or the opposite: a fixed extruder and mobile base. Normally, movement is along the x, y axis and the top border is incremented once the surface to be manufactured is finished. The movement is precise to a few micrometers and it is actually the average diameter of the polymer leaving the extruder that defines the resolution; it is typically around 100–300 μm. For low values of this diameter, problems with blockage, overheating and so on may come into play, as the pressures will become more critical (obviously, precision has an impact on the cost of a machine). Attempts at modeling viscoelastic flows applied to this kind of process have been made by Liravi, Darleux and Toyserkani [LIR 17].

Sher [SHE 15a] shows that "home" machines are accessible, with very competitive prices and very acceptable creation qualities for some FDM machines. At present, it is possible to find FDM (or FFF, fused filament fabrication) machines, whose principle is presented in Figure 3.20, at costs of around 400€, even a little less "if we roll up our sleeves and apply some elbow grease" (see Figure 3.21). Similarly, processes (for toys) use a polymer dissolved in a very volatile solvent. After deposition, the material solidifies through the evaporation of the solvent.

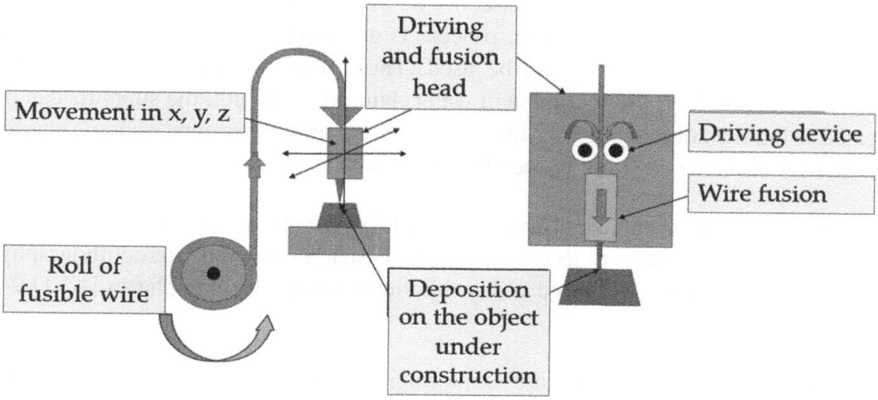

Figure 3.20. *Principle of the FDM or FFF process (wire fusion) according to Castel [CAS 14]*

Figure 3.21. *"Orca" 3D printers with melted wires (Photo credit: GSI Clic Clac and Laurent Dupont ERPI Laboratory, ENSGSI, Lorraine University, France, 2015 – www.lf2l.fr and Morel and Leroux [MOR 16])*

For these systems, the fusion of a wire (generally into a polymer), allows additive manufacturing if the support on which the addition of material takes place is sufficiently solid, hence a slight optimization between the movement velocity, the temperature of the wire leaving the extruder and the shape of the object. As there is not always a desire for high precision, research on optimal conditions is not essential; generally, the creation of a part poses no problem. Modeling the flow of the polymer in the presence of charges has been the object of modeling of the flow of complex fluids in temperature gradients [HEL 16].

In the previous case, that of stereolithography, it had been shown that supports had to be foreseen so that the object could be built. We find ourselves in the same situation here. The advantage of the process is that a machine with several nozzles containing different materials can be used. Then, if one of these is soluble in a particular solvent (e.g. PVA, soluble in water) but others are not, the supports can be created using the soluble polymer (as long as its fusion temperature is close, but not necessarily the same, as those of the other materials).

Moreover, beyond this significant aspect, it is possible to use colored materials to create polychromatic objects (which is easier than with stereolithography, requiring several tanks of colored resin with intermediary cleaning phases [AND 94, AND 92].

Finally, some researchers are exploiting the possibility of re-melting the polymers to re-use them after extrusion, even to use domestic materials with fusible polymers from domestic waste to create objects with the FDM process [3DM 15, MAR 17, CAN 14, GAR 16, WOR 17].

With pastes, it is possible to expand the applications of the process, as shown in Figure 3.22, from Martel [MAR 17], for manufacturing ceramic parts (Ecole Nationale Supérieure des Céramiques in Limoges). This kind of process is also used for BTP printing. The finished part is then dried and classically baked.

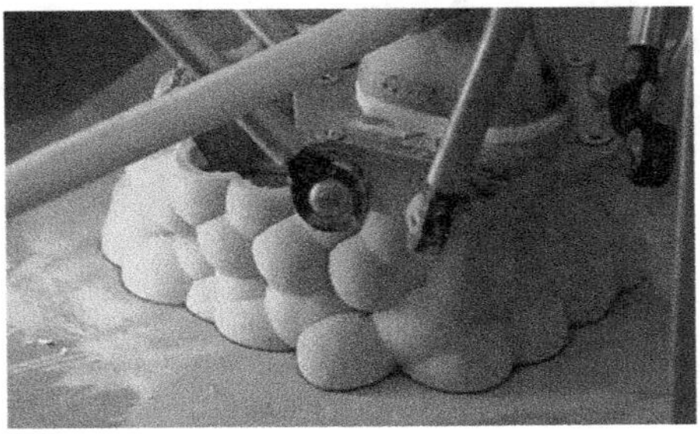

Figure 3.22. *Creation of parts using pastes; the case of ceramics*

3.3.1. *FDM or FFF materials*

Plastic wire, a real consumer good, is generally bought in spools of wire. It unwinds during the 3D printing of the object in an FDM process. Among the plastics used in 3D printing through FDM or FFF (fused filament fabrication) technology, according to Filimprimante 3D [FIL 15], two kinds of wire set themselves apart, the PLA wire and the ABS wire, as they are often used in home applications and "open-source" 3D printers, like the RepRap.

According to this same source, other more technical wires are on the market, like the MoldLay, the PEEK filament or the PCTPE, nylon and its Nylon Glue Complete Tauman, the ColorFabb (which offers a wide variety of wooden filaments (CorkFill) (see e.g. [MAK 15, IMP 14])).

The spools of wire, whether they are PLA or ABS, are of two diameters that have become "standard": 1.75 and 3 mm. The diameter used is a function of the technical specificities of the 3D printer. ABS is a polymer characterized by strong shock and temperature variance resistances. PLA, which comes from cornstarch, is biodegradable. It is a bio-plastic. Its use does not require the presence of a heating plate. It will be less resistant to water. On the contrary, the 3D object can be painted, sanded down and varnished. At present, it is even possible to acquire colored PLA filaments that can artfully display changes in their color intensity [JUL 17].

According to Primante 3D [PRI 15], for an FDM 3D printer (that uses the deposition of melted matter), it may be useful to plan a support that allows the part to be held on the platform during printing so that it remains stable. "This is presented in the form of a grate that allows better attachment, or a sort of scaffolding in order to support certain elements such as the arm of a figurine or the base of a fuselage. The most commonly used materials are PVA and HIPS (high-impact polystyrene), which can both be dissolved in water or d-limonene within 24 h. The latter is used with ABS, which has the same extrusion and platform temperatures; as such, this material is only relevant to double-extruder 3D printers."

Over time, new filaments are appearing on the market. This is the case with the soft wire, for example, proposed by Ninja Flex. This filament, with its astounding properties, allows the production of deformable 3D objects. The objects produced using "soft PLA" re-assume their original shape after deformation. In a recent paper, Quan *et al.* [QUA 15] proposed the use of charged fiber polymers; the orientation of these polymers may be an elegant means of reinforcing the mechanical qualities of manufactured objects.

Thus, all sorts and shapes of materials can be used as long as they are fusible, even polymers from waste wires [GAR 16] or metals and alloys with specific equipment [CHE 17a]. In this range of fusible materials, melted glass has been used as a proof of concept (but with a surface condition that could still be improved). It is this property that has already been exploited to create glass parts according to FDM processes in which the melting wire is replaced by the fusion of a glass wire (change in the injector's deposition temperature) or through the flow of a small melted glass wire that solidifies on the surface being constructed. According to Zaffagi [ZAF 15], to print 3D objects in glass with perfect transparency, the members of the "Mediated Matter Group" [MIT] have developed a manufacturing technique using the extrusion of melted glass. 3ders.org [3DE 14], on the contrary, presents objects created in brass.

To consider the anisotropies and tensions in this type of object created through wire fusion, Shaffer *et al.* [SHA 14] proposed post-treatment with ionizing radiation that should reinforce the cohesion among the melted polymeric wires. Furthermore, if different types of polymers can be used, some of them allowing biological applications, we point out the recent refinement of charged wires in graphene, with very significant mechanical and especially electromagnetic properties, to conduct electricity, serve as an electromagnetic screen, move toward creating sensors, and so on [FRY 13, GRA 15].

NOTE 1.– Processes using, for example, still fluid plaster or concrete are related to those presented above. Adhesion is chemical instead of physical (solidification), but the general principle is the same [DDD 15, JAC 16].

NOTE 2.– Turco *et al.* [TUR 17], for polymers used in additive manufacturing, have developed a simulation of the behavior of 3D structures in networks for different applications.

3.3.2. *Adhesion*

The network of fusible material placed beyond its transition temperature (T_c) comes to solidify on the part under construction situated at a temperature below T_c. This process then involves deposition of fluid, anchoring of this fluid on the solid and heat transfer between the addition and the support allowing T_c to be exceeded in the interface to allow adhesion considered satisfactory by the user.

Figure 3.23 illustrates the physicochemical aspects in question. Thus, for the addition to be able to adhere to the support (part being created), there must be an anchoring of the two parts. Let us not forget that anchorability is a characteristic that is not specific to the surface of the solid, but to a slightly more subtle combination that brings three different energies or tensions into play. The more the addition spreads, the more the surface is considered anchorable. The notion of anchorage is more than a narrowing linked to the connection angle (or contact angle) as presented in this figure. This anchoring angle depends on the surface tension. Cassie's law indicates that if the contact angle is between 10° and 90°, the liquid is considered to be anchorable [MAR 11, DEZ 00].

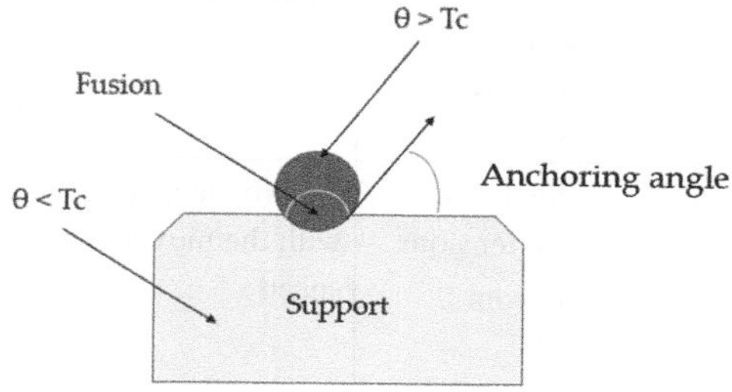

Figure 3.23. *Physicochemical aspects involved in the process*

If the quality of the support and the addition's surface in terms of property is not sufficient, the surface tension of the materials can limit the anchorability. Yet, the extruder, mis-regulated, can lead to the partial pyrolysis of the fusible polymer with the formation of particles that may be responsible for losses in anchorability. (As a note, Merlo and Mazzoni [MER 15], concerning the risks associated with the fusion processes of polymers in the shape of wires, showed that they are likely to expose users to toxic gases and to certain nanoparticles resulting from this process. As we are dealing with smaller, low-cost machines and thus ones with home applications, it is difficult to control the quality of the air in homes and thus to estimate the risks taken by users who rarely have glove boxes and/or appropriate ventilation).

The second associated aspect, the heat transfer: if the addition has a too small size, its temperature can drop too quickly before it can melt a superficial part of the support. It is possible to place the 3D system in a controlled thermal environment to limit this phenomenon and increase the output temperature of the extruder (with the risks presented above). In any case, a limitation in terms of resolution, approximately 0.1 mm, is brought about.

3.3.3. Synthesis

Table 3.3, partially from Koslow [KOS 15d], quoting 3D Matter [3DM 15], recalls the influence of a certain number of factors on the quality of the parts created using the FDM process. As with stereolithography, deformation effects are also to be seen here.

Parameter	Performance	Visual effect	Other comments
Materials	Must anchor the solid		
Extrusion temperature (above T_c)	Little effect on the mechanical strength Low resolution when the temperature increases	Low quality when the temperature increases	Air pollution problems brought about by the partial pyrolysis of the wire
Movement speed/diameter of the polymer leaving the spinneret	Longitudinal mechanical tensions Deviation from instructions Critical adherence for small diameters	Quality decreases with the movement speed	
Color	No effects observed	No effects observed	Behavior of the dyes to temperature
Surface properties	Coupling of the additions Mechanical strength	Coupling	
Aging	Need for a few days to stabilize the constraints (or second baking), otherwise good stability over time	No effects observed	
Recycling	Need for the "appropriate" plastic material	No effects observed	

Table 3.3. *Synthesis of the criteria of influence on the performance of the FDM/FFF process*

3.4. Sheet or powder gluing process

3.4.1. *Bi-material process (SDL)*

More private and relatively seldom applied, the process presented in Figures 3.24 and 3.25 corresponds to the use of sheets that allow the creation of the object and, if necessary, of its "negative"; this process has the advantage of avoiding the introduction of supports to create an object, but introduces complexities into the process with the unused and not directly recyclable unused matter (unlike FDM processes).

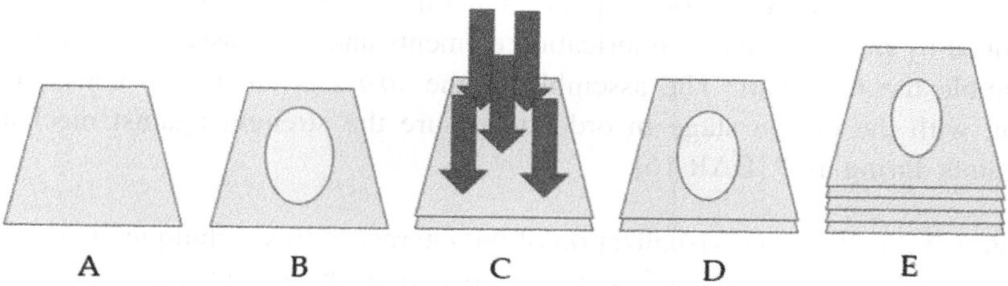

Figure 3.24. *Principle of the SDL (selective deposition lamination) process; the user starts with a sheet (A), then a "glue" in the area corresponding to the surface defined to the corresponding dimension (B), then addition of a new sheet, pressing and gluing, for example, through infrared heading (C), deposition of a new layer of glue (D) and gradual creation of the object (E)*

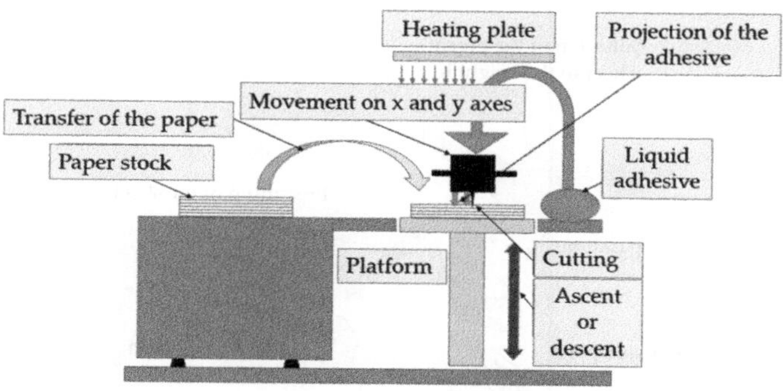

Figure 3.25. *Processing using glued and cut sheets (selective deposition lamination) according to 3D Printing Industry [3DP 15]*

The object created in this way is in a latent form, as the areas not containing linking agent must be eliminated. The process is more complicated than the preceding one but, with glue deposition devices like the "ink jet printer" and very thin sheets, it is possible to achieve precisions of a few dozen micrometers.

NOTE.– Strato-conception (SC):

"The process consists in the automatic decomposition of the object into a series of complementary elementary layers called *strata* in which positioning inserts and reinforcements are placed. Each of these *strata* is directly placed in panoply then manufactured through quick micro-drilling through laser cutting, wire cutting, or all other means of cutting through plate materials. All of these *strata* are then positioned by inserts, clips, or imbrication elements and those assembled in order to reassemble the final part. The assembly of the *strata* is taken into consideration starting with the design stage in order to ensure the strength against mechanical constraints during use" [BAR 16].

Figure 3.26 allows the visualization of the interest of this technique, which could potentially allow the creation of the part's positive at the same time as its negative, with obvious applications in foundry (thus replacing the old, so-called "sandblasting of beaten foundry" processes).

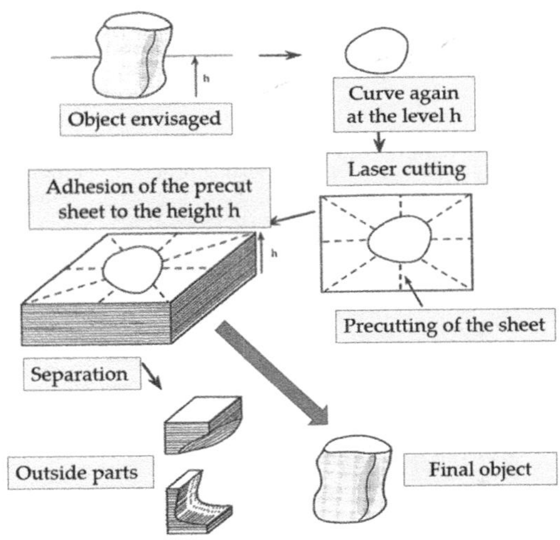

Figure 3.26. *Interest of the strato-conception process*

3.4.2. *Variant using powders: 3DP process*

There are variants of this device; thus, in 1993, a new technology was perfected at MIT (Massachusetts Institute of Technology in Boston, USA). The process is called 3DP (three-dimensional printing) or BJ (binder jetting). In its principle, it is rather close to that of 2D inkjet printers: the process consists in spreading a fine composite powder on a platform. On top of this, the printing head will then place small drops of colored glue (although not in all cases). By combining these drops, the desired color is then achieved to create a 3D reproduction in color. The platform lowers as the layers of powder are attached to obtain the final object. To complete the process, the part must be headed and the excess powder must be blown away. In 1995, Z Corporation (2015) obtained the exclusive rights to the 3DP process from MIT and developed its printers, still with a very industrial aim. 3DP manufacturing is considered fast with a large range of colors. Up to six times less expensive than a stereolithography printer, its price is more attractive despite sometimes inferior precision, mechanical strength and printing quality (except for top-end machine models). Among the limitations, without post-printing treatment, the parts are considered more fragile and their surface is rougher. Figure 3.27, extracted from 3D Systems for Z Corporation (2017), illustrates a specific performance of the process, that of "easy" access to color.

Figure 3.27. *Parts created with the 3DP process (photographs generously provided by 3D Systems)*

In associated processes, it is possible to use successive layers of metallic powder or ceramics that are aggregated by an organic binder projected by a multi-jet printing ramp. The part obtained is then sintered in a high-temperature oven. This operation eliminates the binder and aggregates the metallic grains [NOT 16].

NOTE.– It is possible to use polymer powders containing metallic or mineral charges, to melt them according to this process and then to eliminate the polymer and sinter them [MAR 17]. Aluminum, steel and numerous alloys can be used according to this reference.

3.4.3. *Process using a cross-linkable polymer (SIR, for "Soluble/Insoluble Reaction")*

Another possible path that has been explored is to make a polymer insoluble through reticulation brought about by light (or the opposite). These systems are largely used in electronics [SUN 04]. In the presence of UV light, the chemical process of reticulation described in Figure 3.28 leads to insoluble varieties.

It is thus possible to use a soluble material that is made (photo-resist) insoluble through the action of the light. If a multi-photon photochemical process is available, it is then foreseeable to create the casing of an object that will be separated from its support through the dissolution of the original material in an ad hoc solvent. However, unlike with the stereolithography process, which is based on a chain reaction of reticulation, here there is no amplification; the overall quantum yield is thus less than or equal to the unit.

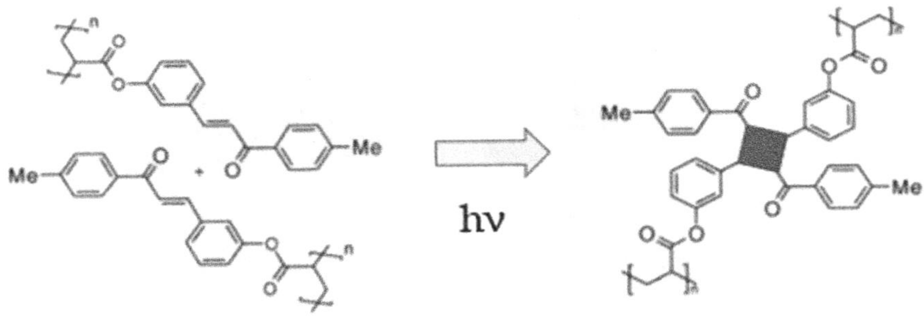

Figure 3.28. *Example of photochemical reticulation (formation of a cycle containing four carbon atoms)*

With a positive resin, in the presence of a ketone, the phenolic compound becomes insoluble. Radiation leads to the formation of an acidic component that is soluble in a base [DAV 11]. Other (photo-resist) materials are classically used in electronics. The resin Novolac (metacrylic polymer with an epoxy group) that is "hardened" with DNQ (diazonaphthoquinone) is one of the most often-used chemicals in electronics. This polymer is called positive photo-resist, because after UV radiation, it is less soluble in a soda bath than the un-radiated polymer [MAR 14].

From a bibliographic standpoint, there are several alternatives to Novolac: Sipley S1813 (methacrylate) [WOO 13]; SU-8 (MicroChem Corp.); SPR-220 (Shipley Company) [DYA 14]; polyurethanes with a possibility of two-photon depolymerization [GAR 14]; nitro-derivatives of polyacetal (photodegradable in UV light) [CHA 13]; acrylamide [GUO 01]; polythiol acrylates [BEL 14]; derivatives of propoxylate diacrylates [LIN 11]; oligo-esters [MAT 09]; commercial resins like XPm-P 15AV [VOI 08] and AZ 9260 or AZ 4562 from Clariant [CON 99] and so on. It has already been possible to carry out additive manufacturing work with the resin SU-8 [DEL 07, ABK 11, ABK 13].

Negative photo-resists have been created (see e.g. a coupling between an acrylic pre-polymer and a metacrylate grafted onto a silica nanoparticle [YUA 14]: the irradiated material is then soluble in a sodium bicarbonate solution) (for other polymers, see: Lin *et al.* [LIN 11]; SU-8 epoxy resins [LIU 14]; modified Novolac [ROY 03], etc.). Femto-ST (2015) uses soluble resins like the commercial resins AZ 9260, S1818 and SJR 5740.

For the classically used polymeric resins, the layers (in 2D photolithography) are made up of an entanglement of macromolecules with comparable sizes. Under the effect of radiation, the chemical transformation is thus translated:

– for a positive resin, by the separation of the macromolecular chains into elements with a smaller mass;

– for a negative resin, by reticulations (as this happens in stereolithography processes).

As indicated in Figure 3.29, it is possible to define a variant to these methods by using a gel that is easily eliminated by dissolution, hardened thanks to light [BAR 11, MUR 06]. In this process, the gel is actually made up of a monomer cooled on the part being constructed (up to –50°C), which does not forbid photochemical polymerization. However, the time required for these operations can become an extreme limitation in an industrial process. Rather than operating in this incremental manner with cooling cycles, it could be advantageous to propose to add a congealed resin to the part being constructed, coming from an extruder and placed on the cold object being constructed. These fine plates can be supported by a removable film that is transparent to UV rays and then taken off after each operation.

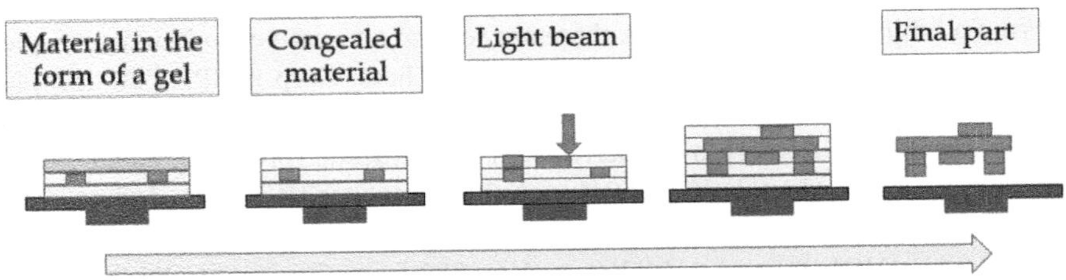

Figure 3.29. *Use of a gel–polymer process*

In different cases, polymers with different masses are obtained, their values depending on the light flow absorbed (and thus the spatial position), which is translated by densities that are a function of the space if there is a decrease or increase in volume on the one hand and solubility (or insolubility) that is a function of the space on the other hand [VAN 06]. These gradients can affect the ability to be soluble (or insoluble) in a solvent.

The process is then defined by the following elements:

1) manufacturing a "solid" resin in sheets containing the resin itself and substances directly or indirectly absorbing light to, after photo-transformation, for example, render it insoluble (see Figure 3.30);

2) radiation on the surface to create points of insolubility according to the spatial instructions imposed to create the casing of the object at the considered dimension or all of the surface at this dimension;

3) dissolution of the non-irradiated resin situated outside the object (and its potential recycling);

4) possible post-curing of the object to solidify it definitively.

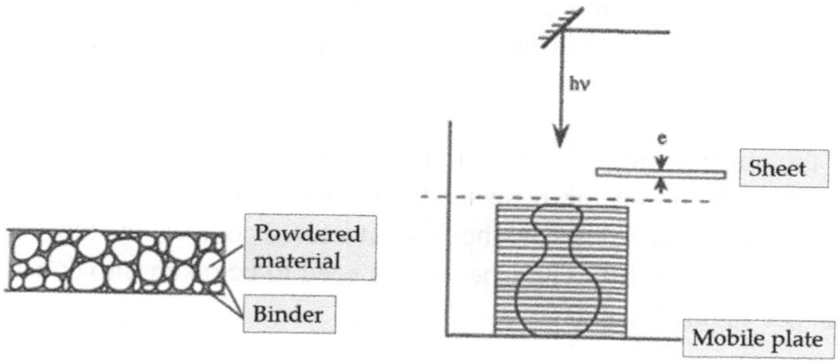

Figure 3.30. *Polymerization of charged resins in sheets*

NOTE.– The problem of dissolution in a solvent:

The low value of the viscosity term for the usual fluids (the solvent in this situation) and the right concordance far from obstacles between the flows in perfect fluids and the real flows led Prandtl to propose the following hypothesis in 1904: in a sufficiently rapid or sufficiently spread out (elevated Reynolds number) flow, the forces of viscosity only play a role in a very small domain, next to the fixed surfaces, which limit flow. On these fixed surfaces, the fluid velocity is zero. Outside the so-called limit layer domain, the speeds of the fluid are those of the flow in perfect fluids that do not take into consideration the forces of viscosity. Thus, beyond the limit layer, the flow is that of the perfect fluid, but the actions exercised by the fluid on the obstacles essentially depend on the limit layer [MID 93, ESP 15, GUY 12].

Thus, between the object and the space where the fluid may be considered free, there is an area in which the velocity progressively varies from 0 to the speed of the free fluid. In this area, the speed gradient is elevated and the forces of viscosity are predominant. The limit layer is a function of the Reynolds number, that is, an increase in the Reynolds number leads to a reduction of the forces of viscosity and thus a reduction of the thickness of the layer. Conventionally, its thickness (δ) is defined at the location where the speed reaches 99% of the maximum flow speed. One difficulty, however, is that the wall is not impermeable, but made up of a resin that is at least partially soluble; there is thus ablation of this wall, possibly depending on the geometry of the system. Moreover, in this layer, there will be contact with the "pure" material of a practically null concentration in solvent while beyond the limit layer it will be maximal, actually imposing a significant gradient of the diffusion coefficient material and chemical potential, as a function of the local resin concentration. This then finds itself in a configuration of "solid–liquid reaction" [GRE 11]. This can easily be modeled for simple configurations (infinite spheres or cylinders), but it is much more difficult for complex shapes [SCH 15, EKM 12].

Thus, if the object formed is very slightly soluble in the solvent, waiting too long to eliminate the remains of the resin in the areas that are difficult to reach can be translated by partial ablation of the manufactured object in the more easily regenerated areas. These different aspects thus need to be taken into consideration in this process (see Figure 3.31).

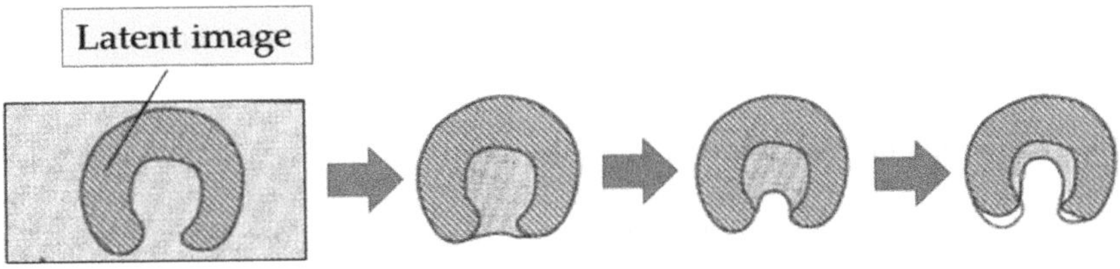

Figure 3.31. *Possible influence of the solubility of the materials on the shape of the final part from the latent image to a revelation of the object as a function of dissolution time (with the unforeseen dissolution of the treated areas)*

Nevertheless, certain successfully tested resins are very insoluble after radiation; the resin Riston® thus could be used in this process without suffering from these kinds of difficulties, as shown in Figure 3.32 of the "handmade" object (using vertical radiation and masks; the film layers being positioned by hand; size of the object on a 5×5 cm^2 square base).

Figure 3.32. *Process involving insolubility of a film polymer: example of creation*

On the basis of machine development, the following project was proposed [AND 94] without creation in terms of a prototype having taken place (see Figure 3.33).

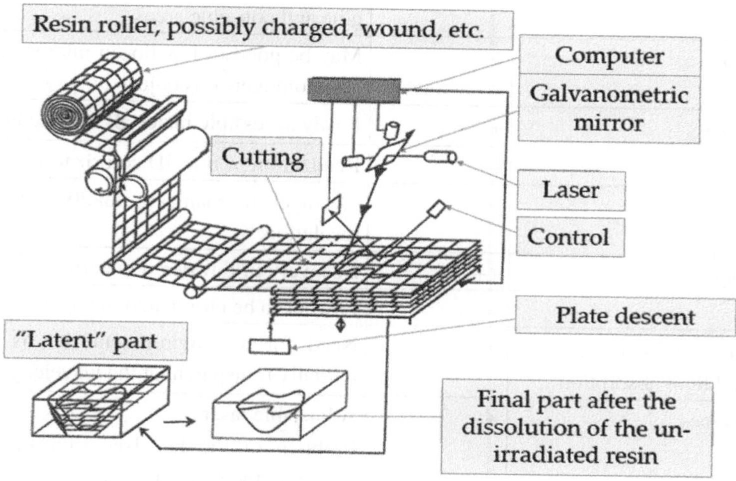

Figure 3.33. *Schema of the principle of a sheet stereolithography machine*

3.4.4. *Synthesis*

Table 3.4 lists all of the elements to be mastered for the development of these processes (classified as + for "easy", = for "doable", - for "existence of barriers" and -- for "barrier point") with comments.

The mastery of these different elements is effectively demonstrated, as seen in the figures representing examples; for all that, the historic aspect of stereolithography with numerous studies on the one hand, the fun and ease of FDM processes on the other hand and finally the strong demand for the creation of (essentially metallic) objects using "good material" have probably not authorized the development of this kind of additive manufacturing.

Theme	Accessibility	Comments
SDL Process: materials and glue deposition	+	Easily accessible materials, classic injectors
Deformation	=	Possible deformation during heating between glue and non-glue areas
Resolution	+	May be good (a few dozen micrometers) but non-homogeneous object
3DP Process	+	Easily accessible materials, classic injectors
Color	+	Realization of fine 3D colorization
Quality	=	Average mechanical quality; satisfactory resolution
Recycling	++	Recovery of powdered material
HSE	=	Powder to be eliminated in the atmosphere
Mono-material process: Localized light absorption	=	Need for an environment that is at least partially transparent at the wavelengths used;
Light sources	+	Adjustable lasers; Problem of the localized division of light energy in the reactionary volume.
Material	=	Choice of polymer materials with low shrinkage (or weak increase in volume) classically used in electronics Possibility of working with films containing a charge
Initial resolution	+-	No foreseen influence of the shape of the latent image on the transfers of matter or heat;
HSE	=	Chemical risk linked to the solvents
Dissolution of the untreated material; final resolution (for processes besides 3DP)	+	Verifying the very weak solubility of the object in the solvent; Examining the effect of the obstacles (the object created) on the kinetics of the dissolution of the resin that was not irradiated;
Supports (for processes besides 3DP)	+++	Useless; the object being constructed is self-supported
Post-curing	+	Classic in stereolithography

Table 3.4. *Elements to be mastered for the development of the process*

3.5. Powder fusion/sintering

The process presented in Figure 3.34 uses powder fusion/sintering [LIP 14, REC 16, FRA 16, DON 09, ZHO 17, NIC 16]. The powder is placed at the desired dimension, compressed and scraped to have a flat and dense thickness. A system to transform the powdered environment resolved on the surface allows fusion according to the trajectory defined by the computer. It can involve infrared, visible light or UV lasers, those from high-energy electron sources and so on. With the surface being transformed, a new layer is placed on the object being constructed, and the object is built slowly. There are multiple names for this: SLM for “selective laser melting”; SLS for “selective laser sintering”; DMLS for “direct metal laser sintering”; EBM for “electron beam melting” and so on.

As will be shown in section 3.5.4, the powder processes transform low-density (ca. 60–70%) porous materials into a dense material capable of reaching porosities below 10^{-3}. From this statement comes a need to define, as precisely as possible, a part that anticipates these effects both in terms of volume reduction and shape. As the layer-by-layer operation also introduces tensions and anisotropies, it is easy to imagine the difficulty achieving this with a computer in the most successful way possible [SCW 16, BIK 16, CON 11, DEH 15, GAN 14, HOD 14, HOY 14, ZHO 15, etc.].

As with the process described in section 3.5.3, in principle it is not necessary when programming to create supports to hold up the part being built. However, we will note that supports/reinforcements may be necessary to maintain the desired shape of the object due to anisotropies introduced by the manufacturing process. After final manufacturing, the object may undergo thermal post-treatment, which allows the structural support elements to be removed. Relative to the three techniques presented above, this process allows the creation of functional metallic parts, which makes it particularly attractive in mechanical industries.

As an example, in this domain, we will cite the joint effort in France of Fives and Michelin [LAM 15] aiming for the industrial creation of metallic parts (and 3D machines dedicated to this) in a co-enterprise “Fives Michelin Additive Solutions”. Furthermore, to illustrate the significant development of this domain, it may be useful to recall the creation of a magazine of interest to machine produces and users (3D Metal Printing Magazine) under the responsibility of the PMFA [PRE 15]. For an example of creation, Figure 3.35, generously provided by Fives, illustrates a set of hydraulic canals using this process.

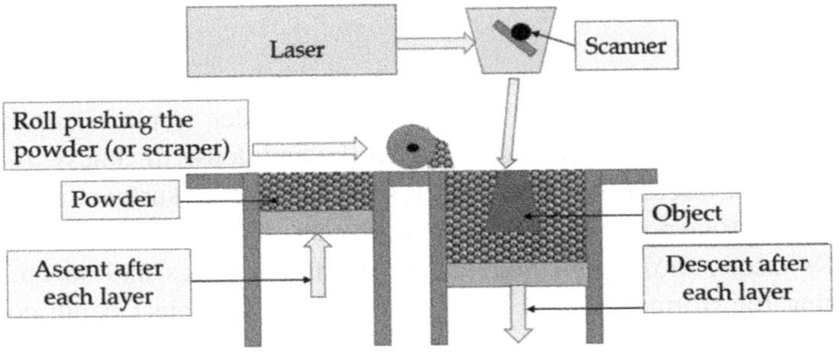

Figure 3.34. *Selective laser sintering (SLS/DMLS)*

Figure 3.35. *Set of pipes created by Fives*

Very recently, the University of Sheffield (UK) has replaced the laser with a set of laser diodes (each element has a strength of only 2.6 W (at 808 nm) and is not capable of melting metallic particles, but the set of diodes allows the fusion of stainless steel). Figure 3.36, from Zavala-Arredondo *et al.* [ZAV 17], illustrates the principle of the process.

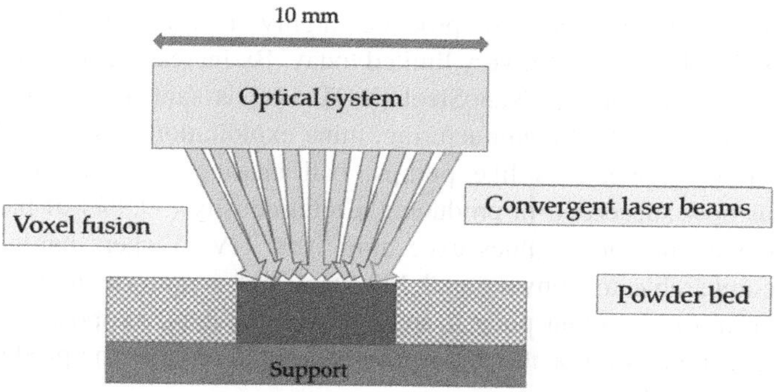

Figure 3.36. *Principle of fusion using laser diodes*

3.5.1. *Materials*

All fusible and wetting materials are, in principle, usable in this process. It can involve polymerized powder materials [ZCO 15], waxes, microparticles [FAN 16] and metals (for metallic materials, see e.g. Farinia Group [FAR 15] for the different commercial powders, SAE Int [GEH 15] for the "scalmalloy" metal alloy aluminum–magnesium–scandium (AlMgSc) used for aeronautics; Kenzari *et al.* [KEN 14] for quasi-crystals; Li *et al.* [LIY 17] for alumina; Primante 3D [PRI 15] and Frazier [FRA 14] for aluminide, which is a mix of polyamide and aluminum; Frazier [FRA 14] for metal alloys and so on). This final material offers the advantage of being able to create both very solid and very flexible parts with significant heat resistance: tool steel is a stainless steel containing martensite, a sub-group of stainless steels characterized by great solidity and hardness, obtained after later thermal treatment. This material is intended, above all, for quick tool manufacturing (cutting, extrusion, etc.) and molds in the domains of aeronautics, astronautics and automobiles; titanium is more often used (automobile, aeronautics and medicine) in titanium alloys like Ti_6Al_4V, more solid than the pure original material and biocompatible; aluminum via the component $AlSi_{10}Mg$ (alloy with magnesium and silicon) is both very light and very solid.

According to Cetim [CET 17], "regularly used in thermal or cold projection, powders nevertheless require very particular attention when used for additive manufacturing. Their characteristics (castability, density, morphology, grain size, etc.) actually depend on the machine and the quality of the part to be manufactured. For some applications, notably in aeronautics, it is necessary to coat the original grains, e.g. with boron carbide coated with cobalt."

The choice of materials to produce highly resistant steel via additive manufacturing is, however, still very limited today. By increasing the significance of steel on the current markets, NanoSteel [NOT 14a] is said to have created cost-efficient wear parts for tool manufacturing, mine exploitation, energy and industrial transportation in applications like pumps, ball bearings and cutting tools. The company has now succeeded in producing a total density (99.9%) of parts with no fissures and with hardness values exceeding 1000 HV (Vickers hardness test), a resistance comparable to conventional M2 steel and a uniform microstructure to wearing. The most important point is that this steel requires no post-treatment like isostatic heat compression or thermal treatment, which reduces the production costs and times considerably.

3.5.2. *Energy sources*

The classical energy sources are lasers. For reasons of yield and cost, CO_2 lasers appear to be excellent candidates for melting materials that are fusible at high temperatures. However, their wavelength is 10.6 µm, which leads to a resolution that is several times this wavelength. If the user wishes to gain more resolution, lasers emitting in visible or UV light can be used (laser diodes in particular, as in the example from Figure 3.36), maybe through impulse stimulation to allow effective fusion. Finally, for maximum resolution applications, electron sources are being used more and more often [OBA 15], allowing 1 µm or a few micrometers to be achieved. Under these conditions, it is the average size of the mineral particles that may become a limiting factor in terms of precision.

3.5.3. *Physicochemical aspects and constraints linked to the process*

First, as a function of the polydispersity of the materials and the regulation of energy flows, there can be the creation of empty spaces in the material making up the object; after all, at present, this technology provides very high quality parts with a precision comparable to industrial needs, particularly for inorganic materials.

3.5.3.1. *Interparticle porosity*

With mono-disperse spheres (simplifying principle), the interparticle porosity (ε) represents the fraction of the interparticle volumes relative to the volume of the whole object considered. If ρ_{app} and ρ_{sol} represent the apparent volumetric mass and that of the massive solid, then:

$$\varepsilon = 1 - \rho_{app}/\rho_{sol}$$

It is about 45%, hence the interest of using a more polydisperse population to gain in this aspect which, obviously, will be translated by a consequent volumetric reduction: the small particles can fill the spaces between the larger ones, which reduces ε. Beyond a certain threshold, this advantage disappears to again find, for an infinite dilution of the large particles, the initial value of ε [MOL 15].

Gradually, it becomes possible to find size divisions so that the space is occupied to the max. However, it is not possible to achieve the compactness of the massive solid with divided solids. The granular materials are stacked upon one another and they block each other to the point that they sometimes result in an arching effect that stops all flow. A vibration can sometimes suffice to break this arch. Maximally compacted, the granular materials, among themselves, maintain a gap representing about 25% of the total volume. Figure 3.37 represents the ideal case of "Apollonian" stacking (Apollonius of Perga having lived in the 2nd Century BC), far from the reality that must consider the morphology of the powders.

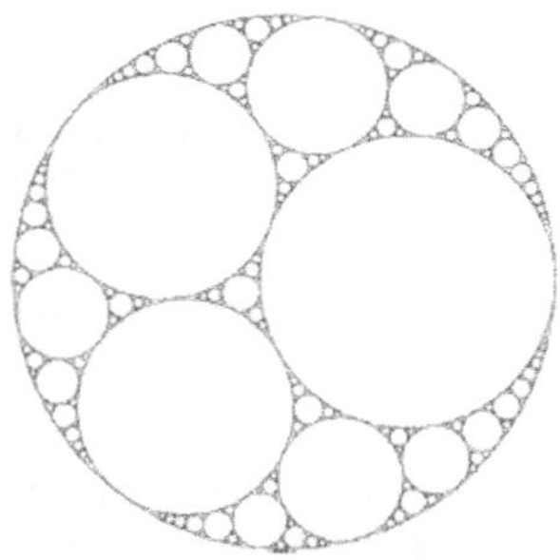

Figure 3.37. *Apollonian stacking*

It is thus understood that there is no way which is likely to easily avoid trapping bubbles in the majority of the melted material because the total compactness theoretically cannot be achieved (hence the interest of working in a vacuum with an electronic source). In any case, the removal aspects belong to the issues that must be dealt with in terms of programming an object's manufacturing (reverse engineering).

3.5.3.2. *Other phenomena*

According to Audrey Loubens in Techniques de l'Ingénieur [LOU 16], "the preparation of DMLS manufacturing involves process elements similar to those of stereolithography and selective laser sintering, as it requires a support system [following in the footsteps of "historic" stereolithography], even if the parts are sintered in a layer of powder (like in SLS). The quick fusion of each layer of metal can produce residual constraints in the part during the DMLS process. The supports holding the parts in place are necessary to avoid all deformation that could take place due to such constraints as each layer cools." Thus, even if the supports are not necessary in principle because the object under construction is "self-supported", the tensions introduced by the process prescribe the presence of supports.

As Lin *et al.* [LIN 14a] indicated, surface fusion, beyond reduction phenomena linked to the porosity of the powder, can lead to different defects: oxidation, undesired residual porosity, internal heterogeneities and so on (see Figure 3.38). Inert gases can be used to avoid part of these unexplored effects (or working in a vacuum when the localized energy source is an electron beam).

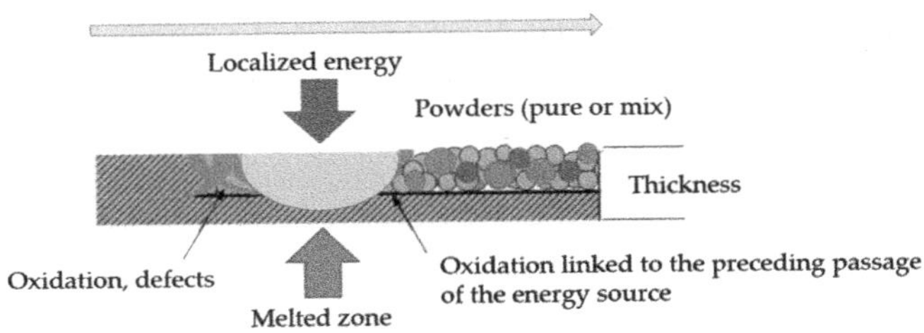

Figure 3.38. *Structural defects introduced by the process*

As a note, Ponche *et al.* [PON 14] proposed a prior "deformation" methodology of the theoretical object in order to consider *a priori* deformations linked to the process. This improvement applied to metallic powders is significant, as it reinforces the interest of the additive manufacturing method for direct applications (and not as a prototype part). However, if the idea is credible, creating an "inverse problem" resolution still remains difficult. It is still often necessary to go through experimental modeling (creating the definitive object using a greatly optimized "sketch").

This situation where local fusion effectively leads to the additive manufacturing of the object, which, under some super-hydrophobic conditions (by analogy with water), cannot be achieved. This is the case with isolated liquid elements, generally spherical, which are supported by highly disorganized surfaces or the opposite (see Figure 3.39). This phenomenon is also reached when the sphere of a liquid like water is mixed with a hydrophobic powder that changes the surface tension and thus isolates the liquid from the support. We then speak of null anchorage and a so-called "fakir" drop (see [AUS 01, LAG 10]). The presence of fine particles of the material to be melted or degradation products of this material (e.g. oxidation) may be likely to resemble conditions of super-hydrophobia, which can lead to prohibitive defects (and possible non-recyclability of the powder materials not involved in the transformation).

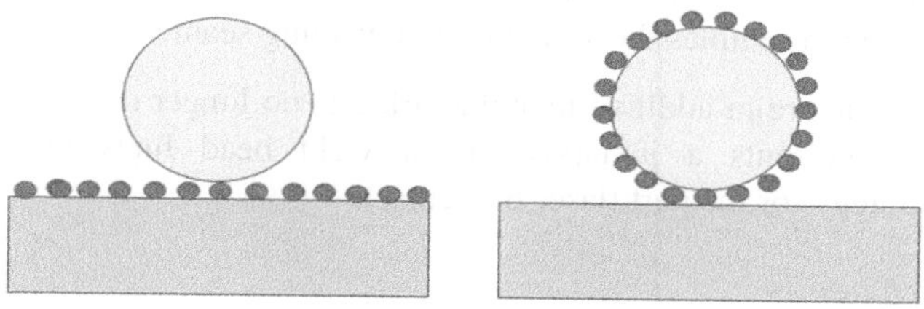

Figure 3.39. *Conditions of "super-hydrophobia"*

Another phenomenon to be considered is the fact that fusion allowed by the energy source is not homogeneous and that solidification can lead to anisotropic structures and tensions in the part created by this process, which then has a "memory" of its manufacturing pathway (see Figure 3.40). It is sometimes necessary, according to the materials used and the function of the object created through additive manufacturing, to effect adapted re-baking.

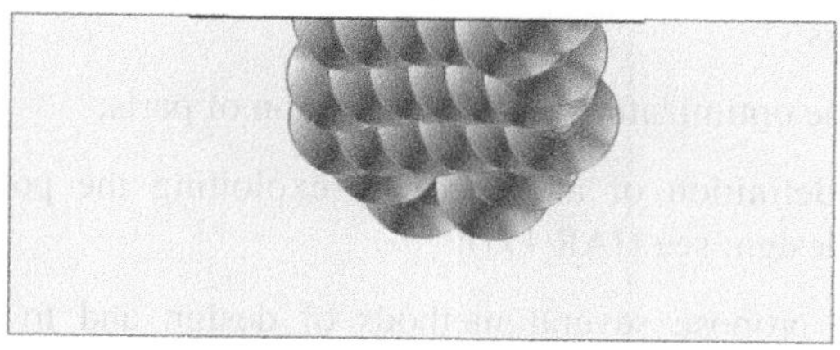

Figure 3.40. *Structural heterogeneities introduced by the process where each treated area has a fusion/solidification gradient with an effect of the object's superior heights on those below*

Thus, according to Stucky [STU 15], a study on steel 316L used in additive manufacturing shows that:

– the static mechanical characteristics at 20°C (Rm, Rp0.2 and extension) are better in additive manufacturing compared to the same foundry nuance;

– the static characteristics at 500°C are still better in additive manufacturing than the characteristics of foundry at 20°C;

– the mechanical characteristics are anisotropic in additive manufacturing and depend on the sampling direction and the test tube;

– the microstructure of the part's case in additive manufacturing is specific (homogeneous) and resembles the structure of a welding seam;

– the microstructure in additive manufacturing is no longer dendritic like that in foundry, but it presents a juxtaposition of weld bead lines that reflect the construction strategy for the part (laser trajectory).

3.5.3.3. *Future needs*

Considering what was presented previously and conclusions from the report provided by Observatoire de la Métallurgie [OBS 17], the development of 3D technology for fusing/sintering metallic powders (metals or alloys) still needs improvements like:

– the creation of parts whose dimensions are consistent with the instructions, created with the "good matter";

– better understanding of the behavior of matter at the macroscopic level, an ability to prove the mechanical resistance of the part over time [CHA 16];

– mastery of the behavioral anisotropies of the objects manufactured brought about by the process;

– research on the optimization of the finalization of parts;

– the optimal definition of an object by exploiting the potential of additive manufacturing (redesign; see [JAR 17]);

– an ability to propose several methods of design and to propose the most adapted technology (with a good knowledge of metallurgy);

– know-how concerning topological optimization, the integration of functionality;

– mastery of the potential recycling of materials;

– and so on.

3.5.4. *Simultaneous contribution of matter and energy*

The process presented in Figure 3.40 uses both an energy source (generally an infrared laser) and fusible powders (even at high temperatures) simultaneously reaching the object being constructed [CAS 14]. The matter injected is melted little by little until the desired object is produced. The powders are projected into the fusion bath created by the interaction of the laser with the deposition substratum, as the different components of Figure 3.41 indicate.

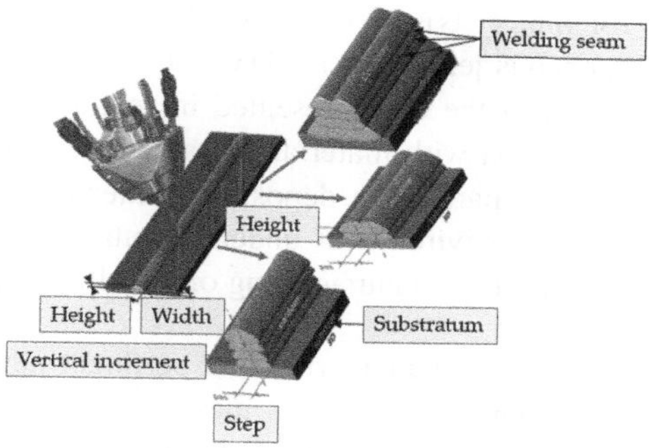

(a) General principle

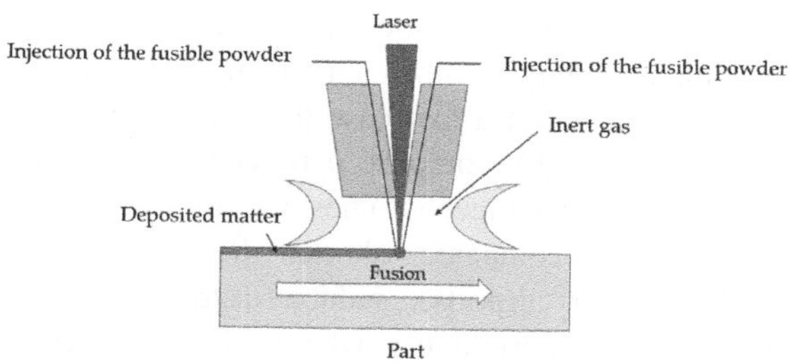

(b) Nozzle function

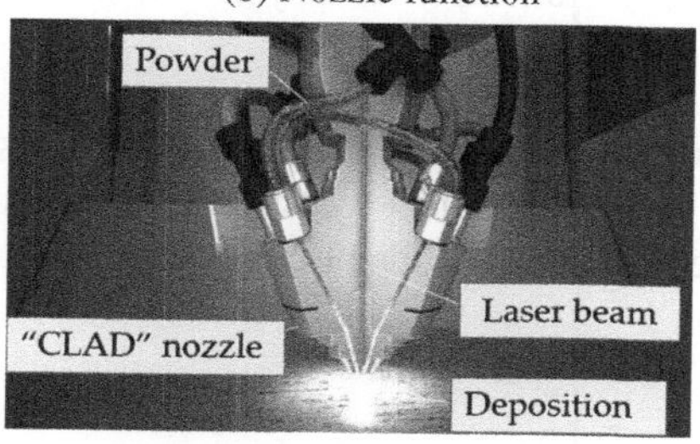

(c) "CLAD ®" nozzle

Figure 3.41. *Localized deposition of matter and fusion by laser: LMD or CLAD process (Figures 3.41(a) and (c) were generously provided by BeAM)*

An interesting variant, the CLAD (Construction Laser Additive Directe) process was improved by the IREPA-Laser in 2012 [IRE 15] in Strasbourg, France, and it melts powders before they come into contact with the fusion bath. It can be called LMD or DED for direct energy deposition. This process uses a fixed nozzle, the object moving across three linear axes, synchronized with two rotating axes and a device using a movement/rotation system for manufacturing on five axes. This technology not only allows complete parts to be created but also the use of pre-fabricated elements on which matter is placed. By pushing the technology into a new application niche, it can be a device for repairing metallic industrial parts (already applied with success for motorists). Metallic powders are injected into the CLAD nozzle to form a homogeneous jet [MOL 14, DMG 14]. In crossing the laser beam, the powders are melted (as in the case presented in section 4.3.2). The result is a dense, homogeneous deposition with material characteristics located between those of foundry parts and forging parts. The deposited matter can be protected locally from oxidation by a machine environment under a neutral gas (see also [VER 14, MAR 15], concerning the additive manufacturing of metallic objects).

The additive manufacturing repair market is a pertinent new application niche. BeAM is one of the few companies that has mastered this process, including the production of equipment based on five-axis machines re-designed to be compatible with the process. With LMD technology, metallic parts become reparable as long as the powder-injection nozzle/laser couple can reach the zone to be treated. Furthermore, it is complementary to other manufacturing technologies: for objects of which 90% can be produced at a low case through traditional methods, BeAM technology, by creating only the complex part, contributes another determinant relative to the panoply of other 3D technologies. It is thus a true revolution that will help constructors and airline companies shorten the repair time considerably (if machines are available near airports), reduce usage costs and thus increase competitiveness [DIA 16]. This originality, where it is the object that is moved, has lately been revisited by two students with a system with 6 degrees of freedom (see [REN 17a]).

Figure 3.42, generously provided by BeAM, illustrates the potential of this technology. The lower part of the object was created with another method and DED (CLAD); matter resolved in space is added to a partially constructed object.

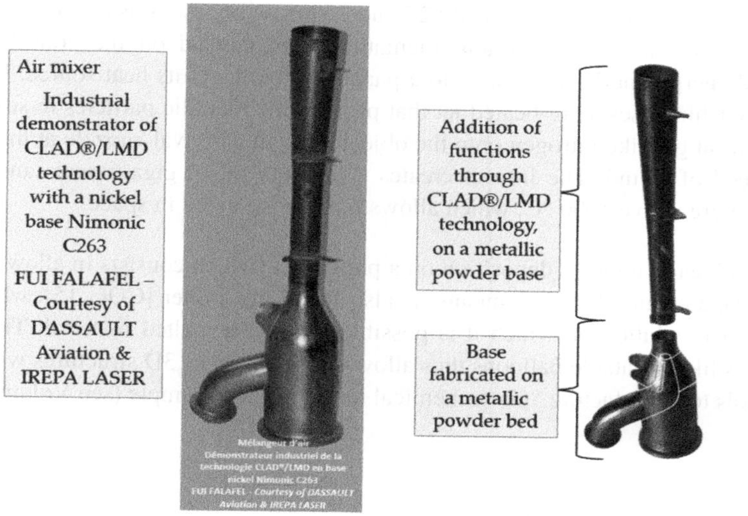

Figure 3.42. *Example of a metallic part created with the DED process*

Equipment has been made by Toshiba using laser metal deposition (LMD) technology associating the deposition of melted metallic powder and an 800 W laser. The laser heats and sinters the powder, allowing the desired object to be achieved layer by layer, like in the CLAD process [3DN 15].

In both cases, the joint choice of the polydispersity of the powders and the operation of the injection nozzles is an element defining the quality of the process.

NOTE.– The idea of the CLAD process has been revisited today by Stratasys, which proposes using a wire printhead (FMD process) associated with an eight-axis robot to adopt an approach of manufacturing in space, making parts with greater dimensions than those offered by current machines [3DN 15].

3.5.5. *MPA process*

The company Hermle in Germany has developed an imminent device using a five-axis milling machine, allowing the same kind of applications. According to Sher [SHE 15b], the process allows the use of powders from metals and alloys with varying natures and sizes (typically 25 and 75 µm) with issues associated with these kinds of depositions with grain orientations that depend on the contribution of thermal energy and heat transfers to a part being built. As its heat source, Hermle's MPA machine uses a pre-heated jet that projects the metallic particles in suspension in a neutral gas like nitrogen onto the object through a "Laval" nozzle at three times the speed of sound. The impact creates a pressure of 10 giga-pascals and a local temperature above 1000 °C, which allows forging localized in space.

This idea to create a deposition on a pre-object (which consists in allowing work on a shape created by other means) is also found in Coulter [COU 15], which uses forms of revolution on which it is possible to deposit melted material. The use of supports like inflatable balloons thus allows the creation of 3D structures with active materials to manufacture "photochemical muscles", for example (see Volume 3).

3.5.6. *Synthesis*

This technology is in a phase of total development because it is (finally) possible to create parts with the appropriate material for an "online" use; furthermore, the possibility of associating the additive process with a five-axis (or more) system to place the object in a good position relative to the nozzle allows the successful exploration of new application niches.

Every mechanical industry has moved towards additive manufacturing, with two primary trends in terms of creating objects with direct metal or polymer part manufacturing:

– Polymer powders allowing better mechanical and thermal resistance. In terms of the resistance of polymer powders, 3D manufactured polymer parts reach a Young module of 11 giga-pascals. Among the most resistant powders are the poly-ether-ether-ketone (PEEK) powders and, more and more, reinforced polyamides (PA 12 nylons) [LES 15]. Charged materials are proposed with fiberglass or carbon elements, even with conductive materials.

– Machines allowing larger part dimensions. The aerospace industry is a good indicator of the state of the art: according to Lescuyer [LES 15], Thales manufactured the largest space parts ever created in Europe with powder bed additive manufacturing (aluminum antenna supports, dimensions $44.6 \times 20.2 \times 39$ cm^3) using a direct laser sintering technology.

The true novelty is the emergence of two segments of use: the re-manufacturing and the repair of detached parts for material in operation (rail in particular; see CLAD process).

Table 3.5 lists all of the elements to be mastered for the development of these processes (classified as + for "easy", = for "doable", - for "existence of barriers" and -- for "barrier point") with comments.

Theme: Fusion of powders/LMD	Accessibility	Comments
Materials	+	Large number of fusible organic (polymers) and inorganic (ceramics, metals, glass, etc.) materials
Multi-materials	+	The mix may be homogeneous
Color	--	No
Sources	+	Classic: lasers, electron beams
Mechanical strength	+	Good, even very good; possibility to re-bake
Manufacturing risks	=	Super-hydrophobicity with very small particles changing the surface tension
HSE	=	Powders
Recyclability	=	Verifying that the powders not used in the process keep their properties
CLAD process	-	The nozzle must reach the area to be treated (limitation of certain shapes)
Multi-manufacturing	+++	Coupling classical matching methods and additive manufacturing
MPA process	+++	Same as above
Post-treatment	+	Possible and rather classic

Table 3.5. *Elements to be mastered for the development of additive manufacturing processes through powder fusion*

3.6. Conclusion

In this chapter, by partially revisiting a schema published by Rosenberg [ROS 16], the machine and process aspects on the one hand and materials on the other (and their association) were described and discussed. It is clear that current developments in computer science can be associated with this presentation. However, Figure 3.43 recalls that innovation in this domain is not separate from the world. Additive manufacturing is developing because there are current markets (unlike what was happening in 1984) which can have various impacts on society. Numerous interactions and returns considering this context can thus have an effect on the technical aspects presented here.

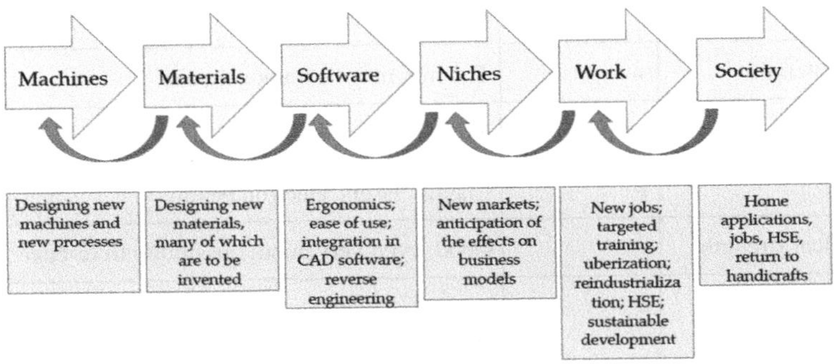

Figure 3.43. *"Virtuous" approach of innovation in additive manufacturing*

As an example of future operations, Zadpoor [ZAD 16] associates the spatial distribution of materials in origami shapes to play on the mechanical properties of the object created with additive manufacturing; he also foresees the creation of structures combining several materials. This example is a great illustration of the necessary trend of the interdisciplinary coupling of process-materials and software.

Several basic technologies have been presented in this chapter; they are diverse even if their principal foundation is based on the concept of adding matter. In an attempt to synthesize these presentations, these different technologies are listed in Table 3.6.

Theme mentioned in this chapter	Abbreviation	Name of the process	Comments
Stereolithography	SLA	Stereolithography	Polymerization of a charged or uncharged resin brought about by light Different irradiation systems
Wire fusion	FDM	Fused deposition modeling	Wire fusion: stratum-by-stratum deposition of a fine layer of melted wire. The matter cools again upon contact with the air and solidifies From the melted polymer to glass to metals
	FFF	Fused filament fabrication	See FDM
Gluing sheets or powders	SDL	Selective deposition lamination	Sheets associated with a glue that can be activated; uses the inkjet system (light-sensitive material or one reactive to UV)
	SC	Strato-conception	Cutting sheets – "mechanical" assembly; layer-by-layer, a laser or a cutting tool removes the contours of the desired object
	3DP	Three-dimensional printing	Powders associated with a glue; uses the ink jet system (light-sensitive material or one reactive to UV) or thermic gluing
	BJ	Binder jetting	See 3DP; particles are placed on the printing plate and linked through the addition of an aggregating, possibly colored liquid

	SIR	Soluble insoluble Reaction	Change in the solubility of an organic material
Fusion/powder sintering	SLS	Selective laser sintering	Sintering brought about by laser on the plane surface
	SLM	Selective laser melting	Fusion brought about by laser on the plane surface
	EBM	Electron beam melting	Fusion brought about by electron beams
	LMD	Laser metal deposition	Deposition of metal simultaneous to its fusion
	CLAD	Construction laser additive Directe	Simultaneous addition of matter and light energy: matter melted during its deposition by a strongly focused heat source (laser)
	MPA	Metal powder application	See CLAD, kinetic energy of the particles

Table 3.6. *Different basic techniques in additive manufacturing*

Process	TP polymer Thermoplastic	TD polymer Heat-hardening	Metal	Ceramics	Multi-materials
SLA	+		+ C	+ C	
FDM/FFF	+		=		
SDL	+	+ (paper)	=	=	=
SC	+	+	+	+	=
3DP/BJ	+	+	=	+	=
SIR	+	+	+ C	+ C	
SLS/SLM	+		+++	+	
EBM/LMD/ CLAD/MPA			+++		

Table 3.7. *Materials – additive manufacturing process relation (C for added charges)*

To attempt a provisional conclusion, in the significant domain of mechanical industries, Table 3.7 presents the different materials that can be used in relation with current additive manufacturing process ([FIM 15, BER 13], quoted by [ITR 15, DE 16] and elements presented in this chapter). This table illustrates multiple possibilities with choices that are today quite board in terms of materials, but with few possibilities to have material gradients, even if studies, presented in Volume 2, seem to prove that it is possible to go in this direction.

Different types of materials used, beyond the "great classics" presented above [CAB 15, LOW 17], display the following primary characteristics:

– metals [ORC 16], for example aluminum, iron, brass, copper, bronze, sterling silver, gold, platinum, titanium and Inconel [NOT 16]. Figure 3.44 presents the more-than-exponential evolution of the use of metallic powders, which reached nearly 800 tons in 2015 [PÉC 17]. The methods of producing these powders revisited by Singh, Ramakrishna and Singh [SIN 17] are numerous (which illustrates a certain difficulty in their manufacturing). Figure 3.45 synthetically presents these methods of producing metallic particles;

– metal–polymer mixed materials [NOT 16, 3DN 16a, LEE 05, MAR 17];

– materials containing graphene, 10 times more resistant than steel [CHA 17];

– polymer: thermoplastic polyurethane (TPU) and poly(lactic acid) (PLA) charged with graphene oxide with excellent mechanical performance [CHE 16a];

– "active" polymers [CHE 17];

– ceramics [COM 14, CHA 03, GON 16]. For traditional ceramics, natural raw materials (loam, clay, kaolin) are used. For technical ceramics, micrometric powders obtained through chemical synthesis are used; this is often referred to as "fine ceramics" [DÉJ 10];

– cement: outer walls of 3D houses [BEN 16];

– clay: inner walls of 3D houses;

– wood: actually a wood paste made up of approximately 50% wood dust and polymeric resin to create objects with a decimeter size;

– stone and marble: as with wood, assembly of stone and resin powders;

– waxes and sands: used in foundry (lost wax and artistic molding);

– paper: specific lamination process;

– glass: still at a very experimental stage;

– ink for tattoos;

– chocolate, chewing gum and various foods (including cellulose according to Bozec, [BOZ 16]);

– and so on.

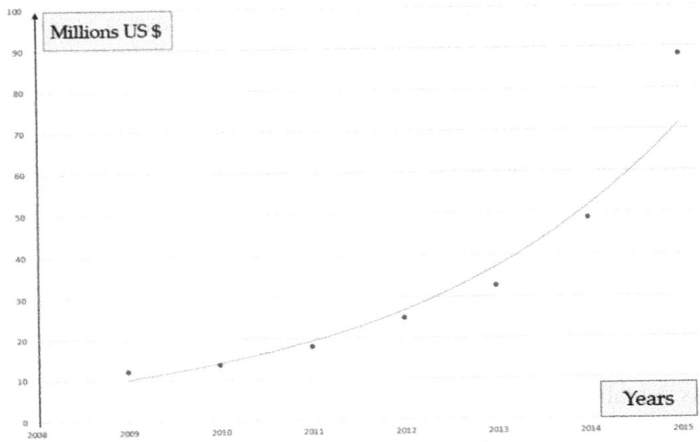

Figure 3.44. *Metallic powders market (exponential curve)*

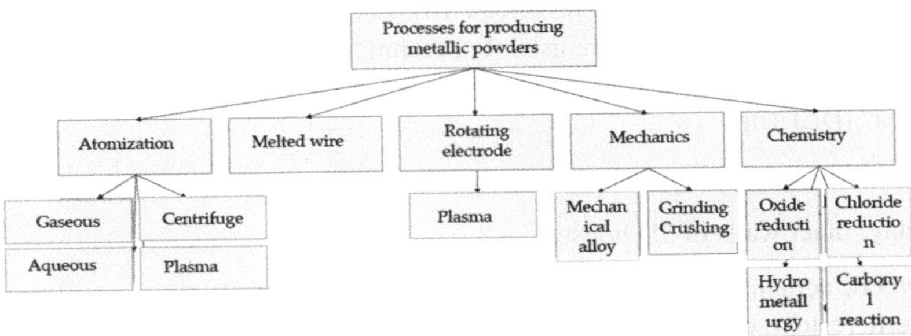

Figure 3.45. *Manufacturing processes with metallic powdered materials*

Berthome *et al.* [BER 16] presented approximate mechanical tolerance and roughness values as a function of the methodology of additive manufacturing used. Table 3.8 presents the data connected to the technologies discussed in this chapter (see also Lieneke *et al.* [LIE 15] for French normative aspects concerning NF EN ISO 286-1, which establishes a codifying system to be used for the sizes of such dimensional entities as a cylinder or two opposite parallel surfaces and which allows the comparison of the 3D technologies with other manufacturing methods with tolerance classes for usage ranging from 5 to 16; additive manufacturing appears to be a perfectly competitive technological set of machining methods).

Technology	Tolerance (μm)	Roughness (μm)
SLA	30	10
FDM/FFF	100	15
SDL	20	15
3DP/BJ	100	15
DMLS	50	8
EBM	200	20
SLS	250	50
FDM	120	8

Table 3.8. *Approximate values of mechanical tolerance and roughness as a function of the additive manufacturing technology*

According to Raynal [RAY 16], a new technique aiming to reduce roughness would allow a smooth and shiny surface to be obtained on 3D objects printed using laser sintering. This finishing process would come into play after printing and increase manufacturing time. “For example, if we take a small plastic the size of a telephone, it would take 1.5 h to print 10. Three more hours would then be added to the process to achieve a fully smooth appearance.” The idea is to produce objects meant for final uses and not as prototypes. “Today, slightly more than 50% of production volumes are dedicated to mass production.”

Thus, in agreement with Wood and Baya [WOO 12], this chapter shows the current trends in the processes-materials paired domain. These elements are gathered in Figure 3.46, which will serve as a backdrop for the presentations in Volume 2.

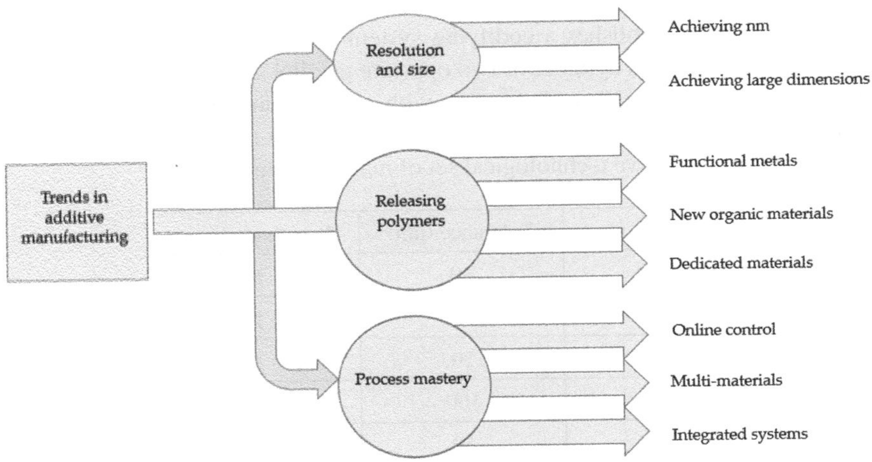

Figure 3.46. *Improvements foreseen for processes-materials in additive manufacturing*

These data are provisional, in a domain where the number of publications and patents is increasing by a few dozen percentage points per year. Despite a significant bibliography, this analysis, which the author hoped to make as exhaustive as possible, is only a small element of a broader set under construction. It must be remembered that this chapter is an update that risks soon becoming outdated. Furthermore, Chapter 1 of Volume 2 already aims to explain the field of incremental novelty with original works, some of which will soon appear on the market, while others are still at the proof-of-concept stage.

We should thus remember that this chapter is a sort of snapshot, already blurred by an incomplete analysis of approximately 60,000 detected publications with a consequent dynamic that reinforces this impression of a certain imprecision, even if the great trends presented here have remained rather stable for some years now. In any case, the current imperfections of the technology presented here in some way constitute a form of demand for further and new research so that the implantation of additive manufacturing in society will continue in a successful manner.

> *"All governance of enterprise supposes having directives in terms of an appetite for risks and it is the administration council's job to give clear directives to the general management, and through it, to all of the collaborators, to explain what appetite for risks is best for the organization". [LOU 17]*

"There is a consensus around the world in terms of scientific policy based on the fact that the future of research rests, among other things, on solving society's stakes with the contribution of numerous disciplines. For the CNRS, as in other places, interdisciplinarity constitutes a true paradox. Although it is one of the most multidisciplinary institutions in existence, it must nevertheless make substantial efforts to stimulate interdisciplinary research. This indicates that there are even barriers within the institution that will have to be removed if we wish to benefit from the wealth and diversity of the science that is produced there". [FUC 16]

3.7. Bibliography

[3DC 16] 3D CERAM, "La stéréo-lithographie appliquée à la céramique", available at: http://3dceram.com/category/3d-ceram/une-technologie-unique/, 2016.

[3DE 14] 3DERS, "ColorFabb announces new brassFill 3D printing filament", available at: http://www.3ders.org/articles/20141219-colorfabb-announces-new-brassfill-3d-printing-filament.html, 2014.

[3DM 15] 3D MATTER, "What is the influence of color, printing speed, extrusion temperature and ageing on my 3D prints?", available at: http://my3dmatter.com/what-is-the-influence-of-color-printing-speed-extrusion-temperature-and-ageing-on-my-3d-prints/, 2015.

[3DN 15] 3D NATIVES, "Toshiba dévoile un premier prototype d'une imprimante 3D métal", available at: http://www.3dnatives.com/toshiba-imprimante-3d-metal-14122015/, 2015.

[3DN 16a] 3D NATIVES, "Stratasys dévoile sa vision du futur de l'impression 3D", available at: http://www.3dnatives.com/stratasys-infinite-build-30082016/, 2016.

[3DN 16b] 3D NATIVES, "Un procédé de jet d'encre métallique dévoilé lors du salon Formnext", available at: http://www.3dnatives.com/xjet-impression-3d-07112016/, 2016.

[3DP 15] 3D PRINTING INDUSTRY, "3D printing processes: the free beginner's guide", available at: http://3dprintingindustry.com/3d-printing-basics-free-beginners-guide/processes/, 2015.

[ABK 11] ABKARIAN M., "La micro-fluidique", available at: https://www.coulomb.univ-montp2.fr/perso/manouk.abkarian/manouk_abkarian_Homepage/Links-files/microflvidiqve.pdf, 2011.

[ABK 13] ABKARIAN M., PROTIÈRE S., ARISTOFF J. *et al.*, "Gravity induced encapsulation of liquids by destabilization of granular rafts", *Nature Communications*, vol. 4, no. 1895, 2013.

[ALL 85] ALLEN N.S., CATALINA F., GREEN P.W. *et al.*, "Photochemistry of thioxanthone - I. Spectroscopic and flash photolysis study on oil soluble structures", *European Polymer Journal*, vol. 21, pp. 841–848, 1985.

[ALL 89] ALLEN N.S., *Photo-polymerisation and Photoimaging Science and Technology*, Elsevier, London, 1989.

[AND 92] ANDRÉ J.C., BRULLÉ Y., BARALDI U. *et al.*, Fabrication d'objets monochromes ou polychromes par stéréo-photo-lithographie à partir de poudres ou de granulés, French Patent, 92-08316, 6 July 1992.

[AND 94a] ANDRÉ J.C., CORBEL S., *Stéréo-photo-lithographie Laser*, Polytechnica, Paris, 1994.

[AND 94b] ANDRÉ J.C., PÉTILLON N., "Vers la photographie tridimensionnelle en couleur", *Proceedings of the 3rd European Conference on Rapid Prototyping*, La Villette, Paris, 5–6 October 1994.

[AND 01] ANDRZEJEWSKA E., "Photo-polymerization kinetics of multifunctional monomers", *Progress in Polymer Science*, vol. 26, pp. 605–665, 2001.

[AND 15] ANDRÉ J.C., "N'est pas Prométhée ou Mary Shelley qui veut: une petite histoire de la stéréo-photo-lithographie…", *Actualité Chimique*, vols 397–398, pp. 99–101, 2015.

[AND 16] ANDRÉ J.C., GALLAIS L., AMRA C., "Procédé pour la réalisation d'un objet tridimensionnel par un processus de photo-polymérisation multi-photonique et dispositif associé", French patent 16-59211, 28 September 2016.

[AND 17] ANDRÉ J.C., "Stéréo-lithographie par photo-polymérisation", *Techniques de l'Ingénieur*, ref. BM7910 V1, p. 19, 2017.

[AUS 01] AUSSILLOUS P., QUÉRÉ D., "Liquid marbles", *Nature*, vol. 411, pp. 924–927, 2001.

[BAL 84] BALLAND D., GUILLARD R., ANDRÉ J.C., "Experimental studies and modelling of light distribution in photopolymerizable composite materials", *Dental Materials Journal*, vol. 3, pp. 93–98, 1984.

[BAR 10] BARTH J., BUBACK M., HESSE P. *et al.*, "Termination and transfer kinetics of butyl acrylate radical polymerization studied via SP-PLP-EPR", *Macromolecules*, vol. 43, pp. 4023–4031, 2010.

[BAR 11a] BARTH J., BUBACK M., RUSSELL G. *et al.*, "Chain length dependent termination in radical polymerization of acrylate", *Macromolecular Chemistry and Physics*, vol. 212, pp. 1366–1378, 2011.

[BAR 11b] BARTOLO P.J. (ed.), *Stereo-lithography; Materials, Processes and Applications*, Springer, New York, 2011.

[BAR 11c] BARTOLO P.J., "Stereolithographic process", in BARTOLO P.J. (ed.), *Stereo-lithography; Materials, Processes and Applications*, Springer, New York, pp. 1–36, 2011.

[BAR 11d] BARTOLO P.J., GIBSON I., "History of stereolithographic process", in BARTOLO P.J. (ed.), *Stereolithography; Materials, Processes and Applications*, Springer, New York, 2011.

[BAR 15] BARLIER C., BERNARD A., *Fabrication additive: du prototypage rapide à l'impression 3D*, Dunod, Paris, 2015.

[BAR 16] BARLIER C., "Strato-conception", available at: http://www.stratoconception.com/ 2016.

[BEL 14] BELBABRA Z., CHERKAOUI Z.M., ALLONAS X., "Photocurable polythiol based (meth)acrylate resins stabilization: new powerful stabilizers and stabilization systems", *Polymer Degradation and Stability*, vol. 110, pp. 296–307, 2014.

[BEN 16] BENSSOUSSAN H., "Interviewing XtreeE: 3D Printing Concrete to Push the limits of Construction", available at: https://www.sculpteo.com/blog/2016/12/07/interviewing-xtreee-3d-printing-concrete-to-push-the-limits-of-construction/, 2016.

[BER 96] BERTSCH A., Micro-stéréo-photo-lithographie par masquage dynamique, Thesis, National Polytechnic Institute of lorraine, Nancy, 1996.

[BER 97a] BERTSCH A., ZISSI S., JEZEQUEL J.Y. *et al.*, "Micro-stereo-photolithography using a liquid crystal display as dynamic mask generator", *Microsystem Technologies*, vol. 3, pp. 42–47, 1997.

[BER 97b] BERTSCH A., JEZEQUEL J.Y., ANDRÉ J.C., "Study of the spatial resolution of a new 3D microfabrication process: the micro-stereo-photolithography using a dynamic mask-generator technique", *Journal of Photochemistry and Photobiology A: Chemistry*, vol. 107, pp. 275–281, 1997.

[BER 11] BERTSCH A., RENAUD P., "Microstereolithography", in BARTOLO P.J. (ed.), *Stereolithography; Materials, Processes and Applications*, Springer, New York, 2011.

[BEU 03] BEUERMANN S., BUBACK M., DAVIS T.P. *et al.*, "Critically evaluated rate constant for free radical polymerization", *Macromolecular Chemistry and Physics*, vol. 204, pp. 1338–1350, 2003.

[BHU 17] BHUSHAN B., CASPERS M., "An overview of additive manufacturing (3D printing) for microfabrication", *Microsystem Technologies*, vol. 23, pp. 1117–1124, 2017.

[BIK 16] BIKAS H., STAVROPOULOS H., CHRYSSOLOURIS G., "Additive manufacturing methods and modelling approaches: a critical review", *International Journal of Advanced Manufacturing Technologies*, vol. 83, pp. 389–406, 2016.

[BOI 15] BOILLAT E., "Techniques de fabrication additive", available at: http://www.swissmem.ch/fr/presentations-fabrication-additive.html, 2015.

[BOU 14] BOUFFARON P., "Impression 3D: Les prémisses d'une nouvelle (r)évolution industrielle", French embassy in the US, available at: http://sf.france-science.org/wp-content/uploads/2014/11/SMM14_025.pdf, 2014.

[BOZ 16] BOZEC H., "'Cellulose' a promising material for 3D Food Printing", available at: http://3dprintingindustry.com/2016/04/14/cellulose-a-promising-material-for-3d-food-printing/, 2016.

[BRA 85] BRAUN C., ANDRÉ J.C., "Industrial Photochemistry. VII. Light distribution in a diffusing medium of titanium dioxide in water", *Journal of Photochemistry*, vol. 28, pp. 13–29, 1985.

[BRE 16] BRETHOME K., PALACIN B., PUECH J. *et al.*, "Les technologies additives pour les applications spatiales", Workshop INSIS CNRS "Fabrication additive pour l'électronique", available at: http://www.cnrs.fr/insis/recherche/docs-evenements/workshop-INSIS_11.01.16_LCarpentier.pdf, 2016.

[CAB 15] CABANIS R., "Guide to unusual 3D printing materials", available at: http://www.sculpteo.com/blog/2015/04/15/unusual-3d-printing-materials/, 2015.

[CAH 14] CAHN J.G., "Impression 3D: perspectives industrielles et/ou utopie sociétale", *Lettre de prospective (CCI Ile-de-France)*, vol. 43, pp. 1–5, 2014.

[CAN 14] CANTON C., "Raw materials for 3D Printing", available at: http://www.think3d.in/raw-materials-for-3d-printing/, 2014.

[CAS 14] CASTEL A., "Technologies duales: Les imprimantes 3D en Chine, Inde et Singapour", Report for the Délégation aux Affaires Stratégiques, 2014.

[CET 17] CETIM, "Les poudres examinées à la loupe", available at: http://www.cetim.fr/fr/Actualites/En-France/A-la-une/Les-poudres-examinees-a-la-loupe, 2017.

[CHA 02] CHARTIER T., CHAPUT C., DOREAU F. *et al.*, "Stereolithography of structural complex ceramic parts", *Journal of Materials Science*, vol. 37, pp. 3141–3147, 2002.

[CHA 03] CHARTIER T., CHAPUT C., "Stereo-lithography as a shaping technique for ceramics", *Journal of the European Ceramic Society*, vol. 1, pp. 29–32, 2003.

[CHA 13] CHATTERJEE S., RAMAKRISHNAN S., "A novel photodegradable hyper-branched polymeric photoresist", *Chemical Communications*, vol. 49, pp. 11041–11013, 2013.

[CHA 14] CHANUT-GUIEU C., GUIEU G., "Créativité, PME et entrepreneuriat: des zones d'ombre et de lumière", *Revue internationale P.M.E.: économie et gestion de la petite et moyenne entreprise*, vol. 27, pp. 35–63, 2014.

[CHA 16] CHASTANG V., Etude du comportement mécanique et des mécanismes d'endommagement de pièces métalliques réalisées par fabrication additive, Thesis, University of Lille, 2016.

[CHA 17] CHANDLER D.L., "Researchers design one of the strongest, lightest materials known", available at: http://news.mit.edu/2017/3-d-graphene-strongest-lightest-materials-0106, 2017.

[CHE 93] CHEN H., CORBEL S., JEZEQUEL J.Y. *et al.*, "Laser induced paste agglomeration: a new process for rapid prototyping", *Journal de Physique III*, vol. 3, pp. 2193–2196, 1993.

[CHE 15] CHECK C., CHARTOFF R., CHANG S., "Inkjet printing of 3D nano-composites formed by photopolymerization of an acrylate monomer", *Reactive & Functional Polymers*, vol. 97, pp. 116–122, 2015.

[CHE 16a] CHEN O., DACULA MANGADLAO J., WALLAT J.D. *et al.*, "3D printing biocompatible polyurethane/poly (lactic acid)/graphene oxide nanocomposites: anisotropic properties", ACS Applied Materials & Interfaces Online, 2016.

[CHE 16b] CHEN W., ZHANG X., XINY S. *et al.*, Synthesis of Filigrees for Digital Fabrication and MARTÍNEZ J., DUMAS J., LEFEBVRE S., Procedural Voronoi Foams for Additive Manufacturing, SIGGRAPH '16 Technical Paper, Anaheim, July 24–28, 2016, available at: http://www.cnrs.fr/ins2i/spip.php?article2229, 2016.

[CHE 17a] CHEN M., GU Y., SINGH A. *et al.*, "Living additive manufacturing: transformation of parent gels into diversely functionalized daughter gels made possible by visible light photoredox catalysis", *ACS Central Science*, vol. 3, no. 2, pp. 124–134, 2017.

[CHE 17b] CHEN W., THORNLEY L., COE H.G. *et al.*, "Direct metal writing: controlling the rheology through microstructure", *Applied Physics Letters*, vol. 110, 094104, 2017.

[CHE 17c] CHEN L., HE Y., YANG Y. *et al.*, "The research status and development trend of additive manufacturing technology", *The International Journal of Advanced Manufacturing Technology*, vol. 89, pp. 3651–3660, 2017.

[COM 14] COMPTON B.G., LEWIS J.A., "3D-printing of lightweight cellular composites", *Advanced Materials*, vol. 26, pp. 5930–5935, 2014.

[CON 99] CONÉDÉRA V., LE GOFF B., FABRE N., "Potentialities of a new positive photoresist for the realization of thick molds", *Journal of Micromechanics and Microengineering*, vol. 9, pp. 173–175, 1999.

[CON 11] CONTUZZI N., CAMPANELLI S., LUDOVICO A.D., "3D finite element analysis in the selective laser melting process", *International Journal of Simulation Modelling*, vol. 10, pp. 113–121, 2011.

[COR 03] CORCIONE C.E., MALLEZZOLI A.G., "Photopolymerization kinetics of an epoxy-based resin for stereolithography", *Journal of Thermal Analysis and Calorimetry*, vol. 72, pp. 687–693, 2003.

[COU 15] COULTER F., "3D printing on inflated substrates and dielectric elastomer artificial muscles", available at: http://3dprint.com/73830/3d-printing-on-inflatable/, 2015.

[CUR 16] CURRAN C., "The role materials play in powering the 3D printing revolution", available at: http://usblogs.pwc.com/emerging-technology/the-role-materials-play-in-powering-the-3d-printing-revolution/, 2016.

[DAU 79] DAUMAS M., *Histoire générale des techniques*, PUF, Paris, 1979.

[DAV 11] DAVIS F.J., MITCHELL G.R., "Polymeric materials and rapid manufacturing", in BARTOLO P.J. (ed.), *Stereolithography; Materials, Processes and Applications*, Springer, New York, 2011.

[DDD 15] DDD NEWS, "L'impression 3D et l'architecture", available at: http://dddfactory.fr/ddd-news/, 2015.

[DEC 02] DECKER C., "Kinetic study and new applications of UV radiation curing", *Macromolecular Rapid Communications*, vol. 23, pp. 1067–1093, 2002.

[DEC 16] DE CAUSSIN D.R., "Design and testing of a top mask projection ceramic stereolithography system for ceramic part manufacturing" Thesis, California Polytechnic State University, 2016.

[DEG 05] DE GENNES P.G., BROCHART-WYART F., QUÉRÉ D., "Gouttes, bulles, perles et ondes", Belin, Paris, 2005.

[DEH 15] DEHOFF R.R., KIRKA M.M., SAMES W.J. *et al.*, "Site specific control of crystallographic grain orientation through electron beam additive manufacturing", *Materials Science and Technology*, vol. 31, pp. 931–938, 2015.

[DEH 16] DEHU E., "L'impression 3D révolutionne déjà l'industrie et l'artisanat: la preuve en images", available at: http://www.wedemain.fr/L-impression-3D-revolutionne-deja-l-industrie-et-l-artisanat-la-preuve-en-images_a1756.html, 2016.

[DÉJ 10] DÉJOU J., "Les céramiques", available at: http://campus.cerimes.fr/odontologie/enseignement/chap17/site/html/cours.pdf, 2010.

[DEL 07] DEL CAMPO A., GREINER C., "SU-8: a photoresist for high-aspect-ratio and 3D sub-micron lithography", *Journal of Micromechanics and Microengineering*, vol. 17, pp. R81–R95, 2007.

[DEL 16] DELAIRE J., PIARD J., MÉALLET-RENAULT R. *et al.*, *Photophysique et photochimie; des fondemants aux applications*, EDP Sciences, Paris, 2016.

[DEW 16] DE WARGNY M., "FDM vs. SLA vs. SLS: Battle of the 3D technologies", available at: https://www.sculpteo.com/blog/2016/08/31/fdm-vs-sla-vs-sls-battle-of-the-3d-technologies/, 2016.

[DEZ 00] DEZELLUS O., Contribution à l'étude des mécanismes de mouillage réactif, Thesis, Grenoble Institute of Technology, available at: https://tel.archives-ouvertes.fr/tel-00009773/document, 2000.

[DIA 16] DIALLO M., "Émeric d'Arcimoles, président de BeAM: "La fabrication additive offre un faible coût d'utilisation", available at: http://www.journal-aviation.com/actualites/32834-emeric-d-arcimoles-president-de-beam-la-fabrication-additive-offre-un-faible-cout-d-utilisation, 2016.

[DMG 14] DMG/MORI, "LASERTEC 65 Fabrication additive: Production de pièces finies, construites couche par couche", available at: http://www.dmgmori.com/webspecial/journal_2014_1/fr-FR/lasertec-65.htm, 2014.

[DON 09] DONG L., MAKRADI A., AHZI S. *et al.*, "Three-dimensional transient finite element analysis of the selective laser sintering process", *Journal of Materials Processing Technology*, vol. 209, pp. 700–706, 2009.

[DOU 04] DOUA J., BEUERMANN S., BUBACK M. *et al.*, "Critically evaluated rate coefficient for free radical polymerization", *Macromolecular Chemistry and Physics*, vol. 205, pp. 2151–2160, 2004.

[DUB 90] DUBARLE D., "Une nouvelle science, la cybernétique – vers une machine à gouverner", *Culture Technique*, vol. 21, pp. 47–49, 1990.

[DUF 05] DUFAUD O., LE GALL H., CORBEL S., "Application of stereolithography to chemical engineering: from macro to micro", *Chemical Engineering Research and Design*, vol. 83, no. A2, pp. 133–138, 2005.

[DUP 17] DUPONT L., "Lorraine Fab Living Lab", available at: https://laboratoire-erpi.wikispaces.com/Lorraine+Fab+Living+Lab?responseToken=021644860a63f191bab4570bf70bdd6e9, 2017.

[DYA 14] DY A., COSMANESCU A., SLUKA J. *et al.*, "Fabricating microfluidic valve master molds in SU-8 photoresist", *Journal of Micromechanics and Microengineering*, vol. 24, 057001, 2014.

[ECK 16] ECKEL Z.C., ZHOU C., MARTIN J.H. *et al.*, "Additive manufacturing of polymer-derived ceramics", *Science*, vol. 351, pp. 58–62, 2016.

[EKM 12] EKMEKYAPAN A., KUNKUL A., YUCEER M. *et al.*, "Modeling of the size of spherical solid particles in turbulent flow", *Journal of Dispersion Science and Technology*, vol. 33, pp. 521–526, 2012.

[EMA 15] EMAMI M.M., BARAZANDEH F., YAGHMAIE F., "An analytical model for scanning-projection based stereo-lithography", *Journal of Materials Processing Technology*, vol. 219, pp. 17–27, 2015.

[ESP 15] ESPCI, "Couches limites", available at: https://cours.espci.fr/site.php?id=2&fileid=306, 2015.

[FAN 16] FANSELOW S., EMAMJOMEH S.E., WIRTH K.E. *et al.*, "Production of spherical wax and polyolefin microparticles by melt emulsification for additive manufacturing", *Chemical Engineering Science*, vol. 141, pp. 282–292, 2016.

[FAR 15] FARINIA GROUP, "Available materials for metal additives manufacturing: characteristics and applications", available at: http://www.farinia.com/additive-manufacturing/3d-materials/characteristics-and-applications-of-available-metals-for-additive-manufacturing, 2015.

[FEM 15] FEMTO-ST, "Les résines photosensibles", available at: http://www.femto-st.fr/fr/Centrale-de-technologie-MIMENTO/Ressources-disponibles/Photolithographie/Les-resines-photosensibles, 2015.

[FEY 98] FEYNMAN R.P., *Vous y Comprenez Quelque Chose Monsieur Feynman?*, Odile Jacob, Paris, 1998.

[FIL 15] FILIMPRIMANTE 3D, "Le filament plastique: consommable de l'imprimante 3D", available at: http://www.filimprimante3d.fr/, 2015.

[FLE 08] FLECK L., *Génèse et développement d'un fait scientifique*, Champs Science, Paris, 2008.

[FRA 14] FRAZIER W.E., "Metal additive manufacturing: a review", *Journal of Materials Engineering and Performance*, vol. 23, pp. 1917–1928, 2014.

[FRA 16a] FRANCE 3, "Une entreprise de Limoges sur le marché chinois", available at: http://france3-regions.francetvinfo.fr/limousin/haute-vienne/limoges/entreprise-limoges-marche-chinois-1112697.html, 2016.

[FRA 16b] FRAUNHOFER, "formnext 2016: Low Cost SLM-Anlage mit Herstellungskosten von unter 20.000€", available at: http://www.ilt.fraunhofer.de/de/presse/pressemitteilungen/pm2016/pressemitteilung_27102016.html, 2016.

[FRY 13] FRYCZKOWSKI R., GORCZOWSKA M., ŚLUSARCZYK C. *et al.*, "The possibility of obtaining graphene/polymer composites from graphene oxide by a one step process", *Composites Science and Technology*, vol. 80, pp. 87–92, 2013.

[FUC 16] FUCHS A., Evaluation scientifique du CNRS, CNRS, Paris, available at: http://www.cnrs.fr/fr/organisme/docs/espacedoc/comite-consultatif-oct-2016.pdf, 2016.

[GAN 14] GANERIWALA R., ZOHDI T.I., "Multiphysics modeling and simulation of selective laser sintering manufacturing processes", *Procedia CIRP*, vol. 14, pp. 299–304, 2014.

[GAR 14] GARCIA-FERNANDEZ L., SPECHT A., DEL CAMPO A., "A polyurethane-based positive photoresist", *Macromolecular Rapid Communications*, vol. 35, pp. 1801–1807, 2014.

[GAR 16] GARMULEWICZ A., HOLWEG M., VELDHUIS H. *et al.*, "Redistributing material supply chains for 3D printing", available at: http://www.ifm.eng.cam.ac.uk/uploads/Research/TEG/Redistributing_material_supply_chains_for_3D_printing.pdf, 2016.

[GEH 15] GEHM R., "High-strength aluminum powder developed for additive manufacturing in aerospace, automotive", *SAE International*, available at: http://articles.sae.org/14175/, 2015.

[GOH 17] GOH G.D., AGARWALA S., GOH G.L. *et al.*, "Additive manufacturing in unmanned aerial vehicles (UAVs): challenges and potential", *Aerospace Science and Technology*, vol. 63, pp. 140–151, 2017.

[GON 16] GONZALEZ J.A., MIRELES J., LIN Y. *et al.*, "Characterization of ceramic components fabricated using binder jetting additive manufacturing technology", *Ceramics International*, vol. 42, pp. 10559–10564, 2016.

[GOR 17] GORGÉ C., "Ensemble de photographies d'objets 3D réalisés par Prodways", Private Communication, 2017.

[GRA 15] GRAPHENE LAB. INC., "Create; print; power up!", available at: http://www.graphene3dlab.com/s/technology.asp#3d-printing, 2015.

[GRA 16] GRAPHENE LAB. INC., "3D printed nanocomposites" available at: http://www.graphene3dlab.com/s/home.asp, 2016.

[GRE 11] GRENMAN H., SALMI T., MURZIN D.Y., "Solid-liquid reaction kinetics – experimental aspects and model development", *Reviews in Chemical Engineering*, vol. 27, pp. 53–77, 2011.

[GRI 96] GRIFFITH M.L., HALLORAN J.W., "Freeform fabrication of ceramics via stereolithography", *Journal of the American Ceramic Society*, vol. 79, pp. 2601–2608, 1996.

[GRO 14] GROSS B.C., ERKAL J.L., LOCKWOOD S.Y. *et al.*, "Evaluation of 3D printing and its potential on biotechnology and the chemical sciences", *Analytical Chemistry*, vol. 86, pp. 3240–3253, 2014.

[GUO 01] GUO Y., MITSUISHI M., MIYASHITA T., "Poly (N-alkyl-methacrylamid) LB films with short-branched alkyl side chains for a self-developed photoresist", *Macromolecules*, vol. 34, pp. 3548–3551, 2001.

[GUO 14] GUO N., LEU M.C., "Additive manufacturing: technology, applications and research needs", *Frontiers of Mechanical Engineering*, vol. 8, pp. 215–243, 2014.

[GUY 12] GUYON E., HULIN J.P., PETIT L., *Hydrodynamique physique*, EDP Sciences and CNRS, Paris, 2012.

[HAG 16] HAGUE R., REEVES P., JONES S., "Innovate UK – Mapping UK research and innovation in additive manufacturing", available at: https://www.gov.uk/government/uploads/system/uploads/attachment_data/file/496991/CO307_Mapping_UK_Capability_AM_WL_Print_pdf, 2016.

[HEL 16] HELLER B.P., SMITH D.E., JACK D.A., "Effects of extrudate swell and nozzle geometry on fiber orientation in Fused Filament Fabrication nozzle flow", *Additive Manufacturing*, vol. 12B, pp. 252–264, 2016.

[HOD 14] HODGE N.E., FERENCZ R.M., SOLBERG J.M., "Implementation of a thermomechanical model for the simulation of selective laser melting", *Computational Mechanics*, vol. 54, pp. 33–51, 2014.

[IMP 14] IMPRIMANTE 3D, "Quels sont les matériaux utilisés en impression 3D? Liquide, en poudre ou en bobine", available at: http://www.lesnumeriques.com/imprimante-3d/quels-sont-materiaux-utilises-en-impression-3d-a1884.html, 2014.

[INN 08] INNERAY D., *Le futur et ses ennemis*, Climat, Paris, 2008.

[IRE 15] IREPA-LASER, "Le procédé CLAD", available at: http://www.beam-machines.fr/innovation/la-technologie.html, 2015.

[IRM 09] IRMOULI Y., GEORGE B., MERLIN A., "Study of the polymerization of acrylic resins by photo-calorimetry: interactions between UV initiators and absorbers", *Journal of Thermal Analysis and Calorimetry*, vol. 96, pp. 911–916, 2009.

[ITR 15] ITRE COMMITTEE, "Open innovation industry, including 3D printing", available at: http://www.europarl.europa.eu/RegData/etudes/STUD/2015/563445/IPOL_STU(2015)563445_EN.pdf, 2015.

[JAC 16] JACKSON B., "3D printing buildings, a summary of additive manufacturing in construction", available at: https://3dprintingindustry.com/news/3d-printing-buildings-summary-additive-manufacturing-construction-97312/, 2016.

[JAR 17] JARED B.H., AGUILO M.A., BEGHINI L.L. *et al.*, "Additive manufacturing: Toward holistic design" *Scripta Materialia*, vol. 135, p. 147, 2017.

[JUL 17] JULIEN, "Du filament PLA multicolore chez UniCoFil", available at: https://www.lesimprimantes3d.fr/unicofil-filament-pla-multicolore-20170208/, 2017.

[KAN 12] KANG H.W., PARK J.H., CHO D.N., "A pixel based solidification model for projection based stereolithography technology", *Sensors & Actuators*, vol. 178, pp. 223–229, 2012.

[KAR 92] KARRER P., CORBEL S., ANDRÉ J.C. *et al.*, "Shrinkage effects in charged photo-polymerizable resins: application to stereo-photolithography", *Journal of Polymer Science*, vol. 30A, pp. 2715–2723, 1992.

[KEL 93] KELLY J.M., MCARDLE C.B., DE MAUNDER M.J., *Photochemistry and Polymeric Systems*, The Royal Society of Chemistry, Cambridge, 1993.

[KEN 14] KENZARI S., BONINA D., DUBOIS J.M. *et al.*, "Additive manufacturing of lightweight, fully Al-based components using quasicrystals", *Journal of Materials Processing Technology*, vol. 214, pp. 3108–3111, 2014.

[KOL 15] KOLLER R., "L'industrie veut aussi sa transformation digitale", available at: http://www.ictjournal.ch/fr-CH/News/2015/12/04/Lindustrie-veut-aussi-sa-transformation-digitale.aspx?pa=1, 2015.

[KOS 15] KOSLOW T., "3D printed micro-fish to explore the ocean of our bodies", available at: http://3dprintingindustry.com/2015/08/31/3d-printed-micro-fish-to-explore-the-oceans-of-our-bodies/? 2015.

[KOU 14] KOURILSKY P., *Le jeu du hasard et de la complexité; la nouvelle science de l'immunologie*, Odile Jacob, Paris, 2014.

[LAG 10] LAGUBEAU G., Interfaces à grains, et autres situations de mouillage nul, Thesis, UPMC, Paris, 2010.

[LAN 15] LANDIS S., "Le CEA-LETI prépare l'industrialisation de la lithographie par impression", available at: http://www.usine-digitale.fr/editorial/le-cea-leti-prepare-l-industrialisation-de-la-lithographie-par-impression.N337639, 2015.

[LAV 15] LAVIGNE C., "J.C. André, A. Le Méhauté, O. de Witte: history of the digital sculpture tools", *What Things May Come: 3D Printing in the Fine Arts and Sciences*, Georgetown, Texas, 2015.

[LEE 05] LEE H.H., CHOU K.S., HUANG K.C., "Inkjet printing of nano-sized silver colloids", *Nanotechnology*, vol. 16, pp. 2436–2441, 2005.

[LEI 11] LEIGH S.J., PURSELL C.P., BOWEN J. *et al.*, "A miniature flow sensor fabricated by micro-stereolithography employing a magnetite/acrylic nanocomposite resin", *Sensors and Actuators*, vol. 168A, pp. 66–71, 2011.

[LER 65] LEROI-GOURHAN A., *Le geste et la parole; la mémoire et les rythmes*, Albin Michel, Paris, 1965.

[LER 83] LEROI-GOURHAN A., *Le fil du temps – ethnologie et préhistoire*, Fayard, Paris, 1983.

[LES 15] LESCUYER T., "Fabrication additive: de nouvelles alliances et des usages plus larges", available at: http://www.innovationreview.eu/articles.php?article_id=2269, 2015.

[LEV 12] LEVET J.L., "L'industrie ne disparaît pas, elle se transforme et ses frontières s'élargissent", available at: http://www.les-industries-technologiques.fr/actualite/regards-croises/jean-louis-levet-lindustrie-ne-disparat-pas-elle-se-transforme-et-ses-frontires-slargissent/, 2012.

[LIE 15] LIENEKE T., ADAM G.A.O., LEUDERS S. *et al.*, "Systematical determination of tolerances for additive manufacturing by measuring linear dimensions", *26th Annual International Solid Freeform Fabrication Symposium*, available at: http://sffsymposium.engr.utexas.edu/sites/default/files/2015/2015-30-Lieneke.pdf, 2015.

[LIN 11] LIN H.M., WU S.Y., CHANG F.C. *et al.*, "Photo-polymerization of photo-curable resins containing polyhedral oligomeric silsesquioxane methacrylate", *Materials Chemistry and Physics*, vol. 131, pp. 393–399, 2011.

[LIN 14] LIN S.L., LIN C.C., LIN D.Y. *et al.*, "Fabrication and analysis of 3D objects using layerwise manufacturing technology", *8th IEEE International Conference on Nano/Micro Engineered and Molecular systems*, pp. 763–766, 2014.

[LIP 14] LIPSON H., KURMAN M., *Impression 3D: la prochaine révolution*, First Interactive, Paris, 2014.

[LIR 17] LIRAVI F., DARLEUX R., TOYSERKANI E., "Additive manufacturing of 3D structures with non-Newtonian highly viscous fluids: finite element modeling and experimental validation", *Additive Manufacturing*, vol. 13, pp. 113–123, 2017.

[LIY 17] LI Y., HU Y., CONG W. *et al.*, "Additive manufacturing of alumina using laser engineered net shaping: effects of deposition variables", *Ceramics International*, vol. 43, no. 10, pp. 7768–7775, 2017.

[LOU 16] LOUBENS A., "Révolution dans les règles de conception: le DMLS", available at: http://www.techniques-ingenieur.fr/actualite/articles/revolution-dans-les-regles-de-conception-le-dmls-necessite-un-nouveau-mode-de-pensee-36705/, 2016.

[LOU 17] LOUISOT J.M., "Optimiser la prise de risques", *Risk-Assur*, vol. 474, pp. 28–37 available at: https://www.riskassur-hebdo.com, 2017.

[LOW 17] LOW Z.X., CHUA Y.T., RAY B.M. *et al.*, "Perspective on 3D printing of separation membranes and comparison to related unconventional fabrication techniques", *Journal of Membrane Science*, vol. 523, pp. 596–613, 2017.

[LUC 15] LUCAS T., "Vidéo: l'impression 3D passe à la polymérisation liquide, façon Terminator 2", available at: http://www.usine-digitale.fr/article/video-l-impression-3d-passe-a-la-polymerisation-liquide-facon-terminator-2.N319562, 2015.

[MAE 98] MAEDER S., GILBERT R.G., "Measurement of transfer constant for butyl-acrylate free radical polymerization", *Macromolecules*, vol. 31, pp. 4410–4418, 1998.

[MAK 15] MAKERSHOP 3D, "Guide d'achat des consomables pour l'impression 3D", available at: http://www.makershop.fr/content/43-guide-achat-filament-resine-impression-3d, 2015.

[MAR 11] MARCHAND A., Mouillage statique et dynamique: Influences géométriques aux échelles moléculaires, Thesis, Paris Diderot University, Paris, available at: http://citeseerx.ist.psu.edu/viewdoc/download?doi=10.1.1.393.6823&rep=rep1&type=pdf, 2011.

[MAR 14] MARTINS J.S., BORGES B.G.A.L., CARPANEZ A.G. *et al.*, "Evaluation of chemical kinetics in positive photoresists using laser desorption ionization", *European Polymer Journal*, vol. 59, pp. 1–7, 2014.

[MAR 15] MARGUERAY T., VOLLARD L., "Fabrication additive métallique: technologies et opportunités", available at: http://www.nae.fr/wp-content/uploads/2015/06/FabAdditiveMetallique_TechnoOpportunites-NAE.pdf, 2015.

[MAR 17] MARKFORGED, "The METAL X; An Industrial Revolution", available at: https://markforged.com/metal-x/, 2017.

[MAR 17] MARTEL A., "TOP 5 des vidéos de la semaine: L'impression 3D en cuisine, interview de la startup française Syos…", available at: http://www.3dnatives.com/top-5-videos-22012017/, 2017.

[MAT 09] MATSUKAWA D., MUKAY T., OKAMURA H. *et al.*, "Photocurable oligo (hemiacetal ester)s having methacrylate side chains", *European Polymer Journal*, vol. 45, pp. 2087–2095, 2009.

[MEI 14] MEIER D., "Regarder le monde de façon interdisciplinaire: l'espace du possible", in V. LEMAY and F. DARBELLAY (eds), *L'interdisciplinarité racontée: chercher hors frontières, vivre l'interdisciplinarité*, Peter Lang, Bern, 2014.

[MER 15] MERLOT F., MAZZONI S., "Gas evolution during FDM 3D printing and health impact", available at: http://www.3dsafety.org/3dsafety/download/mf2015_eng.pdf, 2015.

[MID 93] MIDOUX N., *Mécanique et rhéologie des fluides en génie chimique*, Tech&Doc Lavoisier, Paris, 1993.

[MIQ 00] MIQUEL P.A., *Comment penser le désordre*, Fayard, Paris, 2000.

[MIZ 15] MIZERET J., "Les technologies de fabrication additive pour la créativité, le prototypage, la fabrication", available at: http://www.swissmem.ch/fr/presentations-fabrication-additive.html, 2015.

[MOK 15] MOKBEL H., Vers de nouveaux systèmes amorceurs pour la photopolymérisation radicalaire et/ou cationique dans des conditions plus respectueuses de l'environnement, Thesis, University of Upper Alsace, available at: https://tel.archives-ouvertes.fr/tel-01334572/document, 2015.

[MOL 14] MOLITCH-HOU M., "Mazak adds hybrid metal 3D printer to machine tool lineup", available at: http://3dprintingindustry.com/2014/11/04/mazak-hybrid-metal-3d-printer/, 2014.

[MOL 15] MOLLON G., Mécanique des milieux granulaires, Lecture notes, INSA, Lyon, available at: http://guilhem.mollon.free.fr/Telechargements/Mecanique_des_Materiaux_Granulaires.pdf, 2015.

[MOR 16] MOREL L., LE ROUX S., *Fab Labs: Innovative User*, ISTE, London and John Wiley & Sons, New York, 2016.

[MUR 06] MURAKAMI T., YADA T., KOBAYASHI G., "Positive direct-mask stereo-lithography with multiple layer exposure: layered fabrication with stair step reduction", *Virtual & Physical Prototyping*, vol. 1, pp. 73–81, 2006.

[MUR 17] MURR L.E., JOHNSON W.L., "3D metal droplet printing development and advanced materials additive manufacturing", *Journal of Materials Research and Technology*, vol. 6, pp. 77–89, 2017.

[NEU 05] NEUMANN M.G., MIRANDA Jr. W.G., SCHMITT C.C. *et al.*, "Molar extinction coefficients and the photon absorption efficiency of dental photoinitiators and light curing units", *Journal of Dentistry*, vol. 33, pp. 525–532, 2005.

[NGU 04] NGUYEN-TRI P., DEA: Etude de la polymérisation radicalaire thermique et UV de résines acryliques, Report, University of Upper Alsace, 2004.

[NIC 16] NICKELS L., "Additive manufacturing: a user's guide", *Metal Powder Report*, vol. 71, pp. 100–105, 2016.

[NIK 07] NIKITIN A.N., HUTCHINSON R.A., BUBACK M. *et al.*, "Determination of intramolecular chain transfer and midchain radical propagation rate coefficients for butyle acrylate by pulsed laser polymerization", *Macromolecules*, vol. 40, pp. 8631–8641, 2007.

[NIK 10] NIKITIN A.N., HUTCHINSON R.A., WANG W. *et al.*, "Effect of intramolecular transfer to polymer on stationary free radical polymerisation of alkyl-acrylates, 5 - Consideration of solution polymerisation up to high temperature", *Macromolecular Reaction Engineering*, vol. 4, pp. 691–706, 2010.

[NOT 14] NOTTEAU Y., "Adieu les risques de fissure, NanoSteel a développé un acier dense spécialement conçu pour l'impression 3D", available at: http://www.additiverse.com/actualites/2014-10-01/adieu-les-risques-de-fissure-nanosteel-a-developpe-un-acier-dense-specialement-concu-pour-l-impression-3d, 2014.

[NOT 16] NOTTEAU Y., "Tout ce qu'il faut savoir sur les imprimantes 3D métal!", available at: http://www.additiverse.com/tout-ce-quil-faut-savoir-sur-les-imprimantes-3d-metal/, 2016.

[NUY 13] NUYKEN O., PASK S.D., "Ring-opening polymerization – an introductory review", *Polymers*, vol. 5, pp. 361–403, 2013.

[OBA 15] OBATON A.F., BERNARD A., TAILLANDIER G. *et al.*, "Fabrication additive: Etat de l'art et besoins métrologiques engendrés", *Revue Française de Métrologie*, vol. 37, pp. 21–36, 2015.

[OBA 16] OBATON A.F., "La maîtrise du processus de fabrication additive est un véritable enjeu pour les industriels", available at: http://www.usinenouvelle.com/article/tribune-la-maitrise-du-processus-de-fabrication-additive-est-un-veritable-enjeu-pour-les-industriels.N385082, 2016.

[OBS 17] OBSERVATOIRE DE LA MÉTALLURGIE, "L'élaboration et la transformation des métaux par forge, fonderie et fabrication additive métallique – Rapport Complet, 2017", available at: http://www.observatoire-metallurgie.fr/sites/default/files/uimm_documents/etude_metallurgie_2017.pdf, 2017.

[OEC 16] OECD, "OECD, Science, Technology and Innovation Outlook 2016 – 10 key technology trends for the future", available at: http://www.oecd.org/sti/STIO%2010%20key%20technology%20trends%20for%20the%20future.pdf, 2016.

[ORC 16] ORCUTT M., "One 3-D Printer for 21 Metals", *MIT Technology Review*, available at: https://www.technologyreview.com/s/600876/one-3-d-printer-for-21-metals/, 2016.

[PÉC 17] PÉCOUL N., "Private Communication", 2017.

[PER 05] PERRY M.F., YOUNG G.W., "A mathematical model for photopolymerization from a stationary laser light source", *Macromolecular Theory and Simulations*, vol. 14, pp. 26–39, 2005.

[PEU 97] PEUTZFELD A., "Resin composites in dentistry: the monomer systems", *European Journal of Oral Sciences*, vol. 105, pp. 97–116, 1997.

[PON 14] PONCHE R., KERBRAT O., MOGNOL P. *et al.*, "A novel methodology of design for additive manufacturing applied to additive laser manufacturing process", *Robotics and Computer Integrated Manufacturing*, vol. 30, pp. 389–398, 2014.

[PRE 15a] PRECISION METAL FORMING ASSOCIATION, "Recent news", available at: http://www.pma.org/home/, 2015.

[PRE 15b] PREVÉRAUD J.F., "Les promesses de l'impression 3D métallique", available at: http://www.industrie-techno.com/les-promesses-de-l-impression-3d-metallique.38572, 2015.

[PRI 15] PRIMANTE 3D, "Quels matériaux pour imprimer en 3D", available at: http://www.priximprimante3d.com/materiaux/, 2015.

[PRO 02] PROVIN C., Micro-stéréo-lithographie de céramiques, Thesis, National Polytechnic Institute of Lorraine, Nancy, 2002.

[QUA 15] QUAN Z., WU A., KEEFE M. *et al.*, "Additive manufacturing of multidirectional preforms for composites: opportunities and challenges", *Materials Today*, vol. 18, pp. 503–512, 2015.

[QUI 14] QUICK A.S., ROTHFUSS H., WELLE A. *et al.*, "Fabrication of spatially resolved functionalization of 3D microstructures via multi-photon induced Diels-Alder chemistry", *Advanced Functional Materials*, vol. 24, pp. 3571–3580, 2014.

[RAY 16] RAYNAL J., "Sculpteo voit plus loin que l'impression 3D", available at: http://www.industrie-techno.com/sculpteo-voit-plus-loin-que-l-impression-3d.44292, 2016.

[REC 16] RECROSIO A., "Metal 3D Printing: Which Technologies for Which Applications?", available at: https://www.sculpteo.com/blog/2016/12/14/metal-3d-printing-which-technologies-for-which-applications/?, 2016.

[REN 17a] RENARD M., " 3DCeram accélère le développement de l'impression 3D céramique ", available at: http://www.3dnatives.com/3dceram-spcts-impression-3d-14032017/, 2017.

[REN 17b] RENARD M., " Rencontre exclusive avec les créateurs de l'imprimante 3D à 6 axes", available at: http://www.3dnatives.com/interview-imprimante-3d-a-6-axes-13032017/, 2017.

[ROB 50] ROBERTS D.E., "Heats of polymerization: a summary of published values and their relation to structure", *Journal of Research of the National Bureau of Standards*, vol. 44, Research Paper RP2073, pp. 221–234, 1950.

[ROB 15] ROBINSON A., "The power of 3D printing in manufacturing: an illustrated guide", available at: http://advancedmanufacturinginsight.com/archived-articles/item/infographic-3d-printing-manufacturing, 2015.

[ROS 15] ROSENBERG J., "L'impression 3D, une révolution industrielle mondiale; contexte, enjeux et opportunités", available at: https://www.arts-et-metiers.asso.fr/manifestation_cr/cr_3054.pdf, 2015.

[ROY 03] ROY D., BASU P.K., RAGUNATHAN P. *et al.*, "Photo-induced cross-linking mechanism in azide-novolac negative photoresist: molecular level investigation using NMR spectroscopy", *Magnetic Resonance in Chemistry*, vol. 41, pp. 671–678, 2003.

[SAG 11] SAGER B., ROSEN D., "Simulation methods for stereo-lithography", in BARTOLO P.J. (ed.), *Stereo-lithography; Materials, Processes and Applications*, Springer, New York, 2011.

[SAM 05] SAMORIA G.V., AHRENS C.H., FREDEL M. *et al.*, "Stereo-lithography Somos 7110 resin: mechanical behavior and fractography of parts post-cured by different methods", *Polymer Testing*, vol. 24, pp. 157–162, 2005.

[SAN 07] SANDOVAL J.H., SOTO A.K.F., MURR A.L.E. *et al.*, "Nano-tailoring photo-cross-linkable epoxy resins with multi-walled carbon nanotubes for stereo-lithography layered manufacturing", *Journal of Materials Science*, vol. 42, pp. 156–165, 2007.

[SCA 14] SCALERA F., CORCIONE C.E., MONTAGNA F. *et al.*, "Development and characterization of UV curable epoxy/hydroxyapatite suspensions for stereo-lithography applied to bone tissue engineering", *Ceramics International*, vol. 40A, pp. 15455–15462, 2014.

[SCH 15] SCHAER E., ENSIC, Course, University of Lorraine, Nancy, 2015.

[SCH 16] SCHWALBE M. *Predictive Theoretical and Computational Approaches for Additive Manufacturing: Proceedings of a Workshop*, U.S. National Committee on Theoretical and Applied Mechanics, 2016.

[SHA 14] SHAFFER S., YANG K., VARGAS J. *et al.*, "On reducing anisotropy in 3D printed polymers via ionizing radiation", *Polymer*, vol. 55, pp. 5969–5979, 2014.

[SHE 15a] SHER D., "The top 10 3D printers under $1,000", available at: http://3dprintingindustry.com/2015/06/30/top-10-low-cost-3d-printers/, 2015.

[SHE 15b] SHER D., "Hermle introduces multi-metal 3D printing technology into 5-axis CNC mill", available at: http://3dprintingindustry.com/2015/02/24/hermle-introduces-mpa-multi-metal-technology-5-axes-cnc-mill/, 2015.

[SIE 10] SIEGMANN R., JELICIK A., BEUERMANN S., "Propagation and termination kinetics of PEGylated methacrylate radical polymerization", *Macromolecular Chemistry and Physics*, vol. 211, pp. 546–562, 2010.

[SIN 17] SINGH S., RAMAKRISHNA S., SINGH R., "Material issues in additive manufacturing: a review", *Journal Manufacturing Processes*, vol. 25, pp. 185–200, 2017.

[SOL 16] SOLBERG S., Cost-efficient low-volume production through additive manufacturing, Thesis, Stavanger University, 2016.

[STA 05a] STAMPFL J., LISKA R., "New materials for rapid prototyping applications", *Macromolecular Chemistry and Physics*, vol. 206, pp. 1253–1256, 2005.

[STA 05b] STAMPFL J., LISKA R., "Polymerizable hydrogels for rapid prototyping: chemistry, photo-lithography and mechanical properties", in BARTOLO P.J. (ed.), *Stereo-lithography; Materials, Processes and Applications*, Springer, New York, 2005.

[STA 08] STAMPFL J., BAUDIS S., HELLER C. *et al.*, "Photopolymers with tunable mechanical properties processed by laser-based high-resolution stereolithography", *Journal of Micromechanics and Microengineering*, vol. 18, no. 2, 2008.

[STR 13] STRANO G., LIANG H., EVERSON R.M. *et al.*, "Surface roughness analysis, modelling and prediction in selective laser melting", *Journal of Materials Processing Technology*, vol. 213, pp. 589–597, 2013.

[STU 15] STUCKY M., "Fabrication additive métallique au CTIF", available at: http://souspression.canalblog.com/archives/2015/11/16/32918893.html, 2015.

[SUN 04] SUN H.B., KAWATA S., "Two-photon photo-polymerization and 3D lithographic fabrication", *Avances in Polymer Science*, vol. 170, pp. 169–273, available at: http://nathan.instras.com/MyDocsDB/doc-712.pdf, 2004.

[SUN 05] SUN H.W., FANG N., WU D.M. *et al.*, "Projection micro-stereo-lithography using digital micro-mirror dynamic mask", *Sensors & Actuators*, vol. 121A, pp. 113–120, 2005.

[SUS 10] SUSPERRGUI N., Etude théorique de la polymérisation d'esters par voie organométallique et organique, Thesis, Université de Toulouse, Toulouse, France, 2010.

[TIA 12] TIAN X., ZHANG W., LI D. *et al.*, "Reaction-bonded SiC derived from resin precursors by stereo-lithography", *Ceramics International*, vol. 38, pp. 589–597, 2012.

[TUR 17] TURCO E., GOLASZEWSKI M., GIORGIO I. *et al.*, "Can a Hencky-type model predict the mechanical behaviour of pantographic lattices?", in DELL'ISOLA F., SOFONEA M., STEIGMANN D. (eds), *Mathematical Modelling in Solid Mechanics*, 2017.

[VAE 13] VAEZI M., SEITZ H., YANG S., "A review on 3D micro-additive manufacturing technologies", *International Journal of Advanced Manufacturing Technology*, vol. 67, pp. 1721–1754, 2013.

[VAI 96] VAIRON J.-P., SPASSKY N., "Industrial cationic polymerization: an overview", in MATYJASZEWSKI K. (ed.), *Cationic Polymerizations*, Marcel Dekker, New York, 1996.

[VAN 06] VANG H., "Optimisation des étapes technologiques pour la fabrication de composants de puissance en carbure de silicium", available at: http://csidoc.insa-lyon.fr/these/2006/vang/09_chapitre_2.pdf, 2006.

[VAR 01] VARADAN V.K., JIANG X., VARADAN V.V., *Micro-stereo-lithography and Other Fabrication Techniques for 3D MEMS*, John Wiley & Sons, Chichester, 2001.

[VER 14] VERQUIN B., *Fabrication additive par fusion laser; étude de cas*, Cetim, Paris, 2014.

[VER 15] VERHULST J., Analyse de l'état et de l'avenir du marché de l'impression 3D, Thesis, Université Catholique de Louvain, available at: http://dial.uclouvain.be/memoire/ucl/en/object/thesis%3A2710/datastream/PDF_01/view, 2015.

[VIV 12] VIVERET P., *La cause humaine; du bon usage de la fin du monde*, Editions Les liens qui libèrent, Paris, 2012.

[VOX 15] VOXELJET, "Plastic parts for investment casting and design models", available at: http://www.voxeljet.de/en/services/plastic/, 2015.

[WAN 09] WANG W., NIKITIN A.N., HUTCHINSON RA., "Consideration of macro-monomer reactions in n-butyl acrylate free radical polymerization", *Macromolecular Rapid Communications*, vol. 30, pp. 2022–2027, 2009.

[WAT 91] WATTS D.C., CASH A.J., "Determination of polymerization shrinkage kinetics in visible light-cured materials: methods development", *Dental Materials*, vol. 7, pp. 281–287, 1991.

[WIE 60] WIENER N., "Some moral and technical consequences of automation", *Science*, vol. 131, pp. 1355–1358, 1960.

[WIE 08] WIEDERKEHR F., GOETSCHY O., WUNSCHEL R., "Projet stéréolithographie", Lycée Jean Mermoz, Saint-Louis, 2008.

[WOO 13] WOOD T., BRISSONNEAU V., BRUCKNER J.B. *et al.*, "Optical measurement of exposure depth and refractive index in positive photoresists", *Optics Communications*, vol. 291, pp. 184–192, 2013.

[WOR 17] WORLD ECONOMIC FORUM, "The new plastics economy; rethinking the future of plastics", available at: http://www3.weforum.org/docs/WEF_The_New_Plastics_Economy.pdf, 2017.

[YAG 10] YAGCI Y., JOCKUSCH S., TURRO N.J., "Photoinitiated polymerization: advances, challenges, and opportunities", *Macromolecules*, vol. 43, pp. 6245–6260, 2010.

[YAP 00] YAP A.U., WANG H.B., SIOW K.S. *et al.*, "Polymerization shrinkage of visible-light cured composites", *Operative Dentistry*, vol. 25, pp. 98–103, 2000.

[ZAD 16] ZADPOOR A.A., "Mechanical meta-materials", *Materials Horizons*, vol. 3, pp. 371–381, 2016.

[ZAF 15] ZAFFAGNI M., "De l'imprimante 3D avec du verre en fusion", available at: http://www.futura-sciences.com/magazines/high-tech/infos/actu/d/imprimante-3d-g3dp-impression-3d-verre-fusion-59533/, 2015.

[ZAV 17] ZAVALA-ARREDONDO M., BOONE N., WILLMOTT J. *et al.*, "Laser diode area melting for high speed additive manufacturing of metallic components", *Materials & Design*, vol. 117, pp. 305–315, 2017.

[ZCO 15] Z CORPORATION, "Imprimantes 3D", available at: http://www.zcorp.com/fr/Products/3D-Printers/spage.aspx, 2015.

[ZCO 16] Z CORPORATION, "Z Corporation 3D Printing Technology", available at: https://www.ucy.ac.cy/arch/documents/3d_Printer_Lab/3D_Printing_Technology.pdf, 2016.

[ZHO 15] ZHOU C., YE H., ZHANG F., "A novel low-cost stereo-lithography process based on vector scanning and mask projection for high-accuracy, high speed, high throughput and large area fabrication", *Journal of Computing and Information Science in Engineering*, vol. 15, 011003, 2015.

[ZHO 17] ZHONG Y., RÄNNAR L.E., WIKMAN S. *et al.*, "Additive manufacturing of ITER first wall panel parts by two approaches: selective laser melting and electron beam melting", *Fusion Engineering and Design*, vol. 116, pp. 24–33, 2017.

Conclusion

Perspectives and Considerations

"I would like to propose a simple principle, which I call the 'principle of the cherry and the screwdriver.' Scientists do not have the final perfect method, unveiled in the heavens by some Prometheus who was offering for them to be the sole decoders of the world. Science advances with difficulty by observing events and seeking to detect the regularity of the world. Each problem generates a multitude of theoretical and experimental approaches that are not taken from a reservoir of predetermined certainties, but which are the fruit of inventiveness, imagination, sometimes genius, but still ephemeral methods, trends, derivatives of tradition or zeitgeist. Nature resembles a Sphynx. If we question it, it announces – in the best of cases – that our theories are flawed, but it does not tell us where and how to modify them. Each problem demands the creation of adapted methodological instruments (like a screwdriver) and creates others that must be tackled, one by one (just as we pick cherries). Science is thus a refined form of handicrafts that demands – though this is not true for all trades, and thus there are exceptions – engagement, competence, and a rare ability to change one's ideas about the world when mistakes are perceived. This final point, according to Bertrand Russell, distinguishes men according to their ethics, given that most rather tend to change the world and do not hesitate to shatter it if it does not have the good manners to match their rigid schemata. The essential point for a different vision of the science-society relation remains fully in this crucial knot that consists of allowing people to understand the artisanal aspect of science, by demystifying and offering everyone the possibility to know and truly love it". [LIC 10]

"Technology can allow survival, through the sophistication of models or systems that are not optimal. In other words, it could prevent latent epistemological revolutions in the future. I will call this vice of the technician's vision of the world 'Ptolemy's computer syndrome,' as it concerns much more than the production of knowledge". [MAL 11]

"It is the applications and implications of technoscience that mobilize the public's attention and not the many winding detours that led to it". [CLA 11]

"The sad truth is, however, that it is not the pure savant, but the plumber who reestablishes the lost contact between the world of meaning and appearances and the world of physics". [BOH 61]

"Evolution and innovation require securing the existent. As such, at a certain point, the evolution of a system to increase security is as costly, as every addition of new rules is only slightly efficient, and imperils the economic balance, thereby preventing all innovation". [DEG 13]

"It could indeed be that modern society's tireless drive for innovation and incessant dynamization is what saps its ability for essential innovation and creativity adaptability. In this sense, a very solid form of sclerosis and blockage could appear behind the hyperdynamic surface of delayed modern societies". [ROS 12]

"Modern Man remains starved, but of goods whose existence he/she knew nothing of a few years earlier. As Alfred Sauvy said, he is a hiker who never reaches the horizon". [COH 09]

Kuhn [KUH 72] studied the history of science to better understand how scientific research was structured. A scientific discipline in a recognized domain (thus a stabilized scientific domain) is made up of a group of researchers sharing the same cultures, training and scientific paradigms. According to this author, recalled by Guyot and Rieppi [GUY 14], the disciplinary matrix is made up of heterogeneous elements gathered in four categories:

– symbolic generations, which are commonly employed and often formalized expressions;

– metaphysical paradigms, which are common beliefs;

– values that particularly focus on the criteria for evaluating scientific results, on the rules and methods of organizing work and on acceptable or unacceptable social practices;

– paradigms or examples that are illustrations of typical problems and concrete solutions [...]. These structure how research is carried out.

Additive manufacturing sciences, which correspond to boundary objects implying combined approaches with interdisciplinary origins, can be defined as an engineering science associated with the study of matter transformation (brought about by the addition of energy) to manufacture three-dimensional objects for application or finalized ends. They are based on the acquisition of scientific knowledge allowing these transformations to be described by integrating all of the phenomena and multi-scale and multi-physical processes, as well as their couplings. Several points can be noted in this personal definition:

– it is a science: science is what is known from having learned it, what is held as true, all knowledge, studies of a universal value, characterized by a determined object (domain) and method founded on verifiable objective relations;

– it is a discipline: a discipline refers to knowledge developed by a community of specialists adhering to the same research practices. As long as this community recognizes one another, it naturally tends toward autonomy, through the definition of its borders, the language that makes it up and the theories and techniques that it is led to elaborate and use. However, and we will come back to this, that situation is still emerging;

– it is an engineering science: science of artifacts, objects and systems where knowledge of nature and the intervention of human capacity are associated with a problem-solving view, born indirectly and far away, of functional issues by abstract or concrete means;

– matter is what makes up every body with a tangible reality. It is vast, as it can range from the inert organic and mineral to the living;

– energy is the ability of a system to produce work, a transformation of matter or energy. It is also quite broad, as has been shown in this first volume;

– in connection with this scientific deepening and applications, this definition contributes to technoscience, which highlights the concrete solidarity between material technological developments and theoretical knowledge in the form of constant interactions and positive feedback between scientific discoveries and technical inventions (with the risk of subordination to self-interested application ends and the risk of external piloting of research coming from a "facility or knowledge reservoir", without clear creation of knowledge);

– however, according to Poincaré [POI 11], "the savant must not delay in realizing practical ends. He/She will likely obtain them, but it will be necessary for him/her to obtain them in excess. He/She must never forget that the special object that he/she is studying is only a part of the great whole that must be the only source of his/her activity… Science has had marvelous applications, but the Science that only aims at applications would no longer be Science; it would be cooking"…? Yet, the motor of additive manufacturing is indeed the immensity of its applications;

– modeling allows us to go further and faster. It is a mental construction in which reality is simplified, even reduced to its principal variables (at least those that have been perceived), which can pose the question to contribute robust solutions to the complexity of the world around us. The computer partially replaces the experiment to re-discover the old dream of some to "theorize thought", even palpable reality;

– to avoid limiting the researcher in his/her discipline (in which he/she exercises his/her freedom (his/her autonomy) and his/her activity, allowing him/her to exceed his/her limits of knowledge), is it not necessary to find means of "transgression" that must authorize and support new cultural couplings? How to enrich oneself from others? This is, in principle, what allows an approach on a border object like 3D printing;

– to anticipate the societal demand that may be made of him/her in his/her domain, must this research become engaged, alone or with others, in prospective reflection, only on the potential application visions or, more complexly, on new concepts?;

– how to escape the mundane by developing a bit of creativity: creativity is an ability to easily change one's perceptions of things, the ability to go from one model to another. Demanding creativity of sometime is encouraging him/her to "think outside the box".

At present, knowledge is nearly immeasurable, but what is this knowledge worth applied to additive manufacturing? Concepts help distance ourselves a bit to think "better". Maybe it is necessary to learn how to learn and to organize information around knowledge. This is, moreover, what the European Union is attempting [DUR 17].

Here, a bit chaotically, we can see a few issues, many of which go beyond "simple" additive manufacturing, a difficult scientific concept to explain to an unprepared audience who at least has an approximate idea of the principles of 3D printing (as if there was a competition note-reading in place of an opera!), even if the field has some big names, its gurus, its recognition at the university level, its scholarly journals and learned scientific societies and, above all else, competent industrialists (An attentive reader can find these upon reading this work's bibliography).

In France, beyond this essential aspect of readability (and thus of recognition that remains to be strengthened), daily activity has no risk of remaining stuck in a "feeling of solidarity", a ritual with institutionalized dogmatic models through which knowledge and practices, collectively shared, are staged and played, or are confirmed through self-representation and self-interpretation by the community. Indeed, everything still remains to be done if we wish to escape a situation of hard mentality out of a fear of taking risks. Yet, since the domain of additive manufacturing emerged (1984), our environment has changed considerably: Negatively, a swift transition toward a new world, "new public management", quantitative methods of evaluation, contract management, but convergence, the Internet, network organization, hyper-mobility, the tertiary industry, the silver economy, sustainable development and new forms of production, the relationship to the living, the factory of the future and so on. Recent political evolutions must be integrated into the development of forecasting scenarios, in which 3D printing has its place (present in the surveys in the top 10 emerging technologies), that necessarily have effects on research (particularly with double-digit growth rates). In short, our world of certainties is quite dead with temporalities that are reduced to innovate and be considered as readable actors of public utility. It is this that Volume 2 will try to show in terms of research and incremental innovation activities, and, in Volume 3 particularly, for futures that are largely still to be invented.

In other privileged, less functionalized locations, we witness (without much money) support to the ability offered to evade the frames of action within the fixed system, with the exploitation of its virtuosity and the desire to leave the norms to invent unique processes with pioneering value. Will the possible dogmatization of a domain participate in this inventiveness or will it be linked to a collective will to function in another way with renewed practices? With the international growth of the number of publications and patents, it will be possible to follow this history that is being created and taking shape.

France may be eliminated from the international competition, the country where additive manufacturing saw the light of day. In fact, a fragmentary survey presented in Volume 2 reports on a research unit and the unfavorable sensitivity to risk-taking: Worried about contracts that are in place, the administrative aspects, “fast science”, the response to calls of tender only leaving modest room for creativity, children to be searched for once they leave school, extra hours, classes to be prepared, the transfer of research to doctoral students, the arrival of the French National Evaluation of Academic Research System (HCERES) and so on, all or almost all condition numerous researchers to be reflexively inactive because they are nevertheless busy (while emeritus professors like the author have nothing but that to do because they are no longer signing order forms!).

These few sentences focused on additive manufacturing aim at a triple angle of attack (shaking up and challenging, breadth of questions and, finally, co-construction). They could have been even more harmful to disturb the researcher’s life, so that the informed public (e.g. Fab-Labs) would participate in new relational frameworks in a mobilization of living forces for the development of 3D printing technologies. The author expects only a return on your impressions, your propositions about what is being done and created, in short, to construct meaning with you and maybe new values for this “super-gadget”, which normally brings us together without really knowing how nor why, except that it is magic. But how can we play our cards right in international competition?

With a market of, say 20 billion €/year in 5 years, France holding 5–6% of the world’s economic activity: put simply, this is a market worth more than 1 billion €/year (which remains modest relative to the turnover from a business like the car manufacturer Renault, which is around 50 billion). If the projection on the increase in turnovers remains at 20% for 20 years, the French 3D printing market could approach in 20 years that of the automobile industry in 2016.

However, to do this, it will be necessary to revisit the concept of worth that assumes that an evaluation, considering the complexity, required interdisciplinarity, diversity and the rarity of talents can be carried out justly and pertinently for a “body” of several hundred thousand researchers and teacher-researchers in the public and private domains. An initial axis of reflection is knowing if we belong to an elite group of scouts and how we can judge the societal impact of creativity, the unreasonable and the depth of the associated scientific view for tomorrow, as long as they are given consistency (what is then the legitimacy of the evaluators?); the second is that of the "virtuous followers", who are wise, educated, but conform.

Producers of stabilized knowledge represent the necessary framework for a research unit that needs to base itself on common and effectively shared values; the third, which is not in opposition to the first, plays on the development of disciplinary and/or interdisciplinary links, both internally and externally. It represents another form of intellectual mobility. Between solidly ensured positions, the worth of scouts or openers can only develop if an in-draft is possible, only if it exists for a while, a space of action, human and financial supports to explore the uncertain with its true risk-taking. There is thus probably no single merit, no single "excellence", but diversity; the harmony of skills (not only scientific, but also human) must constitute the richness of a research team, as long as symbolic balances [KUH 72] are removed, as well as non-aggressions and the opaque reality of certain local practices, and as long as there is calm, empathetic management of the tensions between these trends sometimes stemming from the paradoxical injunction to be "readable", "credible", "motors", in short, excellent. It is this governance that must reinforce a collective imagination to have a space of freedom and solidarity action, liberating itself, as much as possible, from a "superior" administrative technocracy that thinks it is piloting research to respond to society's needs (however, it imagines them) and that continues to apply the principles of "meritocracy" adapted to everyone and maybe to nothing. However, it is possibly the price to be paid to continue enterprising, if we know who we are and where we wish to go (and how). For starters, then, let us have a look at the domains being explored in Volumes 2 and 3.

In this same line of thinking, everything leads us to support scientific creativity, that is clear, but it is not urgent because the indicators rethink and quantify the methods of the past, and we can take refuge toward other well-noted (or well-financed) priorities. The space of the "possible", the philosophy of "and if", of the "and why not", collides with local cultures, hidden hierarchies and organizational routines, non-aggressions, tacit rules and evaluation procedures. How often do the lack of time, the short term, the imperatives of production (French National Research Fund Agency (ANR) responses to calls for tender, conferences, publications, etc.) and the hierarchical imperatives (when they exist) and priority management (?) not counteract the newly converted bringers of ideas (when they exist)? The "and if" is opposed to the polite but crippling "yes but", and the spirit of conviviality made of tolerance (or simple disinterest) and possible reciprocity is jeopardized.

Creativity and organization (real, but often different from that which is displayed) thus do not always go hand in hand. To organize a human group is to place a certain number of means, rules, procedures, norms or even values aiming to ensure the effective cooperation of its members. It aims to channel individual actions and to reduce the uncertain and unforeseeable to fashion collectives with an eye for operatory efficiency and a sought-after performance, yet creativity and innovation favor deviation and upset the "established order". It is thus a very difficult art.

The depth of fear for those who are aware of this situation has an effect on their evasion strategies, which are largely subconscious or not expressed. From there, asking the question of knowing if the fear of one day being detected (because the method of control will (finally) have changed) constitutes a motive to block our imagination and risk-taking; there is only one step. Then, we favor the behaviors of powerlessness and self-reflection. One remains infantile in his/her fear of the gendarme and, for fear of exceeding the speed limit, lifts his/her foot from the accelerator when passing in front of the radar.

For an attempt to be made to go beyond these egotistical and childish attitudes, it is necessary for us to re-discover our taste for scientific discussion, for culture in all of its forms, for the imagination, for effort and for the desire to act personally (not through an intermediate party) with true risk-taking. For this, it is necessary to recreate a conscious desire, a collective imaginary, for the energy for this desire is much greater than that of the fear or, in recent times, cowardice. By working in an open system, thermodynamics then allows organizational creativity to be created, because there is double-digit growth with a very large socio-economic demand.

Moreover, innovation is a phenomenon whose complexity, given its sociotechnical, creative, emerging, collective, interactive and multi-disciplinary dimensions, is now largely recognized by all of its decision makers [LER 17]: socio-economic actors (companies, federations, clusters, associations, etc.), institutional actors (state, collectives, financers, etc.) and the different scientific communities that are interested in it (engineering science, design, industrial engineering, management, sociology, psychology, etc.).

The modification of the world landscape, notably with the rise in power of the countries in East Asia (see the national origins of the scientific publications' authors) and the large evolutions in supply and demand constitute the two large trends that will structure the creation, research and innovation activities of tomorrow. Furthermore, the challenges (but also the opportunities) linked to the intensification of the consideration of sustainable development will lead to an increase in the changes and adaptations necessary for all individuals and organizations.

Currently, there is a double stake in additive manufacturing that consists of:

– promoting industrial innovation, which can be translated by the development of new procedures/products, services, organizations, economic models and so on, in order to develop companies' competitiveness;

– as well as accompanying the mutation of all the actors of innovation (companies, institutions, research, businesses, citizens, etc.) toward responsible innovation consisting of a search for balance between economic development, the reduction of environmental impacts and respect for societal and ethical values.

However, to achieve these goals, the working individual and/or citizen must be considered a being who feels, perceives, experiments and evaluates space through his/her body, feelings, senses and emotions. Subjective experiences are a dynamic process through which the individual adds his/her own empirical knowledge to the identity and collective memory of places and processes to give them meaning and to create his/her own vision of his/her environment. Thus, a new paradigm based on the role that would best be given to subjectivity and individual experiences in the design allowed by 3D printing could offer an anthropological and philosophical breakthrough that would have to be explored (independent of this work).

Returning to this civilizational basis implies thinking of additive manufacturing processes as a complex and continuous process, associating knowledge, cultures and different technologies that must be taken into consideration (the average of the averages is not the average). This paragon of a constantly evolving, perpetually changing and thus sustainable, even resilient system would once again place the being, the individual and the personal experience at the heart of additive manufacturing sciences becoming applicational.

Bibliography

[BOH 61] BOHR N., *Atomic Physics and Human Knowledge*, Dover Pub., New York, 1961.

[CLA 11] CLAESENS M., *Allo la science? Analyse critique de la médiascience*, Hermann, Paris, 2011.

[COH 09] COHEN D., *La prospérité du vice; une introduction (inquiète) à l'économie*, Le livre de poche, Paris, 2009.

[DEG 13] DEGOS L., *Eloge de l'erreur*, Le Pommier, Paris, 2013.

[EUR 17] EUROPEAN UNION, "Tackling the skills gap in additive manufacturing", Workshop, Brussels, Belgium, 13 March 2017.

[GUY 14] GUYOT J.L., RIEPPI S., "La prospective en Wallonie: une lecture plurielle", in GUYOT J.L., BRUNET S. (eds), *Construire les futurs – Contributions épistémologiques et méthodologiques à la démarche prospective*, Presses de Namur, Namur, 2014.

[KUH 72] KUHN T.S., *La structure des révolutions scientifiques*, Flammarion, Paris, 1972.

[LER 17] LERU, "Productive interactions: societal impact of academic research in the knowledge society", available at: http://www.leru.org/files/general/LERU_Position_Paper_Societal_Impact.pdf, 2017.

[LIC 10] LICATA I., "Le tournevis et la cerise", *Alliage*, vol. 66, pp. 11–15, 2010.

[MAL 11] MALRIEU J.P., *La science gouvernée*, Rue des gestes, Toulouse, 2011.

[POI 11] POINCARÉ H., "La science et les humanités", Fayard, Paris, 1911.

[ROS 12] ROSA H., *Aliénation et accélération, vers une théorie critique de la modernité tardive*, La Découverte, Paris, 2012.

Index

Other titles from

iSTE

in

Systems and Industrial Engineering – Robotics

2017

ARCHIMÈDE Bernard, VALLESPIR Bruno
Enterprise Interoperability: INTEROP-PGSO Vision

CAMMAN Christelle, FIORE Claude, LIVOLSI Laurent, QUERRO Pascal
Supply Chain Management and Business Performance: The VASC Model

FEYEL Philippe
Robust Control, Optimization with Metaheuristics

MARÉ Jean-Charles
Aerospace Actuators 2: Signal-by-Wire and Power-by-Wire

POPESCU Dumitru, AMIRA Gharbi, STEFANOIU Dan, BORNE Pierre
Process Control Design for Industrial Applications

RÉVEILLAC Jean-Michel
Modeling and Simulation of Logistics Flows 1: Theory and Fundamentals
Modeling and Simulation of Logistics Flows 2: Dashboards, Traffic Planning and Management
Modeling and Simulation of Logistics Flows 3: Discrete and Continuous Flows in 2D/3D

2016

ANDRÉ Michel, SAMARAS Zissis
Energy and Environment
(Research for Innovative Transports Set - Volume 1)

AUBRY Jean-François, BRINZEI Nicolae, MAZOUNI Mohammed-Habib
Systems Dependability Assessment: Benefits of Petri Net Models (Systems Dependability Assessment Set - Volume 1)

BLANQUART Corinne, CLAUSEN Uwe, JACOB Bernard
Towards Innovative Freight and Logistics (Research for Innovative Transports Set - Volume 2)

COHEN Simon, YANNIS George
Traffic Management (Research for Innovative Transports Set - Volume 3)

MARÉ Jean-Charles
Aerospace Actuators 1: Needs, Reliability and Hydraulic Power Solutions

REZG Nidhal, HAJEJ Zied, BOSCHIAN-CAMPANER Valerio
Production and Maintenance Optimization Problems: Logistic Constraints and Leasing Warranty Services

TORRENTI Jean-Michel, LA TORRE Francesca
Materials and Infrastructures 1 (Research for Innovative Transports Set - Volume 5A)
Materials and Infrastructures 2 (Research for Innovative Transports Set - Volume 5B)

WEBER Philippe, SIMON Christophe
Benefits of Bayesian Network Models
(Systems Dependability Assessment Set – Volume 2)

YANNIS George, COHEN Simon
Traffic Safety (Research for Innovative Transports Set - Volume 4)

2015

AUBRY Jean-François, BRINZEI Nicolae
Systems Dependability Assessment: Modeling with Graphs and Finite State Automata

BOULANGER Jean-Louis
CENELEC 50128 and IEC 62279 Standards

BRIFFAUT Jean-Pierre
E-Enabled Operations Management

MISSIKOFF Michele, CANDUCCI Massimo, MAIDEN Neil
Enterprise Innovation

2014

CHETTO Maryline
Real-time Systems Scheduling
Volume 1 – Fundamentals
Volume 2 – Focuses

DAVIM J. Paulo
Machinability of Advanced Materials

ESTAMPE Dominique
Supply Chain Performance and Evaluation Models

FAVRE Bernard
Introduction to Sustainable Transports

GAUTHIER Michaël, ANDREFF Nicolas, DOMBRE Etienne
Intracorporeal Robotics: From Milliscale to Nanoscale

MICOUIN Patrice
Model Based Systems Engineering: Fundamentals and Methods

MILLOT Patrick
Designing Human–Machine Cooperation Systems

NI Zhenjiang, PACORET Céline, BENOSMAN Ryad, RÉGNIER Stéphane
Haptic Feedback Teleoperation of Optical Tweezers

OUSTALOUP Alain
Diversity and Non-integer Differentiation for System Dynamics

REZG Nidhal, DELLAGI Sofien, KHATAD Abdelhakim
Joint Optimization of Maintenance and Production Policies

STEFANOIU Dan, BORNE Pierre, POPESCU Dumitru, FILIP Florin Gh., EL KAMEL Abdelkader
Optimization in Engineering Sciences: Metaheuristics, Stochastic Methods and Decision Support

2013

ALAZARD Daniel
Reverse Engineering in Control Design

ARIOUI Hichem, NEHAOUA Lamri
Driving Simulation

CHADLI Mohammed, COPPIER Hervé
Command-control for Real-time Systems

DAAFOUZ Jamal, TARBOURIECH Sophie, SIGALOTTI Mario
Hybrid Systems with Constraints

FEYEL Philippe
Loop-shaping Robust Control

FLAUS Jean-Marie
Risk Analysis: Socio-technical and Industrial Systems

FRIBOURG Laurent, SOULAT Romain
Control of Switching Systems by Invariance Analysis: Application to Power Electronics

GROSSARD Mathieu, REGNIER Stéphane, CHAILLET Nicolas
Flexible Robotics: Applications to Multiscale Manipulations

GRUNN Emmanuel, PHAM Anh Tuan
Modeling of Complex Systems: Application to Aeronautical Dynamics

HABIB Maki K., DAVIM J. Paulo
Interdisciplinary Mechatronics: Engineering Science and Research Development

HAMMADI Slim, KSOURI Mekki
Multimodal Transport Systems

JARBOUI Bassem, SIARRY Patrick, TEGHEM Jacques
Metaheuristics for Production Scheduling

KIRILLOV Oleg N., PELINOVSKY Dmitry E.
Nonlinear Physical Systems

LE Vu Tuan Hieu, STOICA Cristina, ALAMO Teodoro,
CAMACHO Eduardo F., DUMUR Didier
Zonotopes: From Guaranteed State-estimation to Control

MACHADO Carolina, DAVIM J. Paulo
Management and Engineering Innovation

MORANA Joëlle
Sustainable Supply Chain Management

SANDOU Guillaume
Metaheuristic Optimization for the Design of Automatic Control Laws

STOICAN Florin, OLARU Sorin
Set-theoretic Fault Detection in Multisensor Systems

2012

AÏT-KADI Daoud, CHOUINARD Marc, MARCOTTE Suzanne, RIOPEL Diane
Sustainable Reverse Logistics Network: Engineering and Management

BORNE Pierre, POPESCU Dumitru, FILIP Florin G., STEFANOIU Dan
Optimization in Engineering Sciences: Exact Methods

CHADLI Mohammed, BORNE Pierre
Multiple Models Approach in Automation: Takagi-Sugeno Fuzzy Systems

DAVIM J. Paulo
Lasers in Manufacturing

DECLERCK Philippe
Discrete Event Systems in Dioid Algebra and Conventional Algebra

DOUMIATI Moustapha, CHARARA Ali, VICTORINO Alessandro, LECHNER Daniel
Vehicle Dynamics Estimation using Kalman Filtering: Experimental Validation

GUERRERO José A, LOZANO Rogelio
Flight Formation Control

HAMMADI Slim, KSOURI Mekki
Advanced Mobility and Transport Engineering

MAILLARD Pierre
Competitive Quality Strategies

MATTA Nada, VANDENBOOMGAERDE Yves, ARLAT Jean
Supervision and Safety of Complex Systems

POLER Raul *et al.*
Intelligent Non-hierarchical Manufacturing Networks

TROCCAZ Jocelyne
Medical Robotics

YALAOUI Alice, CHEHADE Hicham, YALAOUI Farouk, AMODEO Lionel
Optimization of Logistics

ZELM Martin *et al.*
Enterprise Interoperability –I-EASA12 Proceedings

2011

CANTOT Pascal, LUZEAUX Dominique
Simulation and Modeling of Systems of Systems

DAVIM J. Paulo
Mechatronics

DAVIM J. Paulo
Wood Machining

GROUS Ammar
Applied Metrology for Manufacturing Engineering

KOLSKI Christophe
Human–Computer Interactions in Transport

LUZEAUX Dominique, RUAULT Jean-René, WIPPLER Jean-Luc
Complex Systems and Systems of Systems Engineering

ZELM Martin, *et al.*
Enterprise Interoperability: IWEI2011 Proceedings

2010

BOTTA-GENOULAZ Valérie, CAMPAGNE Jean-Pierre, LLERENA Daniel, PELLEGRIN Claude
Supply Chain Performance / Collaboration, Alignement and Coordination

BOURLÈS Henri, GODFREY K.C. Kwan
Linear Systems

BOURRIÈRES Jean-Paul
Proceedings of CEISIE'09

CHAILLET Nicolas, REGNIER Stéphane
Microrobotics for Micromanipulation

DAVIM J. Paulo
Sustainable Manufacturing

GIORDANO Max, MATHIEU Luc, VILLENEUVE François
Product Life-Cycle Management / Geometric Variations

LOZANO Rogelio
Unmanned Aerial Vehicles / Embedded Control

LUZEAUX Dominique, RUAULT Jean-René
Systems of Systems

VILLENEUVE François, MATHIEU Luc
Geometric Tolerancing of Products

2009

DIAZ Michel
Petri Nets / Fundamental Models, Verification and Applications

OZEL Tugrul, DAVIM J. Paulo
Intelligent Machining

PITRAT Jacques
Artificial Beings

2008

ARTIGUES Christian, DEMASSEY Sophie, NERON Emmanuel
Resources–Constrained Project Scheduling

BILLAUT Jean-Charles, MOUKRIM Aziz, SANLAVILLE Eric
Flexibility and Robustness in Scheduling

DOCHAIN Denis
Bioprocess Control

LOPEZ Pierre, ROUBELLAT François
Production Scheduling

THIERRY Caroline, THOMAS André, BEL Gérard
Supply Chain Simulation and Management

2007

DE LARMINAT Philippe
Analysis and Control of Linear Systems

DOMBRE Etienne, KHALIL Wisama
Robot Manipulators

LAMNABHI Françoise *et al.*
Taming Heterogeneity and Complexity of Embedded Control

LIMNIOS Nikolaos
Fault Trees

2006

FRENCH COLLEGE OF METROLOGY
Metrology in Industry

NAJIM Kaddour
Control of Continuous Linear Systems

Printed and bound by CPI Group (UK) Ltd, Croydon, CR0 4YY
16/09/2026
14963072-0007